The European Rabbit

The European Rabbit
(Photograph courtesy of J. H. Johns)

The European Rabbit

The history and biology of a successful colonizer

Edited by

Harry V. Thompson and Carolyn M. King

Oxford New York Tokyo

OXFORD UNIVERSITY PRESS

1994

Oxford University Press, Walton Street, Oxford OX2 6DP

Oxford New York Toronto
Delhi Bombay Calcutta Madras Karachi
Kuala Lumpur Singapore Hong Kong Tokyo
Nairobi Dar es Salaam Cape Town
Melbourne Auckland Madrid

and associated companies in
Berlin Ibadan

Oxford is a trade mark of Oxford University Press

Published in the United States
by Oxford University Press Inc., New York

A catalogue record for this book is available from the British Library

Library of Congress Cataloging in Publication Data
The European rabbit: the history and biology of a successful colonizer / edited by Harry V. Thompson and Carolyn M. King.
Includes Index.
1. Oryctolagus cuniculus. I. Thompson, Harry V. II. King, C. M. (Carolyn M.)
QL737.L32E87 599.32'2—dc20 93–10549
ISBN 0–19–857611–0

Typeset by Joshua Associates Ltd, Oxford
Printed in Great Britain by
Butler & Tanner Ltd, Frome, Somerset

In memoriam

Charles Elton, our mentor

Acknowledgements

The processes of production of this book have been long (ten years) and hard by anyone's standards, and not everyone who has put a lot of effort into it over the years is now represented in the published text. We therefore feel a particular gratitude to all the authors of chapters, both those whose efforts eventually bore fruit (listed within) and also those who produced manuscripts which were, for various reasons, eventually withdrawn (J. Sheail, H. G. Lloyd, D. P. Cowan, R. Mykytowycz, and E. R. Hesterman). All have suffered the delays and changes of plan with fortitude and various degrees of good humour.

We thank the editorial staff at Oxford University Press for monumental patience and understanding beyond the call of duty.

The authors of the various chapters have acknowledged their own sources of information and illustrations; we thank J. H. Johns for the frontispiece portrait of the central character of the work.

Contents

Contributors

C. P. Arthur, Office National de la Chasse, Centre expérimental de Saint-Benoist, 78610 Auffargis, France. Present address: Parc National des Pyrénées, 59 route de Pau, 65000 Tarbes, France.

B. D. Cooke, Principal Research Officer, Animal and Plant Control Commission, Adelaide, South Australia.

G. B. Corbet, Little Dumbarnie, Newburn, Upper Largo, Fife KY8 6JQ, Scotland. Curator of Mammals etc. British Museum (Natural History), 1960 to 1988.

Frank Fenner, John Curtin School of Medical Research, Australian National University, Canberra, ACT, Australia.

John E. C. Flux, Hill Road, Belmont, Lower Hutt, New Zealand. Scientist, DSIR Ecology Division, 1960 to 1992. Research Associate, Landcare Research, New Zealand Ltd.

John A. Gibb, Landcare Research New Zealand Ltd, PO Box 31092, Lower Hutt, New Zealand. Director, DSIR Ecology Division, 1965 to 1982.

C. M. King, Royal Society of New Zealand, PO Box 598, Wellington, New Zealand.

K. Myers, Department of Biological Sciences, University of Newcastle, NSW 2308, Australia.

I. Parer, Experimental Research Scientist, CSIRO Division of Wildlife and Ecology, Canberra, ACT, Australia.

P. M. Rogers, Station Biologique de la Tour du Valat, Le Sambuc, 13200 Arles, France. Present address: Mas d'Auphan, Le Sambuc, 13200 Arles, France.

John Ross, MAFF Central Science Laboratory, Tangley Place, Worplesdon, Surrey, England.

R. C. Soriguer, Estación Biológica de Doñana, CSIC, Avenida Maria Luisa s/n, 41013 Sevilla, Spain.

Harry V. Thompson, Rose Cottage, Lower Farm Road, Effingham, Leatherhead, Surrey KT24 5JL, England. Officer-in-charge, Land Pests and Birds Research Laboratory, Tangley Place, Worplesdon, 1959 to 1982.

J. Morgan Williams, MAF Policy, PO Box 8640 Riccarton, Christchurch, New Zealand.

D. Wood, Senior Research Scientist, CSIRO Division of Wildlife and Ecology, Canberra, ACT, Australia.

Editors' introduction

The European rabbit originated on the Iberian Peninsula, and spread naturally, with some human encouragement, to most of continental Europe. In addition, it has been deliberately introduced to many other countries and to hundreds of islands, usually to provide meat for settlers and travellers. Unfortunately, the very character that makes the rabbit such an ideal food species—an almost unrivalled capacity to convert nearly any kind of greenery into large quantities of meat in a short time—also makes it a pest of almost equally unrivalled potential to damage farm crops and to compete with other livestock.

In three of the countries to which it has been introduced (Britain, Australia, and New Zealand), the rabbit has demonstrated that potential convincingly. Largely because of the experience of those three countries and their investment in research, the rabbit has become one of the classic and best-known examples of an invading species, and its most important disease, myxomatosis, the classic and best-known example of a devastating wildlife epizootic. At the same time and, paradoxically, even in parts of the same countries, the rabbit is also greatly valued by hunters as a game animal, by furriers as a source of fine pelts, and by many families as a harmless, cuddly pet.

For centuries in Europe and for many decades overseas, the rabbit has been the subject of disputes between tenants and landlords, while questions of rights of property in rabbits, and their status as an agricultural/sporting asset or liability, have led to extensive and controversial legislation. In the last 40–50 years, several large and expensive programmes of scientific research have been mounted, mostly supported by Government laboratories and mainly concerned with the question of how to control the rabbit as a pest of agriculture. Prominent among those laboratories have been the Australian CSIRO's Wildlife Division (now Division of Wildlife and Ecology) at Canberra; the New Zealand DSIR's Ecology Division (now Landcare Research New Zealand Ltd) at Lower Hutt; and the British MAFF's Land Pests and Birds Laboratory (now Central Science Laboratory) at Tangley Place, near Guildford. Research by these laboratories, in conjunction with regional staff in Britain, the John Curtin School of Medical Research (Australian National University, Canberra), state officials in Australia, and others in New Zealand, scientists in various institutes in all three countries and in Spain, France, USA, Central and South America, plus the support of innumerable farmers, graziers and other countrymen, has contributed to an enormous fund of knowledge about the rabbit in their respective countries.

The problems they tackled were huge and have not been, could not be, entirely resolved. Although the teams working in different countries kept in touch with each other, each had to work in their own particular environment and according to the rules and priorities set by their own governments. Therefore, not all laboratories addressed the same questions, or reached the same conclusions on common themes; but the total picture provided by their combined efforts is impressive and broadly coherent. Now, in the course of various political restructurings of research in each country, all three laboratories have taken on other names; and new generations of rabbit biologists struggle with similar, or consequential, problems.

As in all research, new perceptions must stand on the shoulders of previous studies. With the rabbit, an adaptable and economically important species occupying a wide range of habitats spanning many latitudes, the literature is vast and scattered. The task of proper documentation of the background information behind any new proposal is now almost a project in itself. Yet failure to appreciate the extent of existing knowledge entails the risk of investing scarce research funds in repetition of work already done or not needed.

This book summarizes that historical background, and so provides a firm basis against which new ideas may be assessed. Like most such compilations, it has prised into open print a considerable amount of hitherto unpublished data. Like any review of long-term research, it includes reports of studies that would be impossible under contemporary financial and ethical rules. But its particular advantage, unmatched elsewhere, is that the main chapters are contributed by the very people who were deeply involved in, and/or directing, the key advances in our understanding of rabbits and of myxomatosis during that vital period since the early 1950s. Such a congruence of inside knowledge of rabbits and of the people who worked on them (some of which is personal, and so is not to be found in the literature) has not been assembled before—nor, with respect to the historical period concerned, could be again.

The book is organized as follows. The first two chapters set the scene by describing the origins, evolution and taxonomy of *Oryctolagus cuniculus* (Corbet), and its natural and assisted spread (Flux). The centre of the book is a set of four chapters which describe the biology and control of rabbits in continental Europe (Rogers *et al.*), Britain (Thompson), Australia (Myers *et al.*), and New Zealand (Gibb and Williams). All four chapters cover common subjects, such as reproduction, population dynamics and control operations; but the information is organized geographically rather than thematically so as to bring out the important parallels and contrasts in the behaviour, ecology, and management of rabbits in the very different environments offered by each country. Finally, the last chapter (Fenner and Ross) gives a comprehensive review of myxomatosis, arguably the most important and best-studied epizootic in the history of wildlife management.

All the chapters are necessarily rather concise summaries of the wide fields they cover, but all are fully documented to the primary literature. Inevitably, there are substantial overlaps and differences, even outright contradictions, in attitude to similar problems as described by the different authors. For example, in Australia and New Zealand, political and economic considerations ensured that vigorous campaigns to control rabbit populations preceded any serious attempt to understand their biology. Gibb and Williams have therefore placed their account of control work in New Zealand at the beginning of their chapter; Myers *et al.* follow the more usual convention of discussing biology first and management in Australia last. Authors also vary in the extent to which they quote and comment on primary data. We consider such variations in approach interesting in themselves, and far too important to be swept away in the name of consistency.

Human attitudes to the rabbit have been ambivalent since very early times, and remain so today—especially in Britain and continental Europe. Some of these incongruities are made very obvious by the way this book places together accounts of the rabbit as seen through very different eyes and against very different backgrounds. Take, as examples, the apparently absurd fact that in much of France, the rabbit is a declared pest and yet is protected for nine months of the year. The man who deliberately brought myxomatosis to France, Dr P. F. Armand Delille, was both taken to court by hunters and also awarded a medal by foresters. In Britain, although farmers, landlords, and the general public were fully aware of the agricultural losses due to rabbit damage, they nevertheless regarded the first outbreak of myxomatosis with horror; the Ministry of Agriculture initially attempted to eradicate it, and Parliament decreed it an offence to spread diseased rabbits around the country. The ancient British laws against poaching of individual rabbits, enacted to protect rabbits as valuable property and game (and under which convicted poachers could be transported to Botany Bay), have remained on the statute books into the modern era of official mass destruction of rabbits by cyanide gassing. In Australia and New Zealand, where rabbits have more opponents and fewer friends, and have never been particularly important as game, it has been easier for authorities to take a more pragmatic view; nevertheless, there are always some voices to be heard in defence of rabbits, if only because of the small but persisting value of rabbit meat and pelts, and the lingering influence of Beatrix Potter.

Equally remarkable is the somewhat full-circle feeling that creeps up on a contemporary scientist comparing the histories of rabbit control presented here. In both northern and southern hemispheres, almost any possible method has been tried, including those that have been discarded elsewhere. For example, in New Zealand, large-scale aerial poisoning has been very effective in the past, but is now less favoured. Runholders in the worst-affected areas, the 'semi-arid' lands of the South Island, look enviously at the record of myxomatosis in reducing rabbit populations in Australia and Britain, and continue to press for the release of the virus and its non-flying vector, the European rabbit flea. At the very same time, Britain, whose own rabbit populations are now rapidly recovering from myxomatosis, is considering the use of poison baiting. There are also instructive contrasts between Britain and Australia in official attitudes to cyanide gassing. On continental Europe, on the other hand, hunters are dissatisfied with the extent of their rabbits' natural recovery from myxomatosis, and they have attempted to introduce the American cottontail rabbit, which is virtually immune to it.

It is salutary to consider how much contemporary policy is influenced by historical accident, for example, the early failures of attempts to introduce myxomatosis both in Britain and Australia, where it later became established, and in New Zealand, where it has not (so far). Policy is also, perhaps inevitably, affected by political considerations, for example the arguments about protecting agricultural production from damage by rabbits in Britain and Europe in the present time of huge food surpluses. Scientists and administrators working in any of the countries covered here may be interested to read of the historical and geographical dimensions of their own local version of 'the rabbit problem'.

In March 1992 the New Zealand MAFF organized a workshop at Haldon Station, in the heart of the most intractably rabbit-prone South Island high country. It invited many people from different organizations with an interest in rabbits to debate the relationships between rabbits and their predators, and inevitably also many wider questions of rabbit management in the present and future. Among the participants were three of the people then still struggling to get this book finished. One of the most common recurring themes of discussion, repeated both by invited speakers and by contributors from the floor, was the need for a published summary of existing information on the rabbit, to identify gaps in our knowledge and to test and refine new ideas. The Haldon workshop was only the latest in a long series of meetings which have been held to discuss the rabbit problem at different times and places since the 1950s. We hope that this book will help, not only by providing the convenient reviews requested, and outlining some of the policies that have shaped rabbit research in the past, but also by collecting and listing the key references to the literature on rabbits which is the basis of the advances of the future. The new generation of rabbit biologists works with very different tools and theories from those of their predecessors, but, in research as in evolution, the footprints of history provide the key to understanding the present.

1

Taxonomy and origins

G. B. Corbet

1.1 Introduction

The European rabbit, *Oryctolagus cuniculus*, must be one of the most familiar and best known of all mammalian species, both as a wild animal and in its domesticated forms. It has been studied as a resource, as a pest, and as a tool in medical research; it is kept as a family pet and has a firm place in popular folklore. The resultant literature is enormous—so much so that the very volume of information is liable to get in the way of understanding the rabbit as a species.

The first step in understanding any species of animal is to define it—to recognize it as an entity by delimiting its boundaries in space and by distinguishing it, in all its variability, from the huge diversity of more or less related neighbours with which it shares the available living space of the planet. Once we have thus placed the rabbit in its taxonomic context, we have a key, albeit a very rusty one, that can help unlock its past, both historical and palaeontological. In turn, an understanding of the conditions in which a species evolved its present attributes is a prerequisite for a proper understanding of the way in which it interacts with its contemporary environment, and for intelligent speculation on its future potential.

1.2 Taxonomy

The European rabbit is one of about 40 living species in the family Leporidae, which comprises all the rabbits and hares (Table 1.1). In view of the very close similarity between it and some other rabbits in the Americas, Africa, and Asia it is surprising how little confusion there has been over its recognition as a discrete species since it was named as *Lepus cuniculus* by Linnaeus in 1758, combining the classical Latin names *cuniculus* (rabbit) and *lepus* (hare). Besides two species of hare (the northern *L. timidus* and the African *L. capensis*), the genus *Lepus* of Linnaeus included only one other rabbit, *L. brasiliensis* (now in *Sylvilagus*) from South America. Even when the various North American cottontail rabbits came to scientific attention in the nineteenth century they were recognized, correctly, as being specifically distinct from the European rabbit.

Only two forms now included in *Oryctolagus cuniculus* were originally described as separate species. These are the North African rabbit, named by Loche (1858) as *Cuniculus algirus*, and those on the island of Porto Santo, Madeira, named by Haeckel (1874) as *Lepus huxleyi*. In doing so Haeckel was simply going one step beyond Darwin (1868) who, in *The variation of animals and plants under domestication*, used the Porto Santo rabbits as an example of divergence in form since their recorded introduction to the island in 1418, and commented that *if* their history had not been known most naturalists would consider them a distinct species (on the basis of their small size and reddish colour).

The relationship of the European rabbit to other species of rabbits and hares, as reflected in the European rabbit's generic status, appears on the face of it to be quite clear, since it has for most of the last century been accepted as the sole species of its genus *Oryctolagus*. However, the situation is not as clear-cut as appears at first sight. It was first separated

Table 1.1 Classification of the lagomorphs

Order Lagomorpha
Family Ochotonidae
Ochotona. *c*.20 species of pikas; northern Asia, western North America.
Family Leporidae
Pentalagus furnessi. Ryukyu rabbit; Ryukyu Islands, Japan.
Pronolagus (3 spp.). Red rock rabbits; South and East Africa.
Bunolagus monticularis. Riverine rabbit; South Africa (closely related to *Pronolagus*).
Caprolagus hispidus. Hispid rabbit; northeast India etc.
Nesolagus netscheri. Sumatran rabbit; Sumatra.
Romerolagus diazi. Volcano rabbit; Mexico.
Brachylagus idahoensis. Pygmy rabbit; western USA.
Sylvilagus (12 spp.). Cottontail rabbits; North, Central and South America.
Poelagus marjorita. Central African rabbit; central Africa.
Oryctolagus cuniculus. European rabbit; western Europe etc.
Lepus (*c*.20 spp.). Hares; Holarctic, Indomalayan and Afrotropical regions (except tropical forest).

generically from *Lepus* by Meyer (1790) using the name *Cuniculus*. Although this name was used intermittently throughout the nineteenth century, the situation was confused by its prior use by Brisson (1762) for the unrelated South American paca (a rodent), a use that has continued until recently although it is invalid. The name *Oryctolagus*, meaning 'digging hare', was coined by Lilljeborg (1874) as a subgenus of *Lepus*, containing only the European rabbit, but solely as a means of emphasizing the differences between it and the European hares. This distinction is indeed quite clear, being based upon the much greater adaptation to running in the hares, involving not only the legs and feet but also other features such as the very wide nasal passage, reflected in the wide 'mesopterygoid fossa' or internal nares in the skull (Figs. 1.1, 1.2). There are also substantial differences in development and behaviour, such as the altricial young and the social and burrowing behaviour of the rabbit.

On the other side of the Atlantic a similar development took place; seven species of cottontail rabbits were originally described in the genus *Lepus* as they were discovered between 1837 and 1858. These were subsequently separated in the genera *Sylvilagus* and *Hydrolagus* by Gray (1867) in order to distinguish them from the American hares (*Lepus*), but without any reference to the European rabbit. (*Hydrolagus* has subsequently been combined with *Sylvilagus* by most authors.)

In the same fashion, in India a rabbit was described as *Lepus hispidus* by Pearson (1839) before being segregated in a genus of its own, *Caprolagus*, by Blyth (1845) to emphasize its difference from the local hares, again without comparison with the European or American rabbits. This was repeated in central Africa when *Lepus marjorita* from Uganda was described by St Leger (1929) who later (1932) erected a new genus *Poelagus* to stress that it was a rabbit and not a hare, but without explicitly describing any characters by which it could be distinguished from *Oryctolagus* or *Sylvilagus*. Other, more distinctive, monospecific genera of rabbits were described from the Ryukyu Islands, Japan (*Pentalagus*), Sumatra (*Nesolagus*), Mexico (*Romerolagus*), and western USA (*Brachylagus*) (Table 1.1), although the last has by some been included in the genus *Sylvilagus*.

The first and almost the only person to make comparisons on a world-wide basis was Lyon (1904) in a revision of the entire family Leporidae. Although he upheld *Oryctolagus*, *Caprolagus*, and *Sylvilagus* as separate genera, the diagnostic differences he described were subtle. In particular the differences between *O. cuniculus* and the best known species of *Sylvilagus*, the eastern cottontail, *S. floridanus*, for

Fig. 1.1 (a) Brown hare, and (b) rabbit. (Reproduced with permission from Corbet and Harris (1991).)

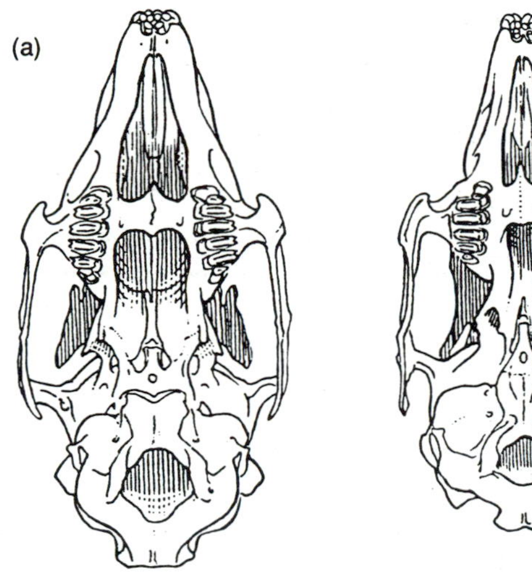

Fig. 1.2 Skulls of (a) brown hare, and (b) rabbit.(Reproduced with permission from Corbet and Harris (1991).)

example the slightly thicker bones of the forearm, especially the ulna, in *Oryctolagus*, related to digging, were no greater than those between the various species of *Sylvilagus*. A later revision by Gureev (1964) was equally unhelpful in differentiating between these genera. In a more recent review of the subject (Corbet 1983) I tabulated all the character-differences between all the species of rabbits and representatives of all the species-groups of hares and concluded that the only justification for retaining the genus *Oryctolagus* distinct from *Sylvilagus* lay in the distinctive social and burrowing behaviour of *Oryctolagus*, which appears to be unique in the family.

Attempts at hybridization between *Oryctolagus* and species of *Sylvilagus* have resulted in some fertilization but no births (Gray 1972). These experiments are of limited value in assessing relationships until comparable data are available for crosses between species of *Sylvilagus* and involving other genera such as *Poelagus*.

The nearest relatives of *Oryctolagus* then appear to be the various species of *Sylvilagus* in the Americas and *Poelagus marjorita* in Africa. It is likely that studies of chromosome banding, isozymes, and DNA will help to resolve these relationships. Although a start has been made (Robinson *et al.* 1983; Robinson and Osterhoff 1983), some of the critical species, for example *Poelagus marjorita*, remain to be examined. Meanwhile there is little point in upsetting the current generic classification. Inclusion of the little-known central African rabbit in either *Oryctolagus* or *Sylvilagus* would cause little confusion; but inclusion of the European rabbit in *Sylvilagus*, as would be required by the International Code of Zoological Nomenclature if these genera were to be combined, would be counterproductive with respect to the use of taxonomy and nomenclature to facilitate understanding of relationships, and the efficient storage and retrieval of information.

1.3 Regional variation

Any original pattern of geographical variation concerning the European rabbit has almost certainly been destroyed, or at least obscured, by the twin human interferences of translocation and domestication. Although several subspecies have been named and described (Table 1.2), no comprehensive description of geographical variation has been undertaken and it is unlikely that orthodox subspecific nomenclature would be a helpful or appropriate way of expressing any geographical variation that exists. Many island populations, such as those on Madeira and several of the Mediterranean

Table 1.2 Synonymy of the genus *Dryctolagus* and the species *O. cuniculus*

Oryctolagus Lilljeborg, 1874. Type species *Lepus cuniculus* L.
Cuniculus Meyer, 1790. Type species *C. campestris* Meyer, in *Lepus cuniculus* L.

Oryctolagus cuniculus (L.)
Lepus cuniculus Linnaeus, 1758. Germany.
Cuniculus algirus Loche, 1848. Algeria (a *nomen nudum* but likely to have been validated by a later author).
Cuniculus fodiens Gray, 1867. Objective synonym of *L. cuniculus* L.
Lepus huxleyi Haeckel, 1874. Porto Santo, Madeira (small, reddish).
O.c. cnossos Bate, 1906. Dhia, off Candia, Crete (pale, greyish).
O.c. brachyotis Trouessart, 1917. Riège, Camargue, southern France (black with short ears).
O.c. oreas Cabrera, 1922. Xauen, Spanish Morocco (blackish).
O.c. habetensis Cabrera, 1923. Dar Amezuk, Anyera, Spanish Morocco (dark brown).
O.c. borkumensis Harrison, 1952. Island of Borkum, east Friesian Islands, Germany (pale).

islands, are characterized by small body size. That on the island of Borkum, off the German North Sea coast, has been given subspecific rank on the basis of pale pelage, but these rabbits are in fact no paler than many Mediterranean animals. Any differences between the African and Iberian populations have apparently not been described from adequate samples.

The names in Table 1.2 are those that are potentially available for subspecies if the existence of sufficiently discrete and diagnosable subspecies could be demonstrated. The characters given in parentheses in Table 1.2 are simply those that were used to justify the original name—there appears to be no information on the geographical extent or variability of these forms. A plethora of other names has been proposed, for example by Hochstrasser (1969), but these are of no standing in orthodox zoological nomenclature because of the absence of diagnoses or because they refer to infra-subspecific groups, and they appear to serve no useful purpose.

Considering that the normal 'agouti' pelage of the wild rabbit is presumably adapted to provide optimum camouflage against predators, especially birds of prey, it is surprising how frequently black morphs are found—even allowing for the fact that some may have originated from domestic stock. The form *brachyotis* in the Camargue region of southern France consists of populations that are predominantly black, and those in Western Sahara also appear to be very dark (Cabrera 1923). In Britain many island populations, for example on Fair Isle and Iona in Scotland, and several mainland populations include a larger proportion of black individuals than is found in most other species of mammals, although in some places a similar pattern may be seen in the water vole, *Arvicola terrestris* (without the complication of domestication).

1.4 Origin and evolution

The differences between the living species of rabbits and hares are subtle, even though we have whole animals for comparison. Since most fossil and subfossil finds consist of isolated teeth or small fragments of skull or other bones, the difficulties of confidently distinguishing species in the fossil record are acute. The problem is exacerbated by the burrowing ability of the rabbit and the consequent difficulty of recognizing remains that have thereby been intruded into earlier strata.

The firmest foundation for an assessment of the fossil history of the rabbit is the abundance of records in Iberia dating from the Late and Middle Pleistocene. At Gibraltar, for example, it is well represented in deep layers of cave deposits that can be dated to the last glaciation, i.e. around 10 or 20 thousand years ago (Zeuner and Sutcliffe 1964). Other records in southern Spain extend back in the Middle Pleistocene before the Mindel glaciation (Lopez-Martinez 1977), and there are records from southern France dated to around 2–300 000 years ago, and later (Donard 1981). The northernmost prehistoric record is that at Swanscombe, Kent, England, at about the same period (Sutcliffe 1964; Mayhew 1975).

Other Middle Pleistocene finds from Spain have been allocated to a separate species of *Oryctolagus*, *O. layensis* (Lopez-Martinez 1977) and likewise *O. lacosti* (Pomel 1854) has been described from the Lower Pleistocene of France. These specific distinctions, based primarily upon the relative sizes of teeth and small differences in the shape of the palate, must be considered doubtful, but they point to an early origin of *Oryctolagus* in south-western Europe. There appear to be no fossil records of either *Oryctolagus* or *Sylvilagus* from anywhere in Asia to bridge the enormous geographical gap and to help explain the relationship between these two genera.

North-western Africa is likely to have been isolated from Iberia throughout (and since) the Pleistocene. In spite of records to the contrary (Kurtén 1968) there appears to be no good evidence of the rabbit in northwest Africa during the Pleistocene, and the earliest records appear to be of Neolithic or even later date. It is therefore quite possible that the rabbit in northwest Africa could owe its presence there to transport by man from Europe.

Whether Britain was joined to the continent at the time of the Swanscombe rabbits or not, these finds can be considered part of a continental mid-Pleistocene population, and there is no further evidence of rabbits in Britain until they were introduced in Norman times, i.e. in the twelfth century AD. A much-quoted record from a Mesolithic archaeological site at Thatcham, Berkshire, in southern England (King 1962) was undoubtedly *O. cuniculus* but has recently been shown by radiocarbon dating to be less than 500 years old (Gowlett *et al.* 1987), rather than the 7500 BC as suggested by the context. Few if any post-glacial finds from anywhere in Europe can be confidently dated by their stratigraphy, and it is only very recently that it has been practicable to carry out radiocarbon dating on specimens as small as a single rabbit bone. However this is now possible and there is a prospect of refining our knowledge of the historical distribution of the species.

An estimate of the time of divergence of *Oryctolagus* and *Lepus* at 2.43 million years was based upon enzyme polymorphism (Harth 1987). Although this seems a plausible date, it is rendered almost meaningless by the corresponding figure of 144 000 years for the divergence of domestic and wild rabbits, which is clearly grossly exaggerated.

1.5 Habitat and zoogeography

Most of the characteristics, of structure, behaviour, and physiology that distinguish a species from its relatives have evolved over an immense period of time to be adaptive to a particular environment. Therefore if we can study a species in the habitat in which it evolved, we are likely to get a better insight into the adaptive significance of its characteristics, and therefore into its whole life-style, than if we study it only in a recent 'man-made' habitat. However the recognition of ancestral habitats in southern Europe is especially difficult in view of the very long history of human interference.

It is likely that, under natural conditions, rabbits and hares occupied separate habitats, with much opportunity for contact at the edges but little actual overlap of home ranges. In Spain today, for example in the Coto Doñana, rabbits tend to live in or close to scrub, while the hares occupy the open grassland (Rau *et al.* 1985). Elsewhere in Europe, hares are found in open forest where the ground cover is low, for example *Lepus europaeus* in Czechoslovakia and Poland and *L. timidus* in Scandinavia, but not where there is a dense shrub layer, for instance of bramble as in many English woods. The importance of scrub for rabbits under 'natural' conditions is somewhat obscured by their ability to live with minimal cover in areas, as in much of Britain, where aerial predators have been grossly reduced through persecution.

One can only speculate as to the factors that originally restricted the rabbit to south-western Europe. Competitive exclusion is likely to have played a part, since lowland Iberia is deficient in other herbivorous mammals compared with central and south-eastern Europe. For example there are no hamsters (*Cricetus/Mesocricetus*), no ground squirrels (*Spermophilus*), no bank voles (*Clethrionomys*) and no mole-rats (*Spalax*). During the last glaciation the presence of rabbits in Iberia and their absence from the other refugia in Italy and the Balkans may well have led to the differentiation of the hares, with the Iberian hares (*Lepus granatensis*) remaining closer to the plains-adapted *L. capensis* of Africa. In Italy and/or the Balkans the absence of rabbits permitted evolution of the more versatile *L. europaeus*, which was then able to spread northwards, westwards, and eastwards as agricultural development broke the monotony of the forests.

1.6 Summary

The European rabbit, *Oryctolagus cuniculus*, has always been defined in relation to the European hares of the genus *Lepus*, from which it is clearly distinct in lacking the hares' adaptations to fast running on open ground. Differences between it and the American rabbits (*Sylvilagus* spp.) and the African rabbit (*Poelagus*) are more subtle, and have never been adequately described.

O. cuniculus is well documented from southwest Europe, from the Middle Pleistocene onwards; it is not known in northwest Africa prior to the Neolithic period, and therefore could have been introduced there by man; in Britain it is unknown between the Middle Pleistocene and its introduction as a domestic animal in the twelfth century AD. The presence of rabbits in Iberia but not elsewhere in southern Europe during the Pleistocene glaciations is likely to have been instrumental in stimulating the speciation of *Lepus* in Europe.

References

Bate, D. M. A. (1906). On the mammals of Crete. In *Proceedings of the Zoological Society of London* (1905), pp. 315–23.

Blyth, E. (1845). Description of *Caprolagus*, a new genus of leporine Mammalia. *Journal of the Asiatic Society of Bengal*, **14**, 247–9.

Brisson, M. J. (1762). *Regnum animale* . . . Lugduni Batavorum.

Cabrera, A. (1922). Una excursión de dos meses por Yebala. *Boletin de la Sociedad Espanola de Historia Natural*, **22**, 101–12.

Cabrera, A. (1923). Sobre dos conejos de Marruecos. *Boletin de la Sociedad Espanola de Historia Natural*, **23**, 356–67.

Corbet, G. B. (1983). A review of classification in the family Leporidae. *Acta Zoologica Fennica*, **174**, 11–15.

Darwin, C. (1868). *The variation of animals and plants under domestication*. Murray, London.

Donard, E. (1981). '*Oryctolagus cuniculus*' dans quelques gisements quaternaires français. *Quaternaria*, **23**, 145–57.

Gowlett, J. A. J., Hedges, R. E. M., Law, I. A., and Perry, C. (1987). Radiocarbon dates from the Oxford AMS system: archaeometry date-list 5. *Archaeometry*, No. 29, 125–55.

Gray, A. P. (1972). *Mammalian hybrids: a checklist with bibliography* (2nd edn). Commonwealth Agricultural Bureaux, Slough.

Gray, J. E. (1867). Notes on the skulls of hares (Leporidae) and picas (Lagomyidae) in the British Museum. *Annals and Magazine of Natural History*, **3** (20), 219–25.

Gureev, A. A. (1964). Zaitzeobrazniye (Lagomorpha). *Fauna SSSR*, **3** (10), 1–276.

Haeckel, E. H. P. A. (1874). *Histoire de la création des êtres organisés d'après les lois naturelles* . . . Paris.

Harrison, D. L. (1952). A new subspecies of the

rabbit (*Oryctolagus cuniculus*) from Borkum Island in North-West Germany. *Annals and Magazine of Natural History*, **12** (5), 676–8.

Harth, G. B. (1987). Biochemical differentiation between the wild rabbit (*Oryctolagus cuniculus*), the domestic rabbit and the brown hare (*Lepus europaeus* Pallas). *Zeitschrift für zoologische Systematik und Evolution-forschung*, **25**, 309–16.

Hochstrasser, G. (1969). Zur Frage der Hauskaninchen-Nomenclature. *Säugetierkundliche Mitteilungen*, **17**, 106–14.

King, J. E. (1962). Report on animal bones. In: Excavations at the Maglemosian sites at Thatcham, Berkshire, England (ed. J. Wymer), pp. 355–61. *Proceedings of the Prehistoric Society*, **28**, 329–61.

Kurtén, B. (1968). *Pleistocene mammals of Europe*. Weidenfeld & Nicolson, London.

Lilljeborg, W. (1874). *Sveriges och Norges ryggradsdjur, 1: daggdjuren*. Schultz, Uppsala.

Linnaeus, C. (1758). *Systema naturae, regnum animale* (10th edn). Uppsala.

Loche, V. (1858). *Catalogue des mammifères et des oiseax observés en Algérie* . . . Paris.

Lopez-Martinez, N. (1977). Nuevos lagomorfos . . . del neogeno y cuaternario Espanol. *Trabajos Neogeno-Cuaternario*, **8**, 7–45.

Lyon, M. W. (1904). Classification of the hares and their allies. *Smithsonian miscellaneous Collections*, **45**, 321–447.

Mayhew, D. F. (1975). *The Quaternary history of some British rodents and lagomorphs*. Unpublished PhD thesis. University of Cambridge.

Meyer, F. A. A. (1790). Ueber ein neues Säugthiergeschlecht. *Magazin für Thiergechichte*, **1** (1, 6), 46–55.

Pearson, J. T. (1839). In: List of Mammalia and birds collected in Assam by John McClelland, Esq. . . . (by T. Horsfield). *Proceedings of the Zoological Society of London*, 146–67.

Pomel, N. A. (1854). *Catalogue méthodique et descriptif des vertébrés fossiles* . . . No. 23. Paris.

Rau, J. R., Beltran, J. F., and Delibes, M. (1985). Habitat segregation between rabbits and hares in Coto Doñana, SW Spain. *Abstracts, International Theriological Congress IV*, No. 0514.

Robinson, T. J., Elder, F. F. B., and Chapman, J. A. (1983). Evolution of chromosomal variation in cottontails, genus *Sylvilagus* . . . *Cytogenetics and Cell Genetics*, **35**, 216–22.

Robinson, T. J. and Osterhoff, D. R. (1983). Protein variation and its systematic implications for the South African Leporidae. *Animal Blood Groups and Biochemical Genetics*, **14**, 139–49.

St Leger, J. (1929). An interesting collection of mammals, with a remarkable new species of hare from Uganda. *Annals and Magazine of Natural History*, **10** (4), 290–4.

St Leger, J. (1932). A new genus for the Uganda hare (*Lepus marjorita*). In *Proceedings of the Zoological Society of London*, 119–23.

Sutcliffe, A. J. (1964). The mammalian fauna. In: The Swanscombe skull. A survey of research on a Pleistocene site (ed. C. D. Ovey), pp. 85–112. *Occasional Papers of the Royal Anthropological Institute of Great Britain and Ireland*, No. 20.

Trouessart, M. E-L. (1917). Le lapin de Porto Santo et le lapin nègre de la Camargue. *Bulletin du Museum d'Histoire Naturelle, Paris*, 366–73.

Zeuner, F. E. and Sutcliffe, A. (1964). Appendix I. Preliminary report on the Mammalia of Gorham's cave, Gibraltar. In: The excavation of Gorham's Cave, Gibraltar, 1951–54 (ed. J. d'A. Waechter), pp. 213–16. *Bulletin of the Institute of Archaeology*, **4**, 189–221, pl. xiii.

2

World distribution

John E. C. Flux

2.1 Early distribution

Like most pests and game animals, rabbits (which are both) now have a distribution strongly influenced by man. The association extends back for thousands of years: bones near Nice show signs of slaughter by man as long as 120 000 years ago (Pages 1980). Many fossils from the upper Pliocene and Pleistocene of southern Europe from Britain to Italy, however, are suspect because the rabbits may have died after burrowing into older strata. Zeuner (1963) accepts only the upper Pleistocene records from Gibraltar and an early post-glacial record from southern France. Corbet (1986) considers the mid-Pleistocene specimen from Kent consistent with similar finds in France and Spain, pointing to a rather more widespread distribution before the final glaciations. Three thousand years ago they seem to have been restricted to Spain, where Phoenicians recorded them as abundant about 1100 BC, and perhaps southern France (Beaucournu 1980). There is no good evidence that they were indigenous in North Africa (Barrett-Hamilton 1912; Beaucournu 1980; Corbet 1986). However, most genera of leporids except *Lepus* and *Sylvilagus* have remarkably small, isolated distributions and the restriction of *Oryctolagus* to Spain is not unduly surprising.

2.2 Spread in Europe

According to Reumer and Sanders (1984) Neolithic settlers took rabbits to Menorca as early as 1400–1300 BC. Phoenicians, and later the Romans, spread rabbits around the Mediterranean, possibly including North Africa. Polybius records them on Corsica in 204 BC, and Strabo (born 63 BC) describes how a pair liberated on the Balearic Islands resulted in a serious pest problem. They reached Italy at least by AD 230, when Athenaeus recorded them as very abundant on Nisida in the Gulf of Naples. Rabbits were a favourite food of the Romans, who kept them in costly enclosures—some were 4 km in circumference and roofed to make them 'impenetrable to cats, badgers, wolves and eagles' (Barrett-Hamilton 1912). Similar rabbit gardens and warrens were built in France, Britain, and Germany in the Middle Ages, especially on small islands where rabbits could be recaptured easily.

Domesticated breeds of rabbits probably originated in French monasteries between AD 500 and 1000 (Zeuner 1963) but coloured varieties are not mentioned until the mid-sixteenth century. Large size would have been favoured far earlier, and Nachtsheim (1949) suggests that unconscious selection in warrens (in contrast to hutches) could have discouraged tameness because tame individuals would have been caught first. Animals reared in captivity were carried over Europe in advance of any natural spread of true wild stock from Spain that could have followed forest clearance, agricultural development, and overgrazing by domestic stock, that together make a habitat suitable for rabbits. However, there is no record of wild Spanish rabbits ever being introduced, or spreading naturally, into northern Europe; genetically the 'wild' rabbit of the north is more closely related to domestic stock than to Spanish rabbits, while those of south-east France are intermediate (van der Loo 1986). This supports Fitter's (1959) suggestion that the only true wild rabbits were in Spain and some Mediterranean and Atlantic islands; all other 'wild' rabbits were escaped domestic stock which had reverted to wild-type colour but retain a larger body-size.

Mistakes by two eminent scientists confused this issue. Darwin's error in comparing the Porto Santo rabbits (mean weight 741 g) with British (*c.* 1500 g) instead of Spanish or Portuguese stock (mean 932 g—Franca 1913) has been widely recognized. Miller (1912) pointed out, 'As might have been anticipated they prove to be exactly similar to the common Mediterranean form.' Unfortunately Miller then compounded the problem in a footnote: 'That the wild rabbit of Germany was considered by Linnaeus as the typical animal is indicated by his statement: "Habitat in Europea australi".' Germany has been retained in modern checklists although many continental authors rightly claim that Linnaeus obviously knew where rabbits originated, and by 'southern Europe' meant Spain, not Germany. If the type locality is Spain, *cuniculus* has priority over *huxleyi*. What the 'wild' rabbit of northern Europe should be called presents the same taxonomic problems as for other feral domestic animals, but ecologically it is sufficient to recognize that its relationship lies in that direction.

Undoubtedly many captive rabbits would have escaped from enclosures or have been freed to save the trouble of looking after them. Most, of course, would have been quickly eliminated by predators. Thus, the first wild colonies in Germany were not recorded until 1423, although domestic rabbits had been kept for almost 300 years before that. This suggests that considerable time is necessary for selection to produce a reversion to wild-type with the anti-predator and other behavioural adaptations needed for free-living existence; but if Fitter (1959) is correct, there seems to have been no tendency to return to the original small body-size even in arid Australia.

The earliest British records are of rabbits on Drake's Island in Plymouth Sound in AD 1135 (Hurrell 1979), in the Scilly Isles in 1176, and on Lundy between 1183 and 1219 (Lever 1985). The first definite mainland record was of a warren at Guildford in 1235, and construction of warrens proceeded north to Scotland in 1264 (Eggeling 1957), the Isle of May by 1329, and the Orkneys by 1497 (Ritchie 1920). These warrens were promoted by landed gentry and monks. Indeed, William of Wykeham in 1387 had to admonish his nuns: 'presume henceforward to bring to church no birds, hounds, rabbits or other frivolous things that promote indiscipline' (Power 1946). In the wild, however, rabbits spread so slowly that they did not reach the north of Scotland until 1793 (Barrett-Hamilton 1912) despite their abundance on many Scottish islands from about 1500 and the thriving common warren at Aberdeen in 1583. Ritchie (1920) considered that the development of root crops for winter feed, the associated increase in sheep numbers, and the destruction of predators combined to make the countryside suitable for rabbits. In the nineteenth century many landowners introduced wild rabbits onto their estates as game, and several Hebridean islands were stocked at that time.

Over much of Europe, also, rabbits spread slowly, assisted by many introductions. In Norway, a few from Shetland (stocked by 1654—Berry and Johnston 1980) were liberated on Fedje Island in 1875 where the population now numbers at least 40, and on nine other islands, including Molen in Oslofjord (where they are now apparently extinct), Edoy in Smola, and Utsira in Rogaland (Myrberget 1984, 1987). Swedish populations originate from a successful liberation in Skane in 1905, and some now reach as far north at Stockholm, occupying Blekings, Halland, Smaland and Gotland (Curry-Lindahl 1975; Andersson *et al.* 1981); there they seem to have reached a climatic limit of low temperatures and snow cover in winter (Fig. 2.1). To the north-east the boundary lies in south-west Lithuania, where 7000 were shot in 1935—8 (Lincke 1943) but their present status is unclear. In Poland until 1850, rabbits were restricted to Silesia. Then they were introduced into many parts of the country, and are now fairly uniformly distributed west of the Vistula, in Mazovia north of Warsaw, and in the Masurian Lake region (Cabon-Raczyńská 1981). Since 1945 there have been sporadic outbreaks of rabbits in west Poland but no permanent expansion in range, which has stayed remarkably static along the −3°C isotherm for the past 40 years (Nowak 1968). A definitive map based on a 10 × 10 km grid is provided by Pucek and Raczyńsky (1983).

Fig. 2.1 Distribution of rabbits in Europe and North Africa. (Based on Van den Brink (1967) with modifications: Sweden (Curry-Lindahl 1975), Poland (Cabon-Raczyńská 1981), Germany (Briedermann 1981), Denmark (Standgaard and Asferg 1980), The Netherlands (Wijngaarden *et al.* 1971), France (Arthur *et al.* 1981), Italy (M. Spagnesi, personal communication 1988), Czechoslovakia (Anděra and Vohralik 1976), Algeria (Kowalski and Rzebik-Kowalska 1991), and Morocco (Aulagnier and Thevenot 1986).)

Rabbits are found over most of Czechoslovakia, up to 450 m a.s.l. in Slovakia and to 500–600 m in the west; in the east where snow cover exceeds 140 days they are absent. Myxomatosis in 1954–6 affected the numbers of rabbits but (as in most countries except Italy—Toschi 1965) had little effect on their distribution (Andera and Vohralik 1976). There are now three somewhat isolated pockets of rabbits in Hungary, where introductions of tame rabbits date from 1749, and the first wild stock was liberated around

Moson in 1779; these apparently spread and were re-released in many places, together with fresh introductions from France and Austria, so that by 1880–1900 hunters could shoot them in most parts of the country. However, by 1957 none remained east of the Tioza (Szunyoghy 1958). In eastern Romania rabbits were introduced in 1905–7, probably from France, about 10 km east of Iasi whence they spread only to the banks of the Prut (Mîndru 1960). In western Romania extensive liberations in Transylvania in the 1880s flourished, but later declined perhaps because of unfavourable winters (Szunyoghy 1958), and according to Vasiliu (1964) only a few remain near Baia Mare.

To the east, in the Ukranian SSR, seven pairs of rabbits were introduced in 1894 or 1895, and some pairs from Austria about 1900. These have established and are widely distributed between the Dniester and the Dnieper, reaching Aleksandrovsk in the east and the Kodyma in the north (Ognev 1966). Their numbers appear to be static despite attempts to assist by controlling predators, constructing burrows, and artificially propagating the rabbits; efforts to acclimatize them more widely are considered worthwhile (Boldenkov 1981). The only other populations in Russia are of domestic stock on eight islands in the Caspian (Aliev 1963; Boback 1970). Introductions of rabbits into Bulgaria before 1944 failed (Dragoev 1978). In Yugoslavia they were restricted to northern Croatia and Dalmatia (Dulić and Tortić 1960), but are now only on Adriatic islands.

The Italian distribution is patchy and was apparently reduced by myxomatosis (Toschi 1965) but curiously rabbits are entirely absent from much of southern Italy and Yugoslavia, and all of Greece. They are present in Sicily and Malta, and on many islands in the Adriatic and Aegean, and are still being widely liberated by sportsmen (M. Spagnesi, personal communication 1982) in Italy. The habitat appears very like that of the 'ancestral home' in Spain, so what limits rabbits in south-eastern Europe? Most of the rabbits for introduction come from Hungary, and this may again support Fitter's theory that the rabbit of northern Europe is a feral domestic animal whose ancestral home is a cage, not Spain.

Over Portugal, Spain, France, Germany, the Netherlands, Britain, and Ireland rabbits are widely distributed in favourable habitats. Thus, all regions of France have rabbits (over 13 million were shot in 1974–5) but they are less abundant where the winter is severe in the north-east, and at high altitudes; elsewhere numbers are highest where the topography is varied and suitable cover of shrubs or hedges is available (Arthur *et al.* 1980). An altitude limit of about 500 m restricts the distribution in Switzerland to small areas near Basel and Valais where they seem to have been introduced about 1860–80 without spreading (Baumann 1949). In Denmark rabbits occupy only the extreme south of Jutland and a dozen small islands, reflecting immigration from Germany and illegal recent introductions respectively (Strandgaard and Asferg 1980). Gamebag totals have remained remarkably constant at about 10 000 a year since 1952, despite eradication attempts and myxomatosis; why they have never spread over Jutland (despite many early liberations—Barrett-Hamilton 1912) is unclear in view of their success farther north in Sweden (Fig. 2.1).

2.3 Spread on other continents

Domestic rabbits, probably silver-greys, arrived in Australia with the first fleet in 1788 but they did not spread beyond the settlements where they had been liberated (Rolls 1969). The classic pattern of successful introduction of rabbits into a new region, initially 'explosive', followed by retraction to the more suitable habitats, began in 1859 at Geelong, Victoria, with the arrival of 20 wild English rabbits. Despite their exponential increase in numbers, the geographic spread of rabbits is linear with time as in other animals (Skellam 1951; Elton 1958). Rabbits crossed Australia at an average rate of 54 km/year (Myers 1971; see also Chapter 5), an exceptional speed matching, for example, that of the European starling in America. Although Myers suggests that trappers liberated rabbits ahead of the main wave, this seems unlikely to have affected the rate much: rabbits had not been easy to establish in the first place, and their rate of spread across the Nullarbor, where no 'help' was available, was 100 km/year (Ratcliffe 1959). The most dramatic figures are given by Strong (1983) for the Northern Territory. There they traversed 450–550 km in 17–20 months at a speed 'within the range of 270 to 390 kilometres per year', settling as far north as 18°S, although today only the south of the Northern Territory holds many rabbits. According to Jarman (1986) rabbits and foxes spread at similar rates from South Australia to the west coast, 'foxes being 17 years behind at Eucla and 14 at Geraldton. Rabbits penetrated the forested south-west of Western Australia slowly, and foxes arrived only 4 years after rabbits at Manjimup.'

The present distribution of rabbits in Australia is curiously uncertain for such a well-studied and economically important animal. Much of the interior of the continent is uninhabited and rarely visited, and the distribution of wild rabbits has varied with time. There are also local variations associated with the erratic rainfall in desert regions. After the initial wave, rabbits reached at least 18°S by about 1900, but probably had already receded by the time of Stead's (1935) map of 'The dark stain, that is the rabbit, spreading over the face of Australia' (Fig. 2.2a). According to Myers and Parker (1965) 'No major shift in regional distribution has occurred since approximately 1910 when the subalpine plains and the coastal valleys were finally reached', and their map (Fig. 2.2b) was reprinted in Myers (1971) and by Cooke (1977). Recently Myers (1983) noted that the rabbit 'continues to probe Australian habitats and is presently making a slow advance northwards along the cool highlands into tropical Queensland', and presented a totally different distribution (Fig. 2.2c). Distribution maps are available for some individual states, for example Letts (1979) for Northern Territory and King (1990) for Western Australia, but they are based on different criteria ('infested land' and presence or absence by postal questionnaire sent to local residents to obtain a general impression) and are hard to compare. Myers *et al.* (Chapter 5) gives the best current summary (Fig. 5.2).

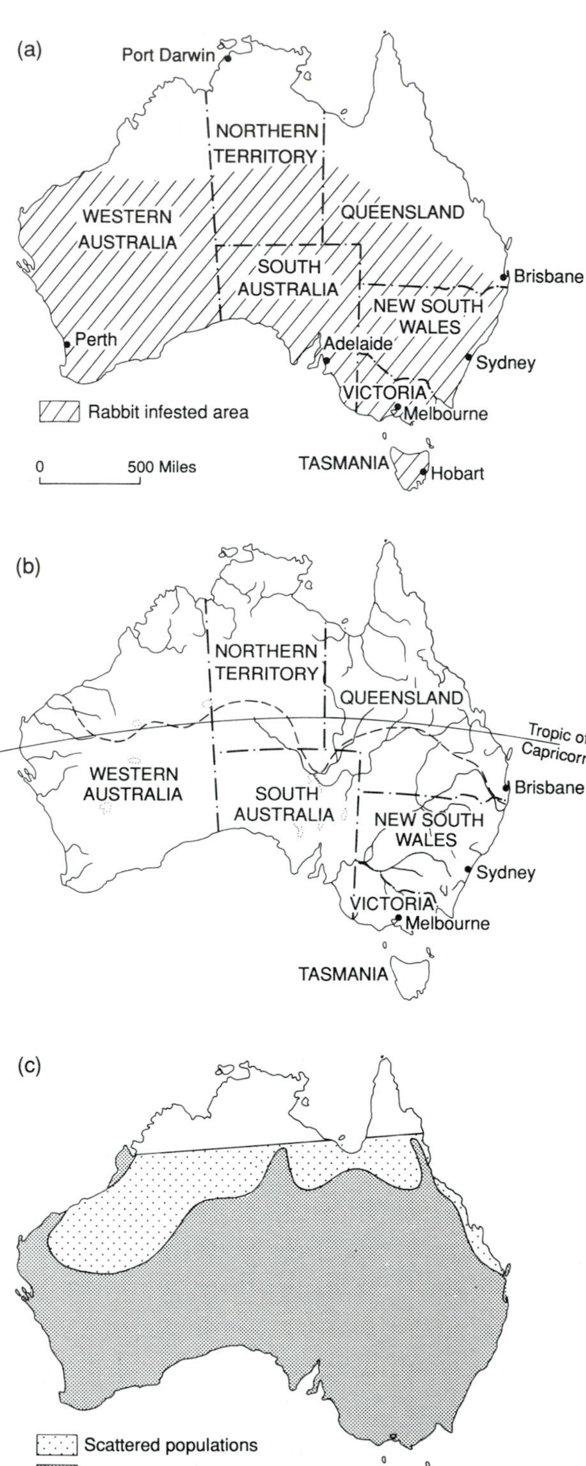

Fig. 2.2 Distribution of rabbits in Australia. (a) Stead (1935); (b) Myers and Parker (1965). Line across continent represents northern limit; (c) Myers (1983).

Rabbits were introduced into South America as early as the mid-eighteenth century (Housse 1953, quoted by Greer 1966) but unsuccessfully—Philippi (1885) remarked that rabbits had not become wild in Chile despite the attempts of French hunters to liberate them, and Lataste (1892) makes it clear in his description of domestic rabbits on an islet in Cauquenes Lake that there were no wild ones in Chile. Missionaries brought domestic rabbits to Isla Grande in 1874, and there was another unsuccessful introduction in 1913 before the first success about 1936 (Arentsen 1954), although they had been present on small islands in the Beagle Channel since about 1880. Numbers on Isla Grande increased rapidly, and one merchant bought 2000 in 1942, 15 000 in 1944, 70 000 and 1948, and 750 000 in 1951 (G. Ferrier, personal communication 1967); Jaksić and Yáñez (1983) state that four rabbits increased to about 30 million in 17 years. Myxomatosis brought in from Germany in 1953 reduced the population by 97 per cent in 3 years, but had become less effective by 1961–4. Today, rabbits are again abundant over central and southern Isla Grande (Bonino and Amaya 1984).

In central Chile, rabbits were not recorded in a detailed survey by Osgood in 1943, but by 1960 were present throughout Malleco Province, to at least 1500 m above sea level (Greer 1966). By 1967 they reached from 32°S to 38°S (G. Ferrier, personal communication 1967), and there is an apparently isolated pocket in the Camarones Valley, Tarapaca, at 19°S (Pine *et al.* 1979). The most alarming extension, however, has been across the Andes into Argentina at Neuquén between 1945 and 1950. Their range has expanded at 15–20 km/year to occupy 45 000 km² by 1984 (Bonino and Amaya 1984) despite the introduction of myxomatosis in 1971 and 1972. Most of Argentina is now available for occupation (Fig. 2.3).

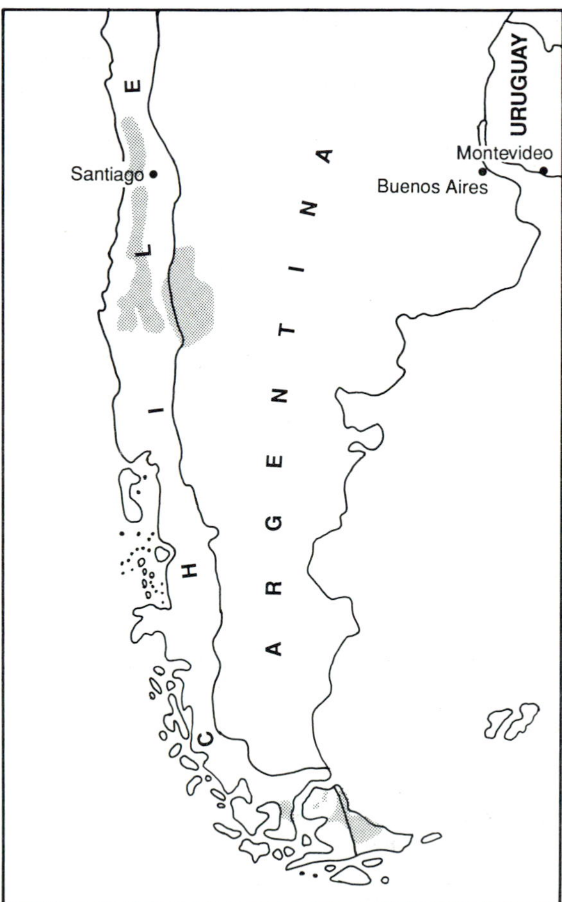

Fig. 2.3 Distribution of European rabbits in South America. (Based on G. Ferrier (personal communication 1967), Pine *et al.* (1979), Bonino and Amaya (1984), and R. H. Taylor (personal communication 1992).)

In North America, rabbits have not become established widely, although they are so abundant on many of the San Juan islands that 50 000 could be harvested annually from 50 square miles (Latham 1955). Fortunately, perhaps because these are domestic stock, liberations of 6–7000 rabbits in Pennsylvania and of many others in Indiana and Nebraska, all failed (Kirkpatrick 1960; Jones 1964). The only mainland populations seem to be in San Clemente Canyon Natural Park, San Diego (W. F. Perrin, personal communication 1982), where there is a small feral stock of wild-type, white, and piebald rabbits, and near Marblemount and Seattle, Washington (W. F. Stevens, personal communication 1980). A population in Idaho is restricted to islands in the Snake River (W. F. Stevens, personal communication 1980).

The final continent, Africa, has an old if not indigenous population of rabbits in Morocco (Heim de Balsac 1936) which from measurements given by Petter and Saint Girons (1972) are *O. huxleyi*. In Algeria, rabbits are numerous in the west but far less common in the east; they are hunted and have been introduced on the Iles Habibas (Kowalski and

Rzebik-Kowalska 1991). The former record for Uganda (Carpenter 1925) is almost certainly a misidentification of *Poelagus marjorita*, which was not described until 1929. *Poelagus* resembles *Oryctolagus* in external appearance very closely, and I have seen it at Masindi where Carpenter reported *Oryctolagus*. Rabbits are still present on seven southern African islands; the original successful introduction on Robben Island (1656–8) was the earliest, but liberations on the mainland were specifically prohibited by 'the perspicacity of the first governor of the Cape, Jan van Riebeeck' (Cooper and Brooke 1982). Later illegal introductions in Natal about 1900 died out because ants attacked the young, according to contemporary observers (Waithman 1981).

In summary, the spread on different continents demonstrates the rabbit's wide-ranging ability to colonize. They are adaptable and can feed on most green vegetation—and even on seaweed, driftwood, and lichens if hard pressed (Cranham 1972). Because rabbits graze closely, a fine turf develops which they can exploit better than their competitors. Elaborate burrows give protection from climatic extremes of heat, cold, and rain, as well as from many predators. Rabbits have a high reproductive rate, and like most colonizers 'are *r*-selected "weeds", potentially capable of specialization' (Berry 1979). Yet compared with other camp-followers of man such as sparrows, starlings, rats, cats, and mice, they have been far less successful in penetrating continents. In crossing Europe, they spread at the slowest rate of 28 colonizing species reviewed by Nowak (1971), and their reputation relies mainly on their incredible performance in Australia.

2.4 The islands

In New Zealand rabbits spread at up to 16 km/year (Wodzicki 1950), a rate much slower than in Australia but similar to that of many other species introduced elsewhere—from sawflies in America to muskrats in Austria (Elton 1958; Nowak 1971; van den Bosch *et al.* 1992). Thus, rates of spread seem to reflect environmental suitability for dispersal rather than reproductive or dispersal ability. House sparrows in America 'expanded at 140 km/year in densely settled areas but at 8 km/year in the deserts of Utah. There were similar differences [7–104 km/year] in Australia' (Blakers *et al.* 1984). Good conditions for adult and juvenile survival are obviously implicated; however, environmental 'suitability' estimated from rates of spread is not necessarily reflected in final population density—areas such as the Nullarbor and Northern Territories, with the fastest recorded rates, now harbour few or variable numbers of rabbits compared with other parts of Australia (or Britain, where rabbits spread at only 1–2 km/year). The explosive phase in New Zealand established rabbit populations on 'all the mountain tops between Manapouri and Te Anau, and through the bush' (Henry 1884). Within 30 years of their introduction about 1865, the rabbits had begun to decline in numbers (Thomson 1922) and pockets of rabbits, isolated on mountain tops or in alpine valleys, are still disappearing (Flux in press) although the basic distribution has been stable now for more than 50 years (Fig. 6.1).

The success of rabbits on small islands is reflected even in geographical names: there are 12 'Rabbit Islands' in New Zealand and 13 in the Falkland group! They now occupy at least 800 islands in all the major oceans, where they were liberated for sport or as food for castaways and arctic foxes (Flux and Fullagar 1983; 1992). Ecologically, their success could be expected: 'On a small island the race for life will have been less severe, and there will have been less modification and less extermination' (Darwin 1878); and Barrett-Hamilton (1912) states, 'It may, however, be successfully introduced on islands, in regions where it could not be established on the adjacent mainland, and this is characteristic of it everywhere throughout its range'. The problem of whether this success (for rabbits) is a function of some physiological attribute or pre-adaptation, a consequence of the lack of ground predators on small islands, or merely a result of the enormous numbers of introductions that have been made over the past 2000 years, is almost impossible to solve: to some extent the factors are interrelated, because only a species with a reputation for success would be persevered with. There is also the problem of the definition of 'success'. Some species such as commensal rodents are found on far more islands than

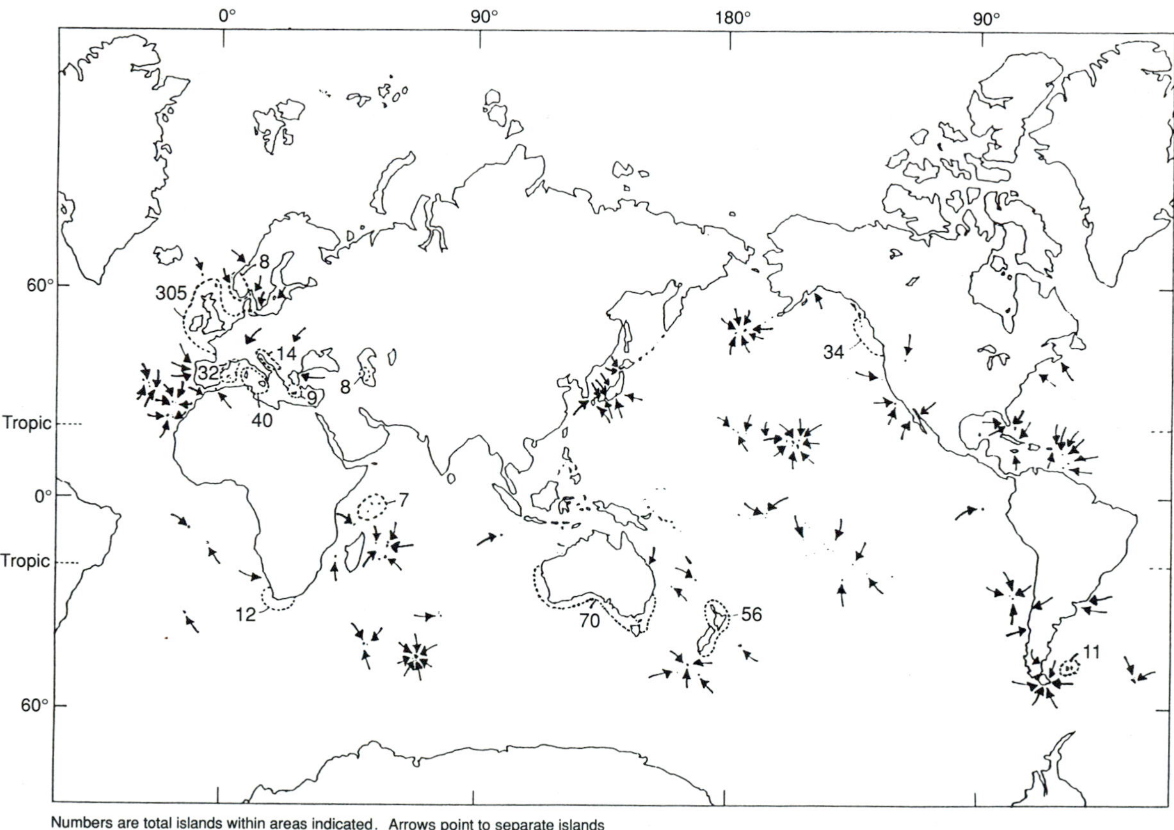

Fig. 2.4 Distribution of rabbits on islands. (Based on Flux and Fullagar (1983, 1992).

rabbits are, and reach them without deliberate human help, but they may not expand into the natural habitat on arrival as dramatically as the rabbit.

Islands on which rabbits survive (Fig. 2.4) range from 59–62°N (Middleton Island, Alaska; Fedje, Norway; the Shetlands, and the Faroes) to 54–5°S (Macquarie Island; islands in the Beagle Channel, South America). The limiting factor seems to be the depth of snow covering their food supply, as in northern and eastern Europe and at high altitudes elsewhere (Australia, New Zealand, Chile). At the other extreme, rabbits can tolerate temperatures of up to 50°C provided they have access to water or green vegetation. They are present in small numbers on Ascension Island (8°S) in the Atlantic, and Chauve-Souris (4°S) in the Indian Ocean. On Phoenix Island (40 ha, 3°S in the Pacific Ocean) they have survived since 1860, but in the absence of fresh water have not reached sufficient numbers to damage the vegetation (Watson 1961). Similarly, on Manana Island (25 ha, 22°N in Hawaii) rabbits introduced just before 1900 survived without markedly affecting the vegetation until a severe drought eliminated them in 1984 (Swenson 1986). On Lisianski Island (240 ha, 26°N in Hawaii), however, rabbits introduced between 1904 and 1909 were abundant by 1913. By 1915 the only vegetation left consisted of a few small patches of tobacco, and most of the rabbits had died. In March 1915 only seven live rabbits were found (and removed) and there was no living vegetation to be seen a year later, so it is practically certain that all the rabbits on Lisianski would have died of starvation (Clapp and Wirtz 1975).

Once the damage caused by rabbits is realized, attempts are often made to rectify the mistake by introducing cats or other predators. Elton (1927) records how a lighthousekeeper on Berlenga Island

(120 ha) brought in cats to control a rabbit plague, which they did and then starved to death. The outcome was similar on South Havra (60 ha, Shetland Islands), Mangere (113 ha, Chatham Group), and on St Helen's (51 ha, Tasmania), although the cats on these islands were said to have died of disease. On Sable Island (2400 ha, Canada) the full treatment was needed: cats eliminated the rabbits, foxes then exterminated the cats, and man killed out the foxes (McLaren 1972).

The outcome on other islands, some equally small, has been different. Rabbits on the Holm of Melby (16 ha, Shetland Islands), 'defied an importation of cats' (Fitter 1959), as did those on Helgoland (60 ha, Germany), Dassen (240 ha, South Africa), Noss (313 ha, Shetland Islands), and many larger islands such as Kerguelen and Macquarie (120 km^2). Other predators that have been tried include weasel, stoat, ferret, mongoose, fox, and genet. The effect of predation varies with the population density of the rabbits, the cover available, the abundance of alternative and seasonal food for the predator, and the presence of diseases, especially myxomatosis; not surprisingly, the result can seldom be predicted.

The first outbreak of myxomatosis in a susceptible rabbit population can result in 99 per cent mortality (Thompson and Worden 1956), so populations on small islands are likely to become extinct. This has been reported for the islands of Eynhallow (110 ha), Churchill (60 ha), St Serfs (50 ha), Citadel (28 ha), and Carnac (17 ha); Mew, Ross, and Roaninish (all 10 ha) and The Skerries (2 ha). On the other hand, myxomatosis failed to remove rabbits from Ceann Iar (70 ha), Montague (49 ha), Cabbage Tree (30 ha), Lighthouse (16 ha), and Sully (3 ha). This last island is especially interesting because the total rabbit population remains at 50 after several bouts of myxomatosis (Ferns 1981)—perhaps a less virulent strain was involved. Rabbits on the Isle of May (57 ha) withstood at least three outbreaks of myxomatosis, in 1955, 1964, and 1969 (Gordon 1970), and are still present. One acre (0.4 ha) of fertile land can support a resident population of 100 rabbits (Simpson 1895), and many offshore islands are extremely fertile because of bird guano deposits; hence the rabbits on a 1 ha island could in theory withstand myxomatosis. Their successful removal by myxomatosis on larger islands will have been the result of lower initial densities, other contributory (additive) factors, or chance. In addition, of course, the area of suitable habitat (rather than that of the island) is critical, and erosion caused by rabbits can often reduce this (Norman 1970).

According to Armstrong (1982) the rabbit 'has proved to be an extremely successful colonizer of small islands'. This claim is hard to support quantitatively. Cheylan (1984) listed the mammal species present on 72 islands in the Mediterranean, and found that *Rattus rattus* was the best colonizer; it is found on most islands larger than 1 ha and on many islands down to 0.3 ha. Rabbits were next best, occupying islands down to 6.5 ha, followed by *Mus musculus* on islands down to 36 ha, and *Apodemus sylvaticus* down to 640 ha. The rabbits, however, had invariably been carefully liberated by man, while all of the rodents would have colonized the islands on their own or as stowaways. A second comparison is between rabbits and hares (*Lepus europaeus*): hares are found on very few offshore islands (none in New Zealand, for example, compared with over 40 islands with rabbits) and would be classed as very poor colonizers. Yet in The Netherlands, where hares are highly regarded as game, they have been introduced to as many islands as have rabbits, with almost exactly the same results (Table 2.1).

The similarity is surprising: of the 49 islands

Table 2.1 Distribution of rabbits and hares on Wadden Sea islands of different sizes, from Laar (1981). (Number of islands on which they have died out, in parentheses.)

Area (ha)	2–49	50–99	100–499	500–999	1000–4999	5000+	Total
No. of islands	5	8	8	9	2	7	49
With hares (ext.)	0	2 (0)	5 (4)	5 (2)	11 (0)	7 (0)	30 (6)
With rabbits (ext.)	2 (2)	1 (1)	4 (0)	3 (1)	11 (3)	7 (1)	28 (8)
With both	0	0	1	1	10	6	18

examined, hares have been established on 30 and have become extinct on 6; rabbits have been established on 28 and have become extinct on 8. There is no obvious difference in distribution or success of the two species with island size, except that they very seldom co-exist on islands smaller than 1000 ha. Hares (unexpectedly, considering their larger size, low population density, and far larger home ranges) are found on the two smallest islands on which lagomorphs still live.

2.5 Conclusion

Overall, the evidence from islands suggests that rabbits are not so much successful colonizers as environmentally tolerant animals which survive well where they are placed. Some indication of their importance relative to other species of colonizers on oceanic islands can be calculated from a review by Fosberg (1983) of the causes of modification in 71 major island groups: 266 were attributable to man (mainly by clearance for agriculture), 18 to goats, 12 to sheep, 7 to cattle, 7 to unspecified introduced mammals, 5 to rats, 5 to insects, and 4 each to pigs, cats, and rabbits. Dogs and parrots scored 2 each, and eight other animals were each mentioned once. Undoubtedly, the main colonizers are humans, and the rabbit is merely one of their lesser associates in the subjugation of the environment.

2.6 Summary

From a relict distribution in Spain 3000 years ago, rabbits have been spread by sailors, sportsmen, and settlers to all continents except Antarctica. The original small wild form, *Oryctolagus cuniculus huxleyi*, was domesticated by the Romans and later by monks; most 'wild' rabbit populations seem to be descended from a large domesticated form which became feral. In Europe they spread slowly, taking several hundred years to reach Scotland, Germany, and eastern Europe, despite the encouragement of sporting landowners. South Africa prohibited the import of rabbits; North America liberated many thousands in unsuccessful attempts to acclimatize them; and South America has plague populations, as yet restricted to relatively small areas. Rabbits crossed Australia at 100–300 km per year, in one of the most dramatic animal invasions ever recorded, to occupy three-quarters of the continent. Liberations of wild and domestic rabbits on more than 800 islands, from Alaska to the equator, provide a fascinating range of experiments in environmental tolerance. The distribution of rabbits remains fluid as they are eliminated from some regions as pests, while hunters continue to liberate them for sport. In comparison with other species, the rabbit is not so much a successful colonizer as a convenient animal to transport.

Acknowledgements

I am very grateful to J. A. Gibb, J. R. Hay, and A. D. Pritchard for helpful comments on the manuscript, and to C. M. King for editorial advice. Drs G. Ferriere, M. Homolka, B. Krystufek, and M. Spagnesi kindly provided maps of the distribution of rabbits in S. America, Czechoslovakia, Yugoslavia, and Italy, respectively.

References

Aliev, F. F. (1963). History and state of acclimatization and reacclimatization of mammals in the Caucasus. In *Acclimatization of Animals in the USSR* (ed. A. I. Janusevic), pp. 31–3. Izvestiya Akademii Nauk Kazakhskoi SSSR.

Andera, M. and Vohralik, V. (1976). Distribution of wild rabbit, *Oryctolagus cuniculus* (Linnaeus, 1758) in Czechoslovakia. *Lynx* (Prague), **18**, 5–18.

Andersson, M., Meurling, P., Dahlbäck, M., Jansson, G., and Borg, B. (1981). Reproductive biology of the wild rabbit in southern Sweden, an area close to the northern limit of its distribution. In *Proceedings of the World Lagomorph Conference* (ed. K. Myers and C. D. MacInnes), pp. 175–81. University of Guelph.

Arentsen, P. S. (1954). Biological control of the rabbit: the spread of the virus *Myxoma cuniculus*, by direct contagion, in the Isla Grande, Tierra del Fuego. *Boletin Ganadero*, **43**, 3–25.

Armstrong, P. (1982). Rabbits (*Oryctolagus cuniculus*) on islands: a case-study of successful colonization. *Journal of Biogeography*, **9**, 353–62.

Arthur, C. P., Chapuis, J. I., Pages, M. V., and Spitz, F. (1980). Enquêtes sur la situation et la répartition écologique du lapin de garenne en France. *Bulletin Mensuel, Office National de la Chasse, Numéro spécial scientifique et technique*, pp. 37–90.

Aulagnier, S. and Thevenot, M. (1986). Catalogue des mammiferes sauvages du Marroc. *Institut Scientifique Charia Ibn Batouta BP. 703 Rabat-Agdal*, p. 59.

Barrett-Hamilton, G. E. H. (1912). *A history of British mammals*, Vol. 2. Gurney & Jackson, London.

Baumann, F. (1949). *Die freilebenden Säugetiere der Schweiz*. Hans Huber, Berne.

Beaucournu, J. C. (1980). Les ectoparasites du lapin de garenne, *Oryctolagus cuniculus*: apports à son histoire. *Bulletin Mensuel, Office National de la Chasse, Numéro spécial scientifique et technique*, pp. 23–35.

Berry, R. J. (1979). The Outer Hebrides: where genes and geography meet. *Proceedings of the Royal Society of Edinburgh*, **77B**, 21–43.

Berry, R. J. and Johnston, J. L. (1980). *The natural history of Shetland*. Collins, London.

Blakers, M., Davies, S. J. J. F., and Reilly, P. N. (1984). *The atlas of Australian birds*. Melbourne University Press.

Boback, A. W. (1970). *Das Wildkaninchen*. Die Neue Brehm-Bücherei Ziemsen, Wittenberg Lutherstadt.

Boldenkov, S. (1981). Forage for hunting animals and game breeding in the Ukranian SSR. *Transactions of the International Congress of Game Biologists*, **12**, 71–3.

Bonino, N. A. and Amaya, J. N. (1984). Distribucion geografica, perjuicios y control del conejo silvestre europeo *Oryctolagus cuniculus* (L.) en la Republica Argentina. *IDIA, INTA*, No. 429–432, 25–50.

Briedermann, L. (1981). Die Jagd in der Deutschen Demokratischen Republik. Jagdinformationen 1–2. *Institut für Forstwissenschaften Eberswalde*.

Cabon-Raczyńská, K. (1981). Lagomorpha. In *Keys to vertebrates of Poland* (ed. Z. Pucek), ppl. 155–64. Polish Scientific Publications, Warsaw.

Carpenter, G. D. H. (1925). Rabbits in Africa. *Nature*, **116**, 677.

Cheylan, G. (1984). Les mammiferes des îles de Provence et de Méditerranée occidentals: un example du peoplement insulaire non équilibré? *Review d'Écologie*, **39**, 37–54.

Clapp, R. B. and Wirtz, W. O. (1975). European rabbit (*Oryctolagus cuniculus*). *Atoll Research Bulletin*, **186**, 150–1.

Cooke, B. D. (1977). Factors limiting the distribution of the wild rabbit in Australia. *Proceedings of the Ecological Society of Australia*, **10**, 113–20.

Cooper, J. and Brooke, R. K. (1982). Past and present distribution of the feral European rabbit *Oryctolagus cuniculus* on southern African offshore islands. *South African Journal of Wildlife Research*, **12**, 71–5.

Corbet, G. B. (1986). Relationships and origins of the European lagomorphs. *Mammal Review*, **16**, 105–10.

Corbet, G. B. and Harris, S. (ed.) (1991). *The handbook of British mammals* (3rd edn). Blackwell Scientific Publications, Oxford.

Cranham, J. (1972). Farne Island rabbits. *Animals*, **14**, 244–5.

Curry-Lindahl, K. (1975). *Djuren i Färg*. Almquist and Wiksell, Stockholm.

Darwin, C. (1878). *The origin of species by means of natural selection* (6th edn). Murray, London.

Dragoev, P. (1978). Introduction of game animals in Bulgaria. *II Congressus Theriologicus Internationalis*, 32.

Dulić, B. and Tortić, M. (1960). Verzeichnis der Säugetiere Jugoslawiens. *Säugetierkundliche Mitteilungen*, **8**, 1–12.

Eggeling, W. J. (1957). The mammals of the Isle of May. *Scottish Naturalist*, **69**, 71–4.

Elton, C. (1927). *Animal ecology*. Sidgwick & Jackson, London.

Elton, C. (1958). *The ecology of invasions by animals and plants*. Methuen, London.

Ferns, P. N. (1981). The mammals of Sully Island. *Transactions of the Cardiff Naturalists' Society*, **99**, 4–14.

Fitter, R. S. R. (1959). *The ark in our midst.* Collins, London.

Flux, J. E. C. (in press). Natural decline to extinction of a New Zealand rabbit population. *Zeitschrift für Säugetierkunde*.

Flux, J. E. C. and Fullagar, P. J. (1992). World distribution of the rabbit *Oryctolagus cuniculus* on islands. *Mammal Review*, **22**, 151–205.

Flux, J. E. C. and Fullagar, P., Jr (1992). World distribution of the rabbit *Oryctolagus cuniculus* on islands. *Mammal Review*, **22**, 151–205.

Fosberg, F. R. (1983). The human factor in the biogeography of oceanic islands. *Compte Rendu des Séances de la Société de Biogéographie*, **59**, 147–90.

Franca, C. (1913). Contribution a l'étude du lapin de Porto Santo (*'Oryctolagus cuniculus Huxleyi'* Haeckel). *Bulletin de la Société Portugaise des Sciences Naturelles*, **6**, 1–14.

Gordon, N. J. (1970). Isle of May Bird Observatory and Field Station report for 1969. *Scottish Birds*, **6**, 129–36.

Greer, J. K. (1966). Mammals of Malleco province Chile. *Publications of the Museum, Michigan State University Biological Series*, *3*, 49–152.

Heim de Balsac, H. (1936). Biogéographie des mammifères et des oiseaux de l'Afrique du Nord. *Bulletin biologique de la France et de la Belgique*, Supplement 21, 1–446.

Henry, R. (1884). Notes from Lake Te Anau. *New Zealand Journal of Science*, **2**, 82–4.

Hurrell, H. G. (1979). The little-known rabbit. *Country-side*, **23** (n.s.), 501–3.

Jaksić, F. M. and Yáñez, J. L. (1983). Rabbit and fox introductions in Tierra del Fuego: history and assessment of the attempts at biological control of rabbit infestation. *Biological Conservation*, **26**, 367–74.

Jarman, P. (1986). The red fox—an exotic, large predator. In *The ecology of exotic animals and plants* (ed. R. L. Kitching), pp. 44–61. Wiley, Brisbane.

Jones, J. K. (1964). Distribution and taxonomy of mammals of Nebraska. *University of Kansas Publications Museum of Natural History*, **16**, 1–356.

King, D. (1990). The distribution of European rabbits (*Oryctolagus cuniculus*), in Western Australia. *Western Australian Naturalist*, **18**, 71–4.

Kirkpatrick, R. D. (1960). The introduction of the San Juan rabbit (*Oryctolagus cuniculus*) in Indiana. *Proceedings of the Indiana Academy of Science*, **69**, 320–4.

Kowalski, K. and Rzebik-Kowalska, B. (1991). *Mammals of Algeria*. Polish Academy of Sciences, Warsaw.

Laar, V. van (1981). The Wadden Sea as a zoogeographical barrier to the dispersal of terrestrial mammals. In *Final Report of the Section 'Terrestrial Fauna' of the Wadden Sea Working Group* (ed. W. J. Wolff), pp. 231–66. Stichting Veth tot Steun aan Waddenonderzoek, Leiden.

Lataste, F. (1892). A propos des lapins domestiques vivant en liberté dans l'îlot de l'étang de Cauquenes (Colchagua). *Actes de la Société scientifique de Chili II. Notes et memoires*, 210–22.

Latham, R. M. (1955). The controversial San Juan rabbit. *Transactions of the Twentieth North American Wildlife Conference* (ed. J. B. Trefathen), pp. 406–11. Wildlife Management Institute, Washington.

Letts, G. A. (1979). *Feral animals in the Northern Territory*. Department of Primary Production, Government of Northern Territory, Darwin.

Lever, C. (1985). *Naturalized mammals of the world*. Longman, London.

Lincke, M. (1943). *Das Wildkaninchen*. Neumann, Neudamm.

McLaren, I. A. (1972). Sable Island: our heritage and responsibility. *Canadian Geographical Journal*, **85**, 108–4.

Miller, G. S. (1912). *Catalogue of the mammals of western Europe*. British Museum, London.

Mîndru, C. (1960). Despre răspîndirea iepurelui de

vizuină (*Oryctolagus cuniculus* L.) în Moldova. *Analele Ştiinţifice ale Universităţii "Al. I. Cuza" din Iaşi (Serie Nouă) Sectiunea 11 (Ştiinţe naturale)*, **6**, 771–4.

Myers, K. (1971). The rabbit in Australia. In *Dynamics of populations* (ed. P. J. Den Boer and G. R. Gradwell), pp. 478–506. Wageningen, The Netherlands.

Myers, K. (1983). Rabbit. In *The Australian Museum complete book of Australian mammals* (ed. R. Strahan), p. 481. Angus & Robertson, London.

Myers, K. and Parker, B. S. (1965). A study of the biology of the wild rabbit in climatically different regions in eastern Australia. *CSIRO Wildlife Research*, **10**, 1–32.

Myrberget, S. (1984). The wild rabbit in Norway. *Fauna*, **37**, 84.

Myrberget, S. (1987). Feral and semi-wild rabbits, goats and sheep in Norway. *Fauna*, **40**, 160–2.

Nachtsheim, H. (1949). *Vom Wildtier zum Haustier*. Paul Parey, Berlin.

Norman, F. I. (1970). Ecological effects of rabbit reduction on Rabbit Island, Wilsons Promontory, Victoria. *Proceedings of the Royal Society of Victoria*, **83**, 235–52.

Nowak, E. (1968). Distribution, variations in numbers and importance of the wild rabbit (*Oryctolagus cuniculus* Linnaeus, 1758) in Poland. *Acta Theriologica*, **13**, 75–98.

Nowak, E. (1971). The range of expansion of animals and its causes. *Zeszyty Naukowe Nr 3*, Warsaw.

Ognev, S. I. (1966). *Mammals of the USSR and adjacent countries*, Vol. 4. Israel Program for Scientific Translations, Jerusalem.

Pages, M.-V. (1980). Essai de reconsitution de l'histoire du lapin de garenne en Europe. *Bulletin Mensuel, Office National de la Chasse, Numéro spécial scientifique et technique*, pp. 13–21.

Petter, F. and Saint Girons, M. C. (1972). Les lagomorphes du Maroc. *Bulletin de la Société des Sciences Naturelles et Physiques du Maroc*, **52**, 121–9.

Philippi, R. A. (1885). Sobre los animales introducidos en Chile desde su conquista por los Españoles. *Annales de la Universidad de Chile*, 1885, 323.

Pine, R. H., Miller, S. D., and Schamberger, M. L. (1979). Contributions to the mammalogy of Chile. *Mammalia*, **43**, 339–76.

Power, E. (1946). *Medieval people*. Methuen, London.

Pucek, Z. and Raczyński, J. (ed.) (1983). *Atlas of Polish mammals*. Polish Scientific Publishers, Warsaw.

Ratcliffe, F. N. (1959). The rabbit in Australia. In *Biogeography and ecology in Australia* (ed. A. Keast, R. L. Crocker, and C. S. Christian), pp. 545–64. Junk, Den Haag.

Reumer, J. W. F. and Sanders, E. A. C. (1984). Changes in the vertebrate fauna of Menorca in prehistoric and classical times. *Zeitschrift für Säugetierkunde*, **49**, 321–5.

Ritchie, J. (1920). *The influence of man on animal life in Scotland*. Cambridge University Press.

Rolls, E. C. (1969). *They all ran wild*. Angus & Robertson, London.

Simpson, J. (1895). *The wild rabbit in a new aspect*. Blackwood, London.

Skellam, J. G. (1951). Random dispersal in theoretical populations. *Biometrika*, **38**, 196–218.

Stead, D. G. (1935). *The rabbit in Australia*. Winn, Sydney.

Strandgaard, H. and Asferg, T. (1980). The Danish bag record II. *Danish Review of Game Biology*, **11**, 1–112.

Strong, B. W. (1983). The invasion of the Northern Territory by the wild European rabbit *Oryctolagus cuniculus*. *Northern Territory Conservation Commission Technical Report*, No. 3, 1–20.

Swenson, J. (1986). Is Manana Island now 'Rabbitless Island'? *Elapaio*, **46**, 125–6.

Szunyoghy, J. (1958). The introduction of the rabbit in Hungary. *Annales Historico-Naturales Musei Nationalis Hungarici*, **50**, 349–58.

Thompson, H. V. and Worden, A. N. (1956). *The rabbit*. Collins, London.

Thomson, G. M. (1922). *The naturalisation of animals and plants in New Zealand*. Cambridge University Press.

Toschi, A. (1965). *Fauna d'Italia. Mammalia*. Edizioni Calderini, Bologna.

van den Bosch, F., Hengeveld, R., and Metz, J. A. (1922). Analysing the velocity of animal range expansion. *Journal of Biogeography*, **19**, 135–50.

van den Brink, F. H. (1967). *A field guide to the mammals of Britain and Europe*. Collins, London.

van der Loo, W. (1986). Geographical immunogenetics of European rabbits. *Mammal Review*, **16**, 199.

Vasiliu, G. D. (1964). Das Wildkaninchen in Rumänien. *Zeitschrift für Jagdwissenschaft*, **10**, 89–91.

Waithman, J. (1981). Rabbit control in Australia: history and future requirements. In *Proceedings of the World Lagomorph Conference* (ed. K. Myers and C. D. MacInnes), pp. 880–7. University of Guelph.

Watson, J. S. (1961). Feral rabbit populations on Pacific islands. *Pacific Science*, **15**, 591–3.

Wijngaarden, A. van, Laar, V. van, and Trommel, M. D. M. (1971). De verspreiding van de Nederlandse Zoogdieren. *Lutra*, **13**, 1–41.

Wodzicki, K. A. (1950). Introduced mammals of New Zealand. *NZ Department of Scientific and Industrial Research Bulletin*, No. 98.

Zeuner, F. E. (1963). *A history of domesticated animals*. Hutchinson, London.

3

The rabbit in continental Europe

P. M. Rogers, C. P. Arthur, and R. C. Soriguer

3.1 Palaeontology and history

Lagomorphs have lived on the continents of Europe, Asia, and North America since the late Eocene or early Oligocene. Members of both families of lagomorphs have occupied the Iberian Peninsula during that time. The family Ochotonidae, once represented by some 15 species in 5 genera, became extinct in Europe (with the exception of *Ochotona pusilla* in eastern Russia) at the end of the Pliocene. The Leporidae have been reduced from 7 species in 4 genera to the surviving 3 species in two genera, *Lepus* and *Oryctolagus*.

The late Neogene and Quaternary periods saw many profound ecological changes in Iberia. Late Miocene beds (5–7 million years ago) at Salobreña in the southern Spanish province of Granada have made an important contribution to our understanding of those changes. The deposits reveal a fauna which included many forms of direct African origin—giraffes, antelopes, hyaena, and various murids—lying together with Asiatic forms. The data are too sparse to calculate the rate of faunal change, although it seems certain that the African pre-Pleistocene immigrants must have bridged across Gibraltar or the Alboran sea. One European form, *Prolagus*, must also have made use of that bridge, but in the opposite direction, to colonize North Africa (Table 3.1).

There was a great increase, in both number and kind, of leporids at the end of the Neogene and during the Quaternary, which was probably related to the development of extensive grasslands, the cooling of the climate, and the arrival of Asiatic immigrants.

3.1.1 The origin of *Oryctolagus*

The earliest known specimen of *Oryctolagus* comes from Salobreña. The remains are sparse (one solitary tooth!) and do not permit identification to species (Lopez Martinez 1977*b*). But they show that *Oryctolagus* originated well before the Pleistocene.

There are three plausible hypotheses to explain the ancestry of *Oryctolagus*. The first is that it is derived from the Asiatic *Trischizolagus*, but this is difficult to accept as the two genera co-existed during the Upper Miocene of Salobreña (Granada). The second, which suggests an African origin, and the third, an indigenous origin from *Alilepus*, seem more likely; but the available evidence does not really favour any of the three hypotheses.

In western Europe, *Lepus* co-existed with *Oryctolagus*, although not constantly. *Oryctolagus* can survive in any environment, but only as long as the soil is suitable to dig and hide in. *Lepus* is more widely adaptable, able to live in forest clearings, alpine meadows, or prairie, whatever the soil. Such wide ecological plasticity makes it difficult to deduce much about the habitats available at the time. In Central Europe, *Lepus* co-existed with *Hypolagus* and with *Ochotona*.

3.1.2 Species of *Oryctolagus* in the last 2.5 million years

Lopez Martinez (1977*b*) in an excellent review of Iberian lagomorphs described a new species of *Oryctolagus*. She also reported new data for

Table 3.1 Occurrence of *Prolagus sansaniensis* in western Palearctic deposits

Country	Miocene (lower)	Miocene (middle)	Miocene (upper)	Pleistocene
Spain	X	X	X	X
France	X	X	X	X
Germany	X	X	X	X
Morocco				X
Portugal	X			
Turkey		X	X	
Hungary			X	X
Greece			X	

Source: After Lopez Martinez (1977*a*).

O. lacosti, already known from France (Lopez Martinez *et al.* 1976) and described the oldest known discovery of *O. cuniculus* on the Iberian peninsula, more than half a million years old.

Oryctolagus lacosti, known from Cataluña (Villaniense and Bihariense), had a large body size (between that of *Lepus* and *O. laynensis*), similar to *O. laynensis* in Italy and France. Cranial and dental characteristics mark it as an *Oryctolagus*, although in body size and in the morphology of the premolars (P3) it had similarities with *Lepus*.

Oryctolagus laynensis appeared in Andalucía, Castilla, and Cataluña in the late Miocene, and is apparently a direct ancestor of present day rabbits. It has also been found in the Pliocene either as a 'descendent' of *Trischizolagus*, also known from the upper Miocene in Murcia, or as an African immigrant. It is the oldest known *Oryctolagus*, with cranial and dental characteristics typical of that genus. Its skeleton appeared to be modified for jumping, but it also had the robust muscular insertions, particularly on the humerus and ulna, common in burrowing animals (Donard 1982).

Oryctolagus cuniculus first appeared in Andalucía, in mid-Pleistocene deposits in Cullar Baza, Granada, co-existing with *Lepus*. It has also been found in the province of Malaga, and in southern France (de Lumley-Woodyear 1969, 1971; Chaline 1976; Lopez-Martinez 1977*b*; Pages 1980). Dental peculiarities suggest that both *O. cuniculus* and *O. lacosti* are descended from a common ancestor, *O. laynensis*.

A plot of tooth lengths from Quaternary remains in the Iberian peninsula shows that the earliest species of *Oryctolagus*, although smaller than *Lepus*, were larger than the later *O. cuniculus* (*O. lacosti* then *O. laynensis*; Fig. 3.1a). Over the last 15 000 years the sizes of various osteological variables have also tended to decrease in *O. cuniculus* (Fig. 3.1b); which was a smaller rabbit than either of the other species, perhaps because it always coexisted with *Lepus*. The rabbit from Cullar Baza (Granada) was similar to *O. cuniculus* today.

This is consistent with Donard's (1982) research on the recent history of the rabbit in France. From bones in numerous caves she was able to show that from 500 000 years (Mindel glaciation) to 15 000 years ago (Magdalenian epoch) local populations of rabbits showed substantial variation in body size, following changes in climate. Four geological sub-species are thus defined: *O. c. lunellensis*, a rabbit of moderate size and temperate climate, dating from 300 to 200 000 years ago; *O. c. grenalensis*, a large rabbit of very different climates, dating from 150 to 70 000 years ago; and *O. c. huxleyi*, a small rabbit of hot climates some 40 000 years old, still living on the Canary Islands (where it was introduced long ago, or was possibly there before people) and Madeira; and *O. c. cuniculus*.

3.1.3 Geographical variation in contemporary *Oryctolagus cuniculus* in Europe

Not only has *O. cuniculus* become smaller in the course of its evolution over the last 15 millennia; it also exhibits a clear reduction in body size over its

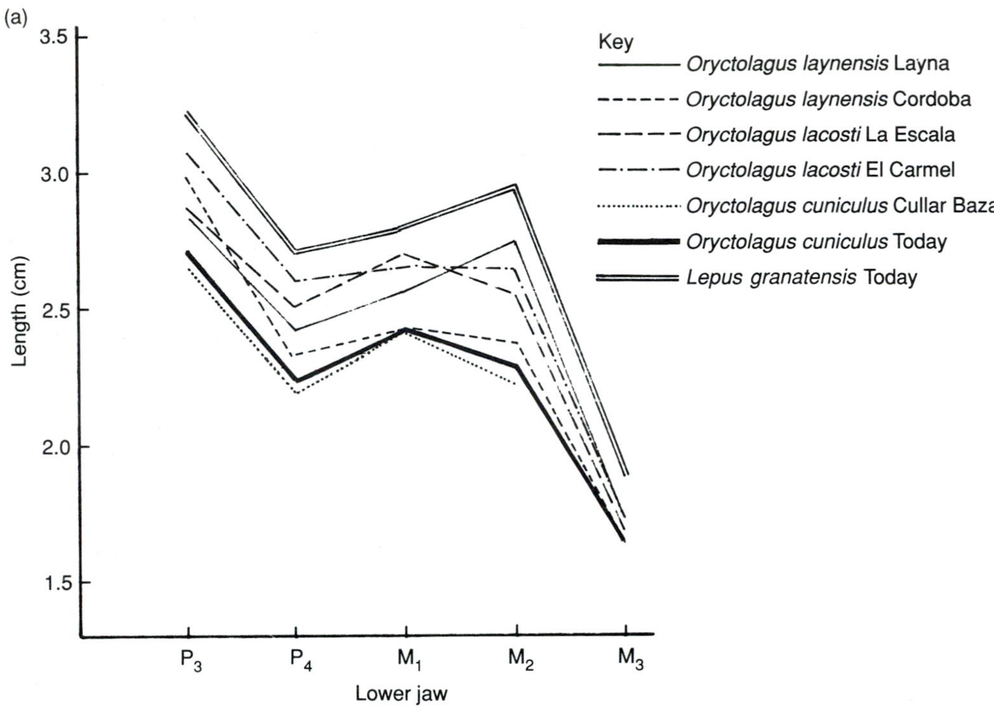

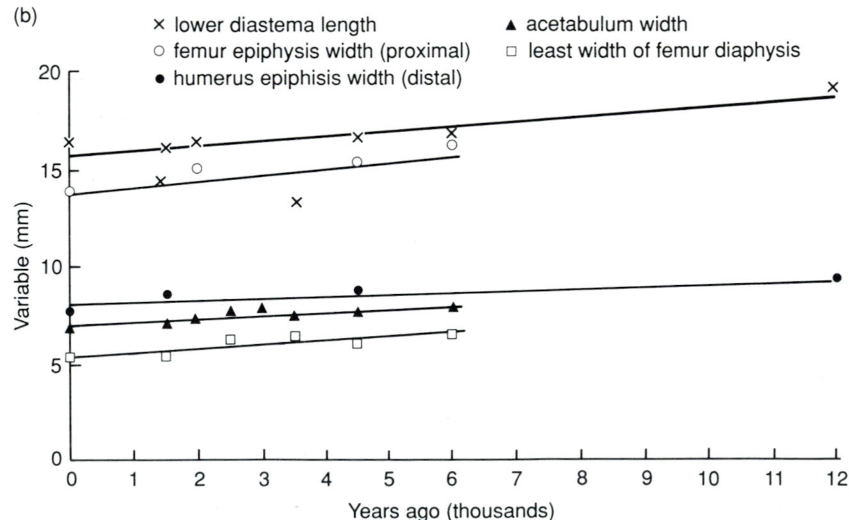

Fig. 3.1 Tooth lengths of lagomorphs on the Iberian peninsula.

present range, from north to south (Table 3.2). The rabbits of North and Central Europe are much larger than their relatives in Mediterranean Europe and North Africa. Even within France or the Iberian peninsula, there is a clear north–south gradient in body size.

Several hypotheses have been suggested to explain this pattern. One associates selection of body size with various ecological influences (climate, predation etc.). The rabbit apparently conforms to one such hypothesis, Bergmann's law, which predicts larger body size in higher latitudes. Soriguer (1981*a*, 1983*b*) has shown that rabbit populations which differ in body size also differ in other respects. The Mediterranean Iberian rabbit grows faster, and as we note later, has smaller litters, breeds younger and dies sooner than its bigger cousins. The final result in terms of the rate of increase 'r' is the same, however, albeit through different strategies. Another explanation suggests the influence of human activities (introductions, 'improvements' etc.) and domestication.

A third possibility is that the northern and southern rabbits represent different evolutionary lines. At the time of the glaciation of Mindel (50 000 BC), *O. cuniculus* had a fairly restricted range. Studies of the systematics of rabbit parasites, principally fleas, suggest that there were then two relict populations of rabbits: one in Spain, where the rabbit was particularly abundant near Gibraltar, and the second in the south of France (Beaucournu 1980*a*, 1980*b*). Indeed while *Spilopsyllus cuniculi*, the most common flea on rabbits, is present throughout the rabbits' range, three other fleas, *Xenopsylla cunicularis*, *Caenopsylla lactaevi*, and *Odontopsyllus quirosi*, are present only in France and Spain (including Morocco for *Xenopsylla*). The morphological differences are clearest between French and Spanish individuals of two of these species of flea, suggesting prolonged separation between the two relict rabbit populations (Beaucournu 1980*a*).

Studies of mitochondrial DNA and different immunoglobulin alleles support this hypothesis (van der Loo *et al.* 1991; Biju-Duval *et al.* 1991). In general there is high polymorphism in southern Spain, decreasing towards the north. Phylogenetic relationships between the rabbit's mtDNA types suggest that the separation of the two relict populations could have begun even before the Mindel glaciation. Domestic rabbits appear most similar to northern wild rabbits, and least similar to those of Andalucía (Richardson *et al.* 1980; Biju-Duval *et al.* 1991; Van der Loo *et al.* 1991; Van der Loo and Arthur in preparation). This suggests that the foundation stock for domestication came from only one (northern) population. It is interesting to note that the mtDNA of French rabbits is more closely related to that of rabbits in the north of Spain than the south. Almost all the alleles of domestic rabbits are present in the wild genotypes. Rabbit populations could have suffered genetic impoverishment the further they have moved from their ancestral home.

Today's wild rabbits still retain traces of their diverse historical origins. Genetic characters distinguish at least three separate rabbit populations in Europe: one in south, central, and west Spain and Portugal, a central one in northern Spain and southern France, and a northern one which includes rabbits in Belgium, England, Scotland, Wales, and northern France. Rabbits in the last group are genetically similar to those now in Australia and New Zealand (Richardson *et al.* 1980).

3.1.4 History of *Oryctolagus cuniculus* in Europe

Donard's (1982) studies showed no rabbits in all deposits of the Holocene (10 000 years ago) and earlier, north of the Loire. No rabbit bones have been found in all the excavations undertaken in the départements (administrative divisions of France) of Côte d'Or, Eure-et-Loire, the north of Charente, in Seine-et-Marne, Essonne, Yonne, Seine-Maritime and Haute-Savoie, whereas the hare (*Lepus*) is relatively well represented. In contrast, rabbit bones have been found in Charente Maritime, a deposit in the south of Charente, in Gironde, Haute Garonne, Hérault, Bouches-du-Rhône, Ain, Dordogne, Corrèze and Ardèche (Fig. 3.2).

Many deposits in Hérault and Bouches-du-Rhône contain rabbit remains, sometimes in abundance and notably of the geological sub-species *lunellensis*. They confirm that the range of the rabbit has extended progressively northwards from a relict population on the coast of the Mediterranean, and that it was only at the start of the historical epoch (20 000 to 1000 BC) that rabbits invaded other lands to the north of the Loire (Donard 1982).

Table 3.2 North–south variation in contemporary rabbit weights (g)

Country		département (France only)	Sex	Whole weight (mean)	Sample size	Maximum weight	Reference
Sweden			F	1670	521	>2000	Andersson *et al*. (1979)
			M	1670	521	>2000	
Holland			F	1500	–	>1700	Wallage-Drees
			M	1530			(pers. comm.)
France[a]	N	Nord	F	1710	45	>2000	Arthur and Guenezan
		(sand dune)	M	1725	25	>2000	(unpublished)
	W	Morbihan[bc]	F	1450	47	–	Arthur (unpublished)
		(heathland)	M	1415	56	–	
		Vendée	F	1450	24	1830	Arthur and Aubineau
		(bocage[f])	M	1350	16	1760	(unpublished)
		Deux-Sèvres	F	1355	17	>2000	Arthur and Aubineau
		(bocage)	M	1370	22	1980	(unpublished)
	Centre	Loiret[bc]	F	1385	29	–	Arthur and Guenezau
		(bocage)	M	1360	23	–	unpublished
	CW	Yvelines	F	1415	67	1830	Arthur and Guenezau
		(farmland)	M	1355	84	1780	(unpublished)
		Yvelines	F	1500	100	1930	Arthur (unpublished)
		(parkland)	M	1510	83	1860	
		Essonne	F	1360	51	1830	Arthur and Angibault
		(woodland)	M	1405	63	1760	(unpublished)
	E	Haute-Saône[bc]	F	1350	37	–	Arthur (unpublished)
		(pasture)	M	1360	28	–	
	CE	Saône-et-Loire[bc]	F	1375	28	–	Arthur (unpublished)
		(pasture)	M	1345	31	–	
	SW	Landes	F	1360	115	1750	Arthur and Avignon
		(fallow/pasture)	M	1430	90	1650	(unpublished)
		Landes	F	1555	15	>2000	Launay (pers. comm.)
		(farmland)	M	1555	8	>2000	
		Haute-Garonne[d]	F	1730	9	>2000	Launay (pers. comm.)
		(farmland)	M	1630	9	>2000	
		Dordogne	F	1345	33	1640	Arthur and Garcia
		(bocage)	M	1310	28	1610	(unpublished)
	SE	Hautes-Alpes[c]	F	1290	12	1370	Arthur and Reudet
		(bocage)	M	1295	18	1450	(unpublished)
		Bouches-du-Rhône	F	1245	41	1580	Arthur and Gaudin
		(pasture)	M	1210	28	1510	(unpublished)
		Bouches-du-Rhône	F	1250	22	–	Vandewalle (pers. comm.)
		(fallow)	M	1235	19	–	
		Bouches-du-Rhône[e]	F	1354	101	–	Rogers (1979)
		(salt-marsh)	M	1301	104	–	

Table 3.2 (*cont.*)

Country	département (France only)	Sex	Whole weight (mean)	Sample size	Maximum weight	Reference
France SE						
	Hérault[c]	F	1230	10	1510	Arthur and Taris
	(garrigue/cultiv.)	M	1250	10	1430	(unpublished)
	Corse	F	1255	12	1430	Arthur and Roux
	(maquis)	M	1320	14	1490	(unpublished)
	Vaucluse	F	1300	45	1590	Arthur and Gaudin
	(fallow)	M	1340	65	1540	(unpublished)
Portugal		F	1017	18	–	Lopez Ribeiro (1981)
		M	1023	11	–	
Spain NE		–	1190	4	–	Soriguer (unpublished)
NE		–	1158	71	–	Mañosa and Real (unpublished)
N		–	1224	97	–	Ceballos (unpublished)
N		–	1274	113	–	Ceballos (unpublished)
NW		–	934	6	–	Soriguer (unpublished)
SE		–	1043	28	–	Marquez (unpublished)
S		–	1011	1044	–	Soriguer (unpublished)
S		–	910	16	1050	Soriguer (unpublished)
SW		–	1092	521	–	Soriguer (1980*c*, 1981*a*)
SW		–	923	18	–	Soriguer (unpublished)
Morocco		–	1039	–	–	Soriguer (unpublished)

[a] All data except Rogers (1979) are from adult rabbits (>9 months) collected by ferreting in January–February.
[b] Data collected by the Service technique of the Fédération Départementale des Chasseurs.
[c] Data collected during the hunting season (September/October) from adult rabbits (>9 months).
[d] Includes domestic × wild crosses.
[e] Data collected in all seasons by night shooting; January–February mean weights were 1237 g (♀♀, n = 22) and 1282 g (♂♂, n = 28).
[f] A landscape of small fields (pasture or crop) surrounded by hedgerows on mounds.

By Roman times rabbits had become a problem in Spain. In the Balearic Islands they caused so much damage that the colonists asked the Emperor Augustus to send a Roman legion to clear the land of them, or to allocate them land elsewhere (according to Strabo, 58 BC to AD 20). Pliny the Elder, in his *Natural History*, tells of the ramparts of Tarragone being undermined by rabbit warrens, and recommended even then using ferrets against them. The rabbit was already used as the symbol of Spain by the poet Catullus, and imprinted on the reverse of some coins by the Emperor Hadrian. The Romans modified the range of the wild rabbit by introducing it to many countries, including to Italy and Corsica from France (Bodson 1978; Zeuner 1963), and to North Africa from Spain (C. Louzis personal communication).

In France, domestication of rabbits began much later, by monks in the Middle Ages. Their motivation was that newborn rabbits (the famous *laurices*) were considered to be aquatic, so were authorized in the Catholic religion for consumption in Lent. From the second to the fifteenth centuries there are numerous

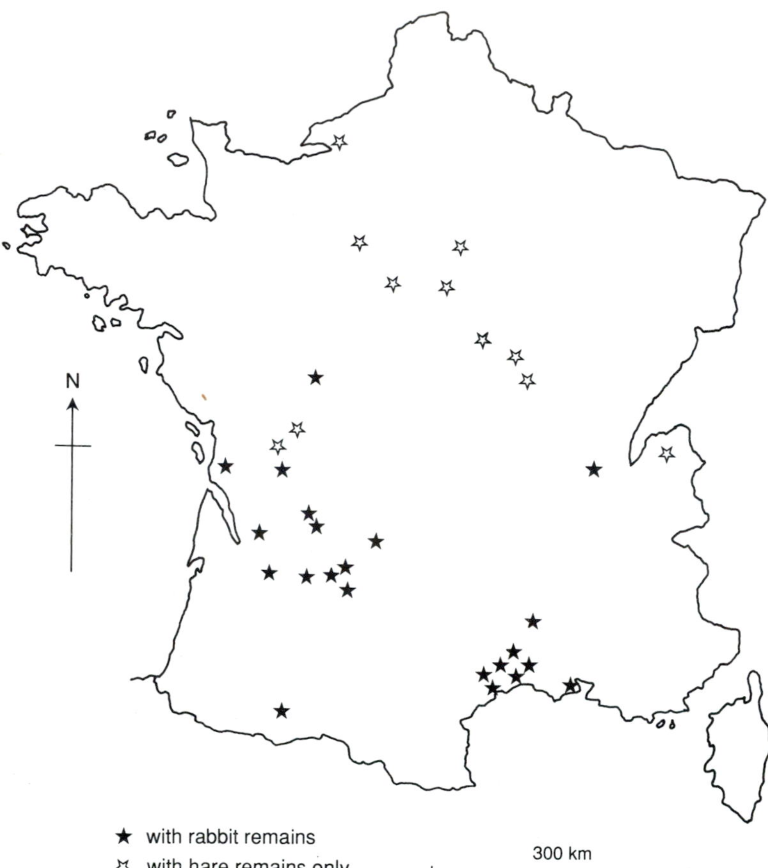

Fig. 3.2 Sites in France with rabbit remains, and with remains of hares but not of rabbits, 300 000 BC to AD 100 (after Donard 1982).

references to rabbits kept by monks and sometimes by peasants, or by the feudal gentry in enclosures of varying extent and degree of security, sometimes made of stone. Rabbits were often sold in markets, as shown in the book *Conejeria de Toledo* (the Rabbitries of Toledo) of the twelfth century. Trade sometimes spread far afield; there are records of the sale of rabbits between the abbeys of Corvey (Germany) and Solignac (France) in 1148, and the despatch of 6000 rabbit pelts from Castile to Devon in 1221 (Delort 1984). However, there is scarcely a mention of rabbits living free in the wild.

Nevertheless the monarchy in France always attempted to restrict 'warren rights', by forbidding the creation of new warrens or the enlargement of old ones, or the re-establishment of old boundaries (there are examples of such rules in the ordonnances of Jean le Bon in 1356 and of Charles VI in 1413). In the seventeenth century Colbert (prime minister to King Louis XIV) ordered the destruction of rabbits in all the royal forests, and undertook to compensate for rabbit damage where the royal servants were unsuccessful in their task. The French Revolution abrogated Colbert's ordonnance, and annulled the exclusive right of the gentry to control warrens, but the rabbit in France nevertheless remained enclosed and controlled. Not until the Second Empire, in 1862, did Napoleon III declare rabbits to be game, and order that they be allowed to range freely in the forest of Compiegne.

At the end of the nineteenth century, numerous bourgeois hunts were established close to the towns, on agricultural areas which were abandoned after the Industrial Revolution and the phylloxera crisis. The rabbit became their principal quarry. In less than 20 years, from 1845 to 1863, the number of rabbits

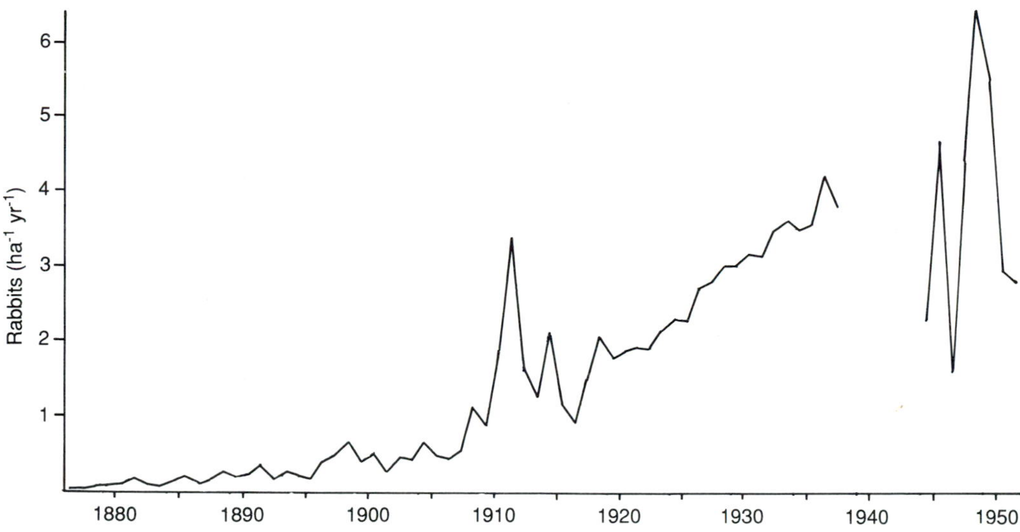

Fig. 3.3 Number of rabbits shot annually per hectare on a private hunt of 720 ha in Loiret, France, 1877–1953 (after Giban 1956).

sold in the markets of Paris rose from 177 000 to nearly 2 million (Gayot 1865). From then on, the extension of the natural range of the rabbit continued progressively. In Sologne, there were so few rabbits in 1880 that it was judged a waste of effort to fence nurseries for reforestation programmes: in 1930, hunters took more than 10 rabbits/hectare in a season from the same area.

At the start of the twentieth century the introduction of winter cereals and the rotation of crops, combined with the massive destruction of all predators (aerial or terrestrial) by gamekeepers, greatly favoured population increases in rabbits. Some gamekeepers even provided food for game (and especially for rabbits) in winter—apples, beetroot, and fodder. Rabbits quickly reached the very high densities recorded from 1920 to 1930 (Fig. 3.3). Bags of over 10 rabbits/hectare were common in northern départements of France from then until 1950–2. In Mediterranean France bags were between 1 and 2 rabbits/hectare (Giban 1956).

Damage to farm crops due to the excessive abundance of rabbits did not generally cause concern in the north of France, but it was nevertheless the reason that, on 14 June 1952, Dr Delille introduced the myxoma virus on Maillebois, his estate in the département of Eure-et-Loire. Because of previous failures to introduce the virus in other countries, and because he did not know the mechanism of transmission, Dr Delille assumed that any disease would be confined to his 300 ha walled enclosure. But the result of his experiment was a little more spectacular than he expected: by summer 1952 the disease had reached 9 départements. By the end of 1953 the whole of France was affected (Fig. 3.4), and isolated cases of myxomatosis were reported from Spain, Belgium, Holland, Germany, and England (Joubert *et al.* 1972).

For hunters, the rabbit population crash in France was catastrophic. In 1953–4 bags were 15 per cent of those before 1952; in 1954–5, 2 per cent, and in 1955–6 about 7 per cent. Some 90 to 98 per cent of the French rabbit population was killed by myxomatosis between 1953 and 1955. The pressure of public outrage was enormous, and legal proceedings were started against Dr Delille. He was convicted of illegally spreading an animal disease in September 1954, but received only a nominal penalty—a one franc fine. On the other hand, in 1956 the Syndicat National des Forestiers Français presented him with a medal 'in recognition of services rendered to agriculture and sylviculture'! In fact the estimated increase in agricultural and sylvicultural production at that time was in the order of 1000 million francs (using 1992 values for the franc; Siriez 1957).

French hunters quickly attempted to limit the

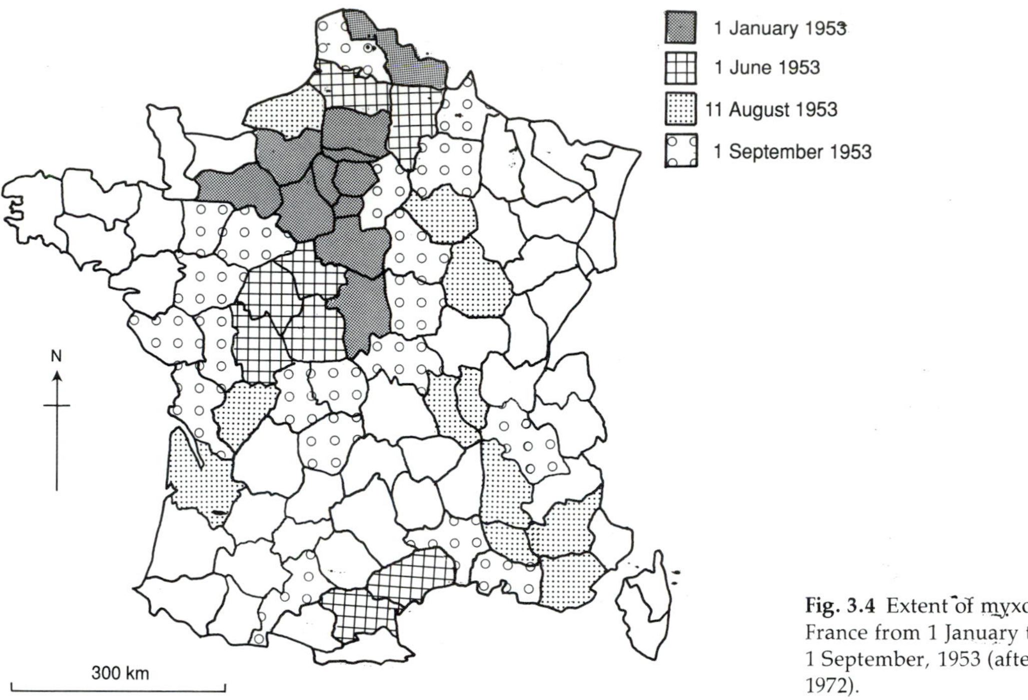

Fig. 3.4 Extent of myxomatosis in France from 1 January to 1 September, 1953 (after Joubert *et al.* 1972).

effects of the myxoma virus, trying to establish 'cordons sanitaires' around affected areas, and even, in 1956, importing Australian rabbits reputed to be resistant to the virus. All their efforts failed. Nevertheless, 1956 saw the start of a small increase in bag size. The first attenuated strains of the virus were found in 1955 (Jacotot *et al.* 1956), and they spread gradually until in 1968 more than half the strains isolated from wild rabbits were much attenuated, grade IIIB–IV (Fenner and Ratcliffe 1965). That proportion seems to have been more or less maintained until 1977–8 (Joubert 1979). Over the last four or five years, however, there has been an increase in the more virulent strains.

3.2 Present distribution and regional variations in numbers

The rabbit is now found over most of western Europe, excluding most of Austria, Italy, and Switzerland, and around the coasts of the western Mediterranean Sea (Fig. 2.2). Its eastern range extends as far as the Crimea. It is also found on the Balearic Islands, Corsica, Sardinia, Sicily, and Crete.

Wild rabbits were introduced in southern Ukraine at the end of the nineteenth century. About 80 years later their range was increased extensively by further introductions to shoots in the Ukraine, Maldavia, and the Pre-Caucasus. In 1979 experiments began on establishing rabbits in Uzbekistan, and later in Lithuania. At each site, 2000–3000 animals were released. Irrigation development in Central Asia provided opportunities for wild rabbits to become locally abundant in desert zones (Skulyatvev 1987).

A detailed picture of the distribution and abundance of the rabbit in France was produced by a 1977 survey of all gamekeepers by the Office National de la Chasse (Arthur *et al.* 1980). It showed, with other data (Arthur and Chapuis 1985), that rabbits are found in every département (Fig. 3.5). They are less abundant and more dispersed in mountain areas (Jura, Alpes, Vosges, centre of the Massif Central, Pyrénées) and the north-eastern départements, perhaps discouraged by a high snowline, thin soils, too

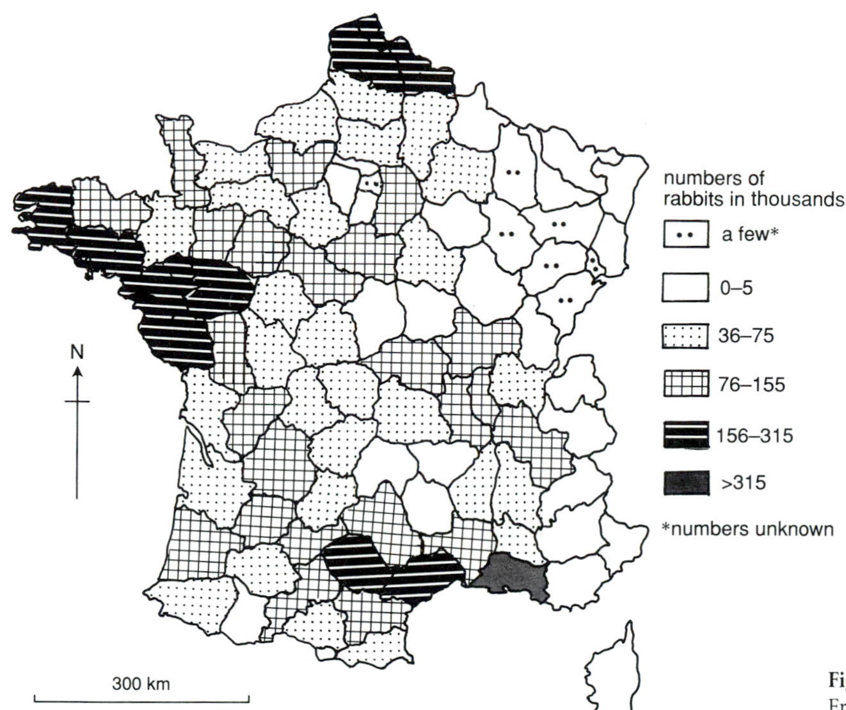

Fig. 3.5 Numbers of rabbits shot in French départements in 1983 (after Arthur and Guenezan 1986).

Table 3.3 Density and size of rabbit warrens in various French landscapes

Region	Landscape	Area sampled (ha)	Warrens per ha	Entrances per warren	Reference
Camargue	woodland	60	2.6	8.6	Pages (1980)
Camargue	salt-marsh	450	1.4	2.8	Rogers (1979)
Midi	vines	40	1.5*	3.0	Pages (1980)
Midi	garrigue†	25	0.2	1.3	Pages (1980)
Brittany	heath	400	2.0	1.0	Chapuis (1979)
Sologne	wood/cultivated	283	0.7**	3.0	Servan (1972)
Ile-de-France	cultivated	200	0.4**	2.2	Panaget (1983)
	wood/cultivated	85	1.3**	2.1	Arthur and Guenezan (unpublished)
	parkland	250	2.5	24.0	Arthur and Guenezan (unpublished)
Vaucluse	fallow/woodland	38	2.6	15.0	Arthur and Guenezan (unpublished)

† Shrub-covered calcarious land.
* Warrens mostly in the brush beneath the vines.
** Warrens mostly in woods and hedges.

much woodland and, in the north-east of France, harsh winters. In contrast, they have high population densities in Brittany, Vendée, along the Rhône valley, and the Mediterranean coast, where there are smaller fields, more uncultivated land (heathland, garrigue—shrub-covered calcarious land) and many hedgerows. Rabbits are also abundant in Nord-Pas-de-Calais, Sologne, and around the cities of Paris and Lyon, where they are encouraged by many private hunts. The pattern of distribution has remained fairly consistent over the years, although abundance has varied.

As might be expected, landscape also determines the distribution and size of warrens. In most of France the density of warrens is less than 1.5 per hectare, each with a maximum of 3 entrances, but may reach 2.5 or more warrens per hectare, some with over 20 entrances (Table 3.3).

Since the introduction of myxomatosis in France four national surveys have been made to monitor rabbit abundance (Giban 1956; ONC 1976; Arthur *et al.* 1980; Arthur and Guenezan 1986). Although the survey methods used differed (survey of some private hunts from 1920 to 1956, survey of gamekeepers in 1977, sample survey of hunters in 1974 and 1983), the general trend is clear enough.

After the population crash in 1952, hunters' bags remained small until 1956; then they slowly increased, more rapidly after 1970. In 1977, more than 15 million rabbits (about a quarter of the pre-myxomatosis figure) were shot. Thereafter data are limited, but only 6.5 million were shot in 1983—a real reduction in total population since 1977 of around 55 per cent. But counting only the area of grassland, cultivated land, hedgerow and woodland edge, the number shot per 100 hectares per year was 36.4 in 1974 and 37.6 in 1977, dropping to 24.4 in 1983; a real reduction of about 35 per cent. However, the pattern of decline varied in different regions of the country (see below).

Variations in abundance of rabbits have been observed since at least 1880, long before the advent of myxomatosis (Middleton 1934; Giban 1956; Tapper 1985). The explanations offered then were usually coccidiosis or climate; also, natural cycles may be inherent in the population dynamics of rabbits. Part of the reason for the rise and fall in the 1970s and 1980s might be that the years 1970–8 were, in France at least, climatically favourable for the rabbit—high temperatures with low precipitation—while the period 1978–83 often saw high spring rainfalls, which tend to drown nestlings in their burrows (Birkan and Pepin 1984). However, since 1970 it seems more likely that myxomatosis has caused periodic changes in the equilibrium between the virulence of the virus and the resistance of the rabbit (Ross 1982).

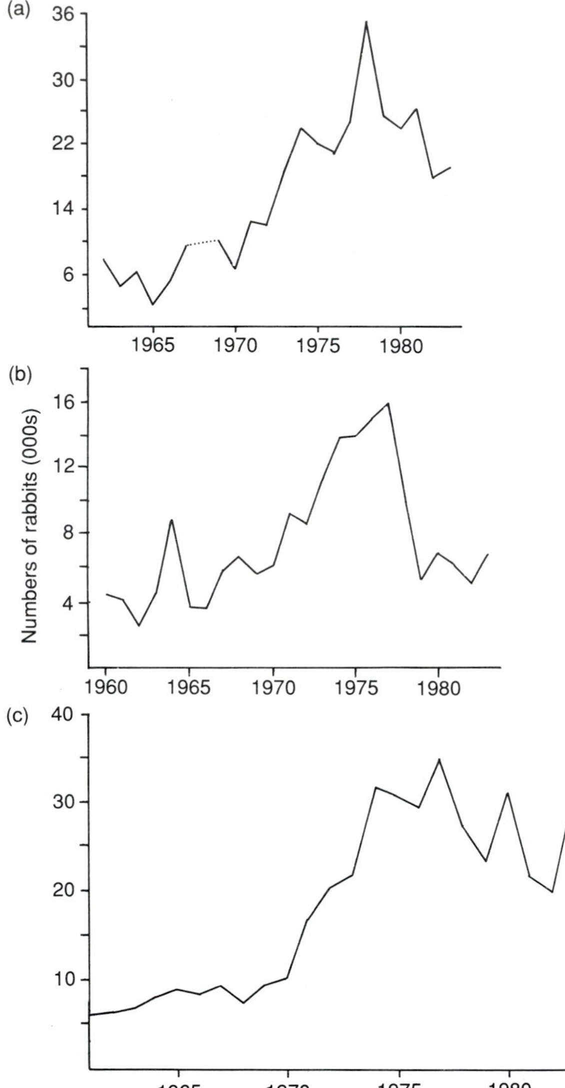

Fig. 3.6 Changes in numbers of rabbits shot in (a) Austria, (b) Germany, and (c) England from 1961 to 1984 (after Tapper 1985 and data from the agricultural ministries of Austria and Germany).

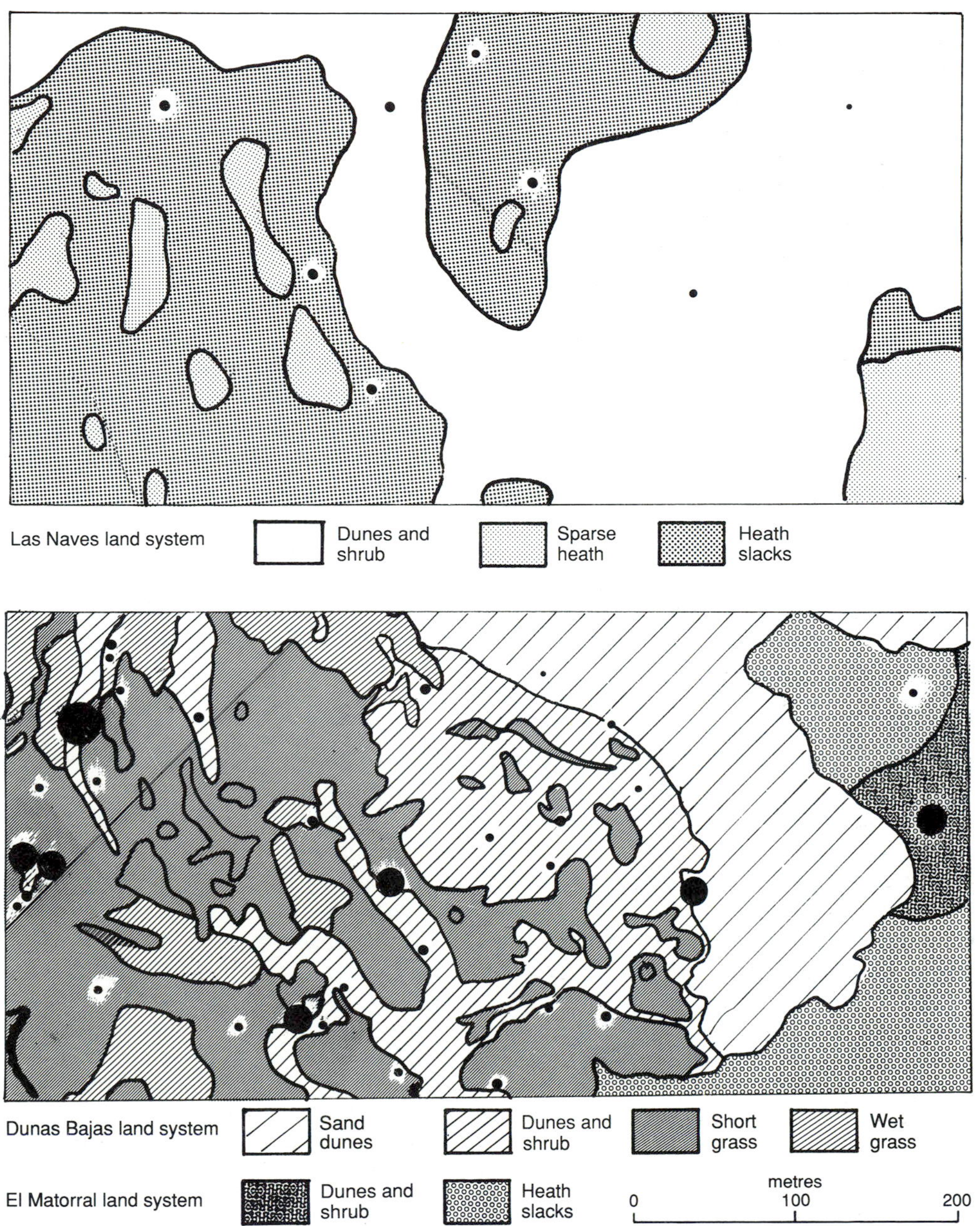

Fig. 3.7 Distribution of warrens (size of dot is proportional to number of active entrances). Favourable (bottom) and unfavourable sites (top) are in Doñana National Park, southern Spain (after Soriguer and Rogers 1981).

The variation in bag size in other European countries (West Germany, Austria, Britain) mirrors this picture; they also report an increase from 1970, peaking towards 1977 or 1978, then a decrease until 1983, varying somewhat from one country to another (Fig. 3.6).

If the numbers shot per unit area decrease more than the total for the area in question, it implies that there has been a real decrease in abundance associated with a decrease in habitat. Habitat loss has been significant in some areas of France; for example, between 1974 and 1983 many hedges in Brittany were destroyed, while in Normandy grassland increased; in Nord-Pas-de-Calais and Alsace, field size and monoculture of maize or beetroot increased, combined with the destruction of hedges; woodland increased in the Massif Central, Burgundy and Franche-Comté, and so did urbanization in Ile-de-France. In the south-west and north-east there was an increase in garrigue, decrease in the area of vineyards and orchards, and increase in built-up areas which must have decreased the availability of favourable habitat. Those changes were all generally unfavourable to rabbits, and were reflected in bag records.

In northern France and in the Massif Central both bag size and the numbers shot per unit area decreased by 60 per cent from 1974–83. In contrast, in the south-west and north-east the total bag decreased by 42 per cent from 1974–83 but the number shot per 100 ha decreased by only 33 per cent. Our conclusion is that there has been a general decline in the northern populations, less marked in unhunted reserves; the southern populations, though fluctuating annually, have been maintained over a decade.

Data from Spain are less detailed—no comparable surveys have been made. It is clear, however, that rabbits have long since been widespread over almost the whole peninsula. In 1956–8 their range covered all of Spain except the north coastal provinces (Muñoz-Goyanes 1960). We have no data from Portugal for those years. A 1988 survey showed much the same picture, showing in addition that rabbits were distributed throughout Portugal. Limited data from hunters in Cordoba Province indicate a marked increase in abundance from the early 1960s to the early 1980s, rising to a plateau towards the end of the decade. More recently numbers declined markedly, a consequence of the combined effects of myxomatosis and viral haemorrhagic disease (see Sections 3.6 and 3.7 below).

On a smaller scale, the local distribution of rabbits in relation to landscape demonstrates their terrain requirements. Detailed surveys in southern Spain, in areas both favourable and unfavourable for rabbits, showed that they thrive best where sand ridges suitable for warren building interdigitate with or abut moist feeding grounds (Rogers and Myers 1979; Fig. 3.7). Similarly in the Camargue, southern France, warrens are associated with levées and other high ground adjacent to lower, moister areas producing vegetation suitable for feeding. Such a juxtaposition favours a high local population density of rabbits (Rogers 1981).

3.3 Status as a game species and as a pest

Between 6 and 13 million rabbits are shot each year in France by almost two million licensed hunters, and an unknown though probably similar number in Spain by 1.5 million licensed hunters. In stark contrast to Australia and New Zealand, the status of rabbits in much of Europe is that of game rather than vermin. In Spain the rabbit is the most important game species; in France it constitutes 40 per cent of the total bag of small plains species, coming second only to thrushes and blackbirds.

The hunting season in Spain is generally from October to mid-January, varying a little between provinces. Recent legal changes will reduce this period to October and November only. If rabbit numbers are particularly high, however, they may be taken from the first week of July. The season in France also varies, from 15 September to 5 January or from 8 October to 15 January, except in those départements in which the rabbit has been declared a pest, in which case the closing date is either 28 February or 31 March. The taking of rabbits out of season is allowed only when there has been damage to orchards, vineyards, or other crops. Similarly, rabbits, but only bucks, may be taken out of season in Germany.

In France, rabbits are usually hunted with small dogs of the Fauve de Bretagne, beagle, or teckel breeds, or with ferrets. Gas or other poisons, however, are prohibited even where the rabbit has been declared a pest. In Spain, traps and snares are allowed in addition to shooting, but under official control in order to avoid killing predators. Traps are also allowed where rabbits are raised and released for hunting stock. The use of ferrets will be illegal in Andalucía after the 1992 season.

Hunters in France go to considerable effort and expense to ensure that there are adequate numbers of rabbits for shooting. Surveys by the Office National de la Chasse (ONC) show that up to half-a-million rabbits are released each year for hunting, at a cost of 30 to 45 million francs. More than half are captive-bred, of which 40 per cent are sold as pure wild-type and 10 per cent as wild-domestic crosses. In addition, many hunting groups have, in the last decade or so, built artificial warrens with tree stumps, stones, and branches, or have raised areas of earth or sand in which rabbits may build their own warrens protected from predators or flooding.

In Spain the rabbit has never been considered a pest in the way it is in France where, in spite of its status as game, it is still the greatest pest after wild boar. By 1952 damage to agriculture and sylviculture in France had reached the estimated 1000 million francs (1992 values, Siriez 1957), although the level has much reduced since, and 88 of France's 90 départements had declared the rabbit a pest. Only two years later, in 1954, rabbits had been so much reduced by myxomatosis that only 54 considered it a pest throughout the département, and 21 in certain parts of the département: 15 départements considered the rabbit only as game. By 1985 only 45 considered the rabbit a pest throughout the département (Fig. 3.8).

Where the rabbit has been declared a pest translocation of wild rabbits, or raising wild-type rabbits or wild-domestic crosses for restocking, is forbidden. Even so, wild rabbits were restocked in over 70 départements (illegally in 25 of them) and raised for restocking, either as pure wild-types or as crosses, in about 50 départements (again illegally in 5 of them). The law allows rabbits to be raised in départements where they are a pest as long as they are not released in those départements, a condition which is not always fulfilled.

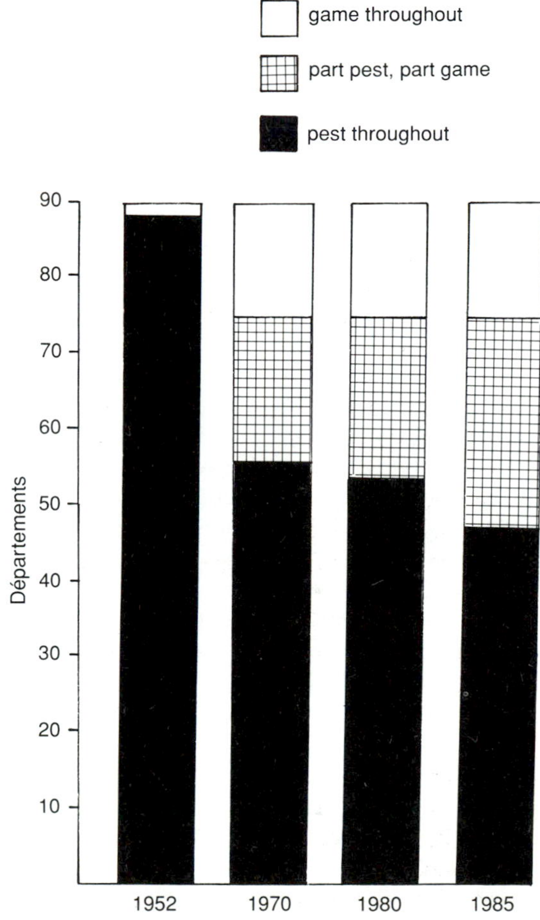

Fig. 3.8 Number of French départements in which the rabbit was declared a pest, 1952–85 (Arthur unpublished).

A 1977 survey by the ONC showed that in those départements having rabbits, they were present, on average, in two out of three communes (the smallest administrative sub-division within a département). Of those with rabbits, only a quarter suffered any damage, mainly in the west of the country, mostly minor, seriously affecting only 7 per cent of cereal crops, and 2 per cent of woodland. Of those communes suffering damage to arable crops, 35 per cent take some form of protective measure, usually repellents; 77 per cent of communes protect plantations, usually with wire netting. The total cost of the damage in 1977 was 30 million francs (1992 values).

To some degree the potential conflict between

farmers and hunters in France (and many farmers are also hunters) is reduced by payment of compensation through insurance policies taken out by the hunting groups. In 1977, 5 per cent of hunting groups made such payments, usually after amicable negotiation between hunter and farmer. Nevertheless each year some cases go before a tribunal which, after expert opinion, defines the extent of damage and rules on the fault of management by the hunters. In 1985 there were 50 such cases, claiming damages totalling 2 to 3 million francs; but in half the cases no payment was awarded, because it was judged that there had been no error of management or unusual damage. It is important to note here that in France it is generally expected that wildlife will feed on agricultural crops to some degree; only when the damage is excessive are there grounds for complaint.

Whatever the cries of despair from farmers, particularly in the north of France, the rabbit is considered the primary game species by hunters of both north and south. The dramatic reduction in populations caused by myxomatosis led to increased hunting pressure on other game, for example hares, partridges, and pheasants, which do not have the ecological plasticity of the rabbit. Their populations too declined rapidly. To compensate, hunters in France, and likewise in Spain, made numerous releases of alternative game species each year; they also attempted to introduce new species, for example *Sylvilagus floridanus* and bobwhite quail, but without success.

Elsewhere in Europe (for example Belgium, Germany, Italy, Spain) hunters reacted similarly to myxomatosis. At least three applications to make legal introductions of *Sylvilagus* in Spain have been denied; the Spanish Ministry of Fisheries and Food (MAPA) prohibited introduction of *Sylvilagus* in 1982. European rabbit populations certainly do recover from epizootics, and with increasing success as time goes on, but in Mediterranean France and Spain the lack of native rabbits has been lamented as much by the hunters (who were sometimes the same farmers who complained of damage!) as by conservationists, albeit for different reasons.

3.4 Rabbit–plant ecology

The diets of rabbits in various parts of Europe have been studied by microscopic analysis of stomach contents or faecal pellets, by chemical analysis both of food eaten and of plants available, by selective exclosures, and by direct observation. The results show considerable differences between sites and, perhaps more significantly, between seasons. There is, nevertheless, a clear general trend showing that rabbits eat mostly grasses and forbs when available, preferring the former, although not always (Table 3.4).

In the Camargue, a deltaic, saline habitat, the diet of rabbits is unlike that documented at other sites studied, except Czechoslovakia and parts of Brittany, in that dicots are the most important items, particularly in summer. Of grasses, only *Bromus* is frequently eaten. The choice of food is also restricted—eight species of plants may account for 86 per cent of the diet throughout the year (Rogers 1979). The staples of winter are *Atriplex*, *Halimione*, and *Arthrocnemum* spp. In spring the diet is more diverse, reflecting a greater choice available; the winter items are supplemented by *Trifolium*, *Medicago*, and *Melilotus* spp., which may account for 40 per cent of the diet. *Bromus* is eaten most often towards the end of spring, and *Agropyron pungens* mainly in summer. Rabbits often cut long stems and other parts of plants, apparently without any intention of eating them (Arthur and Rogers, unpublished observations), although Soriguer has observed them doing it to reach flowers and fruits.

In Doñana National Park in Spain, rabbit population densities are highest in the ecotone between shrubland and marshland. There they eat mostly grasses (67 per cent) and forbs (30 per cent) (Soriguer 1988). Other herbivores (red and fallow deer, wild boar, horses, and cattle) are also present in high densities, and with rabbits consume over 70 per cent of the available plant biomass. If they are excluded from the rabbits' feeding places, then the rabbits' diet changes to forbs. It appears that rabbits in Doñana really eat only what is left by the other herbivores, which is mainly grass (Soriguer 1988). In sand-dune systems, where a herbaceous layer is rare, rabbits eat the stems, leaves and fruits of shrubs (Soriguer 1981*b*; Soriguer and Herrera 1984).

Table 3.4 Analysis of faecal pellets and stomach contents of rabbits in Europe

Country	Sample size[a]	Type[b]	Number of species	Graminae per cent	Dicots per cent	Shrubs per cent	Reference
Holland	–	F	18 (11)	58.0	23.5	18.6	Wallage-Drees (1989)
Czechoslovakia	20	F	43	43.9	41.3	14.8	Homolska (1985)
England	20	F	20 (4)	80.7	19.3	0	Bhadresa (1977)
UK	–	F	20 (<14)	80.0	<15.0	0	Williams *et al.* (1974)
N France I	20	F	–	15.8	74.7	8.1	Chapuis (1979)
N France II	20	F	–	86.2	3.9	6.5	Chapuis (1979)
S France	80	S	33 (9)	15.7	42.8	41.5	Rogers (1979)
S Spain	130	F	14 (5)	66.7	29.6	1.4	Soriguer (1988)

[a] Number of pellets or stomachs.
[b] F = faecal pellet analysis, S = stomach contents analysis.

Rabbits living in scrubland with few forbs and grasses turn to eating woody plants (Table 3.5), selecting different parts of each, often the upper few millimetres. For example, rabbits eat the soft fleshy new stems of *Stauracanthus genistoides*, and 'prune' *Halimium halimifolium* and *Juniperus phoenicea* when eating their seed and fruits. Similar behaviour can be observed in rabbits feeding on *Corema album*, and (in Brittany) on gorse (*Ulex* spp.).

In northern France, variation in diet reflects different landscapes and climates. Those rabbits living on the edge of substantial forests in the Ile-de-France (the region around Paris) eat mainly grasses (including cultivated varieties, for example maize, wheat) from spring to late summer, turning to bramble (*Rubus* spp.) and the bark of trees in winter. There is a greater variety of food plants available in spring, as in southern Europe (Chapuis 1979). In contrast, rabbits living in small wooded plots or hedgerows in cultivated landscapes eat grasses all year round. Again, the cultivated grasses predominate when available: rye, wheat, maize, and particularly barley. As the cereals mature wild grasses replace them in the diet, which are in turn replaced by dicots (bramble, ivy, rape) when grasses are unavailable (Chapuis 1979; Panaget 1983). In the more heathland landscapes of Brittany, rabbits also eat grasses (*Holcus*, *Festuca*, and *Agrostis* spp.) when available, but turn to gorse (*Ulex europaeus*) and heathers (*Calluna vulgaris* and *Erica* spp.) in winter, and their young buds in spring. Gorse may account for 50 per cent of the diet of some rabbits at times; bryophytes are also taken (Chapuis 1979).

The quality of food varies considerably over the year, which has important implications for breeding (see below). Different parts of plants also vary in food value, and rabbits are capable of detecting those differences and selecting the best parts. For example, the ratio of protein to fibre is two or three times higher in the diet than in the standing vegetation (Fig. 3.9; Rogers 1979). Seasonal changes in diet accompany, as might be expected, changes in the nutritive value of plants, and of particular parts of plants. Stomach analysis shows the food of Camargue rabbits in spring to be high in protein (19 to 20 per cent) and low in fibre (*c.* 25 per cent). At the end of spring (in May or June, according to the year) the proportion of protein drops quickly to 12–13 per cent, and fibre rises to over 30 per cent, reflecting a change in the quality of food available. It obliges the rabbits to eat more, and coincides with a reduction in the physical condition of adults (Fig. 3.10; Rogers 1979; Vandewalle 1986).

The choice of plants eaten in the Camargue also depends largely on the balance between salt and water in the available standing crops. For example, the reason that rabbits eat apparently unappetizing grasses in summer (dry *Bromus*, *Agropyron*) may be that *Atriplex* and *Halimione* contain too much salt (Rogers 1979).

The effect of rabbits on vegetation is, at a gross level, predictable—a reduced standing crop. The

Table 3.5 Impact (consumption + pruning etc) of rabbits on woody vegetation in Doñana National Park, southern Spain

Location	Date	Plant density stem ha^{-1}	Plant volume m^3 ha^{-1}	Impact			Species	%
				Total[a]	old[b]	new[c]		
Porquera-Fraile	Oct 1982	11665	8729	25.0	0.0	0.1	*Halimium halimifolium*	0.1
Sabinar-Ojillo	Oct 1992	22835	8539	81.0	3.5	0.4	*Stauracanthus genistoides*	12.5
							Juniperus phoenicea	4.5
Rancho Manuela	Oct 1982	33200	16885	55.5	0.5	0.4	*S. genistoides*	17.5
							G. triacanthos	17.5
							Cistus salvifolius	0.9
							Rosmarinus officinalis	0.4
							H. halimifolium	0.2
							Phillyrea angustifolia	0.4
Sabinar-Cota-32	Jan 1983	46098	11147	41.5	1.7	1.3	*Lavandula stoechas*	17.5
							C. salvifolius	7.6
							H. commutatum	3.5
							C. libanotis	2.4
							J. phoenicea	1.9
							R. officinalis	0.8
Dunas-Cota-32	Jan 1983	4933	2870	19.3	2.4	3.4	*Scrophularia frutescens*	17.5

[a] Total impact of rabbit + others on all parts of plants.
[b] Impact of rabbits on old parts of the plant.
[c] Impact of rabbits on new growth (<1 year old).
Source: Soriguer (1981*b*).

various exclosure experiments that confirmed this expectation, for example by Bassett 1978; Bassett and Rogers 1979; Soriguer 1981*b*, 1983*a*, 1988, must have seemed redundant to many farmers! In early summer in 1976 and 1977 in Doñana National Park rabbits accounted for only 15 to 26 per cent of the total forage offtake (70 per cent) by large herbivores (Soriguer 1983*a*; Soriguer and Herrera 1984), and slightly more in June 1983. In low Mediterranean mountain areas (western Sierra Morena) rabbits consumed 33 per cent of the standing crop (Soriguer 1981*a*). In Mediterranean France (Camargue) comparable figures vary from 1 to 27 per cent (Bassett and Rogers 1979). Of four significant herbivores on one Camargue study area, rabbits were second only to horses as consumers of primary production (Table 3.6), ahead of both coypu and grasshoppers (Duncan 1992).

Herbivores may also cause changes in floristic composition. Results from exclosure experiments to determine how much rabbits do so have not been clear cut, largely because it is difficult to assess the effect of herbivore activity on natural succession. In the Camargue the most common winter foods of rabbits (*Halimione*, *Torilis*, *Limonium*) increased when rabbits were excluded, but few other changes could be attributed directly to the effects of grazing. Much the same conclusion applies in southern Spain (Soriguer unpublished data).

An important consequence of the feeding behaviour of rabbits, particularly their tendency to select certain parts of plants including whole fruits and seeds, is that they play an important role in seed dispersal. This aspect of the relationship between rabbits and vegetation has been poorly documented (see Staniforth and Cavers 1977), but in Table 3.7 we

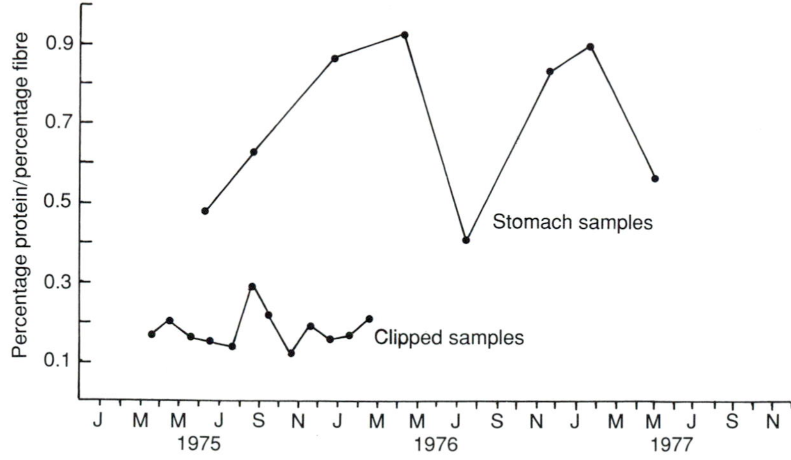

Fig. 3.9 Relationship between season and crude protein/crude fibre ratios in standing vegetation and rabbits' stomach contents (Rogers 1979).

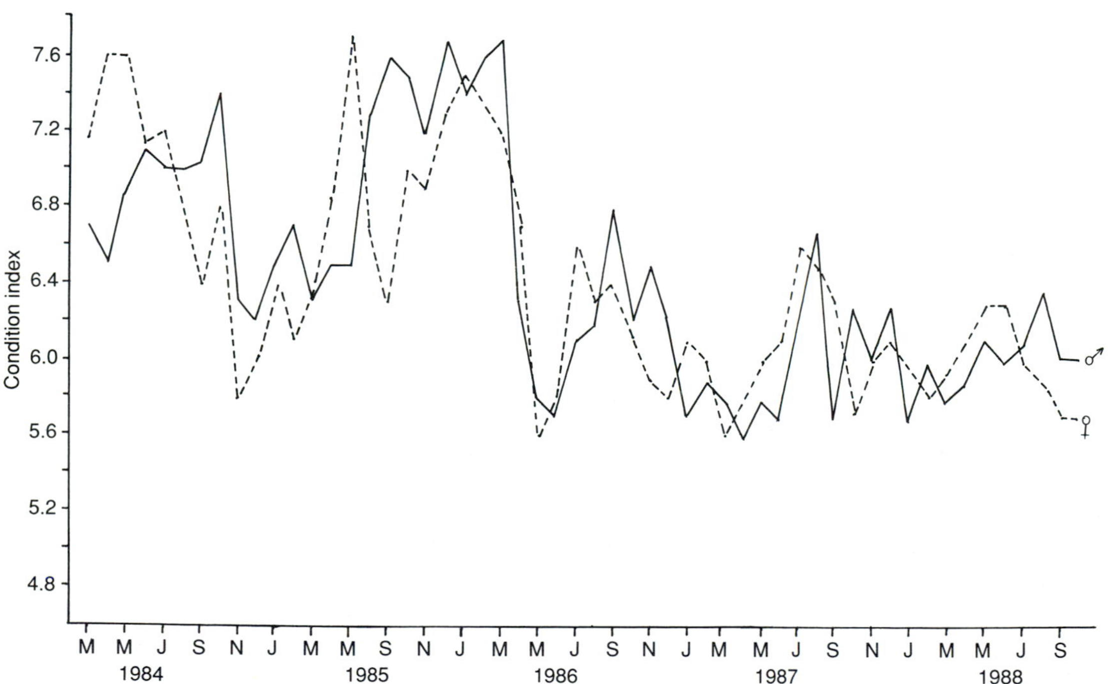

Fig. 3.10 Changes from 1984 to 1988 in indices of physical condition (Bailey 1968) of adult rabbits (>9 months) in the Camargue (Vandewalle unpublished).

Table 3.6 Estimated offtake of herbaceous plants by the main herbivores on a 335 ha Camargue study area (S. France)

Species	Year	Number	Dry matter offtake (t. day^{-1})
Coypu	1977–78	50	0.017
Grasshoppers	1977	–	0.020
Horses	1976	26	0.356
	1977	35	0.466
Rabbits	1976	838	0.087
	1977	670	0.067
	1983	3450	0.352

Data from various sources in Duncan (1992).

list the seeds of 16 species found in the faecal pellets of rabbits in Spain. All were checked and found to germinate normally.

Table 3.7 Reproductive parts of plants found in faecal pellets of rabbits in Spain, illustrating their role in seed dispersal

Species	Part of plant
Corema album	complete fruit
Cistus salvifolius	capsules with seeds
Halimium halimifolium	—ditto—
Lavandula stoechas	flowers and seed
Armeria gaditana	flowers and fruits
Asphodelus aestivus	—ditto—
Juniperus phoenicea	buds, fruit and seed
Rosmarinus officinalis	old flowers and seed
Olea europea	whole ripe fruit
Pistacia lentiscus	ripe fruit
P. terebinthus	—ditto—
Retama sphaerocarpa	seed
Juncus bufonius	—ditto—
Trifolium spp.	—ditto—
Asparagus aphyllus	fruit and seed
Phillyrea augustifolia	—ditto—

Source: Soriguer (1981*b*).

3.5 Population ecology

The population ecology of the rabbit is not very well-known in many countries of Europe. We have used European data when they are available, but have otherwise largely relied on data from Spain and France. Those from Spain are from Andalucía, the main study areas being of evergreen-oak forest (highland) and olive forest and scrub (lowland). Much of the French data comes from three areas where many studies have been done: in the north, Ile-de-France, the area around Paris, the landscape is predominantly agricultural, but with extensive woodland, and the climate temperate; in the south the Vaucluse is an upland area with a Mediterranean climate; and southernmost is the Camargue, a flat, saline, deltaic area on the Mediterranean coast.

3.5.1 Reproduction

Just as there is a latitudinal gradient in rabbit weights, so there are clear trends in the length and timing of the reproductive season (Table 3.8). The larger rabbits of northern France and Sweden start breeding later and continue longer (180 days or more) than the smaller rabbits of the south of France (96 days); in Mediterranean areas the season starts earlier still, but its end is very variable.

Seasonal patterns of reproduction in males are clear throughout Europe, with latitudinal trends. Indices of testicular activity, for example size and position of testes, tend to start rising slightly earlier in the south of Spain (Fig. 3.11) than in the south of France (Fig. 3.12), with variations from year to year, but clearer is a tendency to later decline in testicular indices in the north of France than in the south (April–May) (Fig. 3.13).

Male rabbits can reach sexual maturity at 6 months of age in France, although 8–9 months is the average (Rogers 1979; Arthur 1980); in Sierra Morena (southern Spain), Soriguer (1981*a*) reports 4–5 months.

In the north, gravid females may be found from early March to mid-August or even mid-September;

Table 3.8 Reproductive season of rabbits for various places in Europe

Locality	First pregnancies		Last pregnancies		Length of season (days)
	Month	Week	Month	Week	
Spain SW	November	1	May	3	210
″ S1 (dry)	January	1	March	1	90
″ S2 (irrigated)	October	2	June	2	210
Portugal S	September	3	April	4	240
Spain N	December	2	August	1	270
France S (Camargue)	January	2	May	1	96
″ C (Vaucluse)	January	2	June	1	142
″ N (Paris)	February	1	July	3	192
Holland	March	2	July	3	150
Sweden	March	3	August	2	180

Source: Andersson *et al*. (1979); Arthur (1980); Arthur and Gaudin (unpublished); Lopez Ribeiro (1981); Soriguer (1981*a*); Vandewalle (1989); Wallage-Drees (1983).

Fig. 3.11 Changes in testis weights of rabbits from 3 plots in southern Spain from 1980 to 1982 (Soriguer unpublished). (– – – –) Plot 1, irrigated throughout all years. (• – • – • –) Plot 2, seasonal rainfall only in 1980, irrigated throughout thereafter. (———) Plot 3, no irrigation, seasonal rain only.

in the south of France, from early January to the end of May; and in southern Spain from December or earlier, to March assuming a dry spring, sometimes later, except on irrigated plots or in exceptional years. The proportion of females that are gravid tends to rise throughout the season, as do other indices of fertility in females (ovary weight, numbers of corpora lutea, *in utero* litter size); 95 per cent of births in the Camargue were recorded between the 6th and 22nd week of the year (Table 3.9; Rogers 1979; Vandewalle in preparation).

In southern Spain most females become gravid at 3–4 months, whereas in the south of France, where adult weights are 10–15 per cent more, the equivalent age is up to 6 months. In all regions, rabbits are just capable of reproducing in the year of their birth,

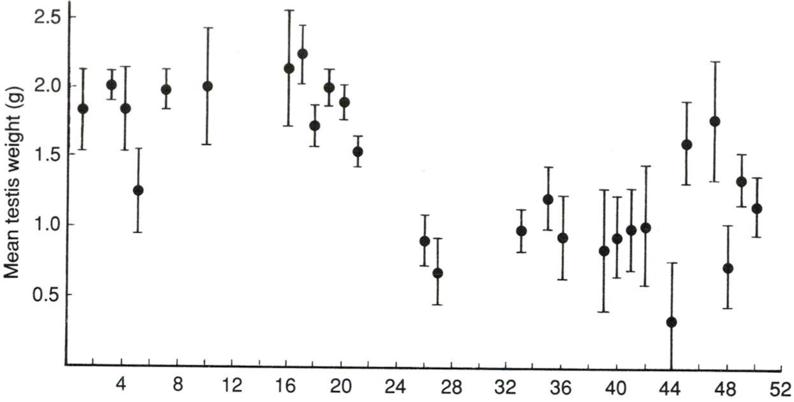

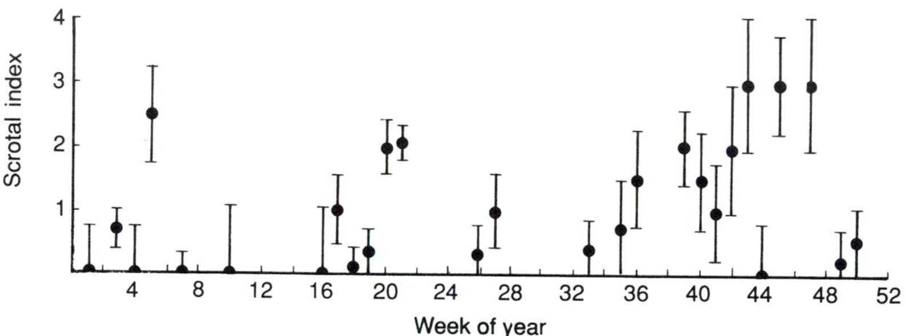

Fig. 3.12 Indices of male fertility in rabbits at La Tour du Valat, southern France, 1975–77 (after Rogers 1979).

though they have more opportunity to do so in the south. However, the contribution made by first-year females to recruitment is questionable, given that they are less fecund than adults, and that they must give birth late in the season when the quality of food is poorer (Rogers 1979).

In general, litter sizes are smaller in the south of Europe than in the north; this may, in turn, be associated with the smaller body size of rabbits in the south (Soriguer 1981*a*, 1983*b*). Average litters are around 3 or 4 in the south of Spain. In the south of France, they average about 5 versus 4.5 in the north (Tables 3.10; 3.11), but the season is shorter and the average female produces only 10 to 13 young per year in the south of France versus about 17 in Ile-de-France.

Whilst the reproductive cycles of both males and females appear to be fundamentally linked to photoperiod, geographical patterns of reproduction suggest that changes in temperature and rainfall can modify both the beginning and the end of the reproductive season. For example, at the beginning of the reproductive season, low temperatures affect both male fertility (Figs. 3.11–3.13) and the weight of the uterus (Fig. 3.14). Conversely, ambient summer temperatures in the Mediterranean are close to the upper limits for spermatogenesis in other lagomorphs (Terroine and Trautmann 1937; Hart *et al.* 1965). In the Camargue, reproduction may start late if the previous winter was unusually wet and cold (Vandewalle personal communication), and stop early in a dry summer. In Ile-de-France, drought reduces pregnancy and lactation rates, whereas in the south of France, early autumn rain stimulates new vegetative growth and a second reproductive season. The difference in reproductive seasons on irrigated and unirrigated plots near Cadiz (southern Spain), shown in Table 3.8, is consistent with the effect of rainfall elsewhere.

Soriguer and Myers (1986) planned a field experiment to test the effects of climate and food on reproduction in rabbits in southern Spain, their presumed ancestral home. They showed that both male and female reproductive cycles can be pre-

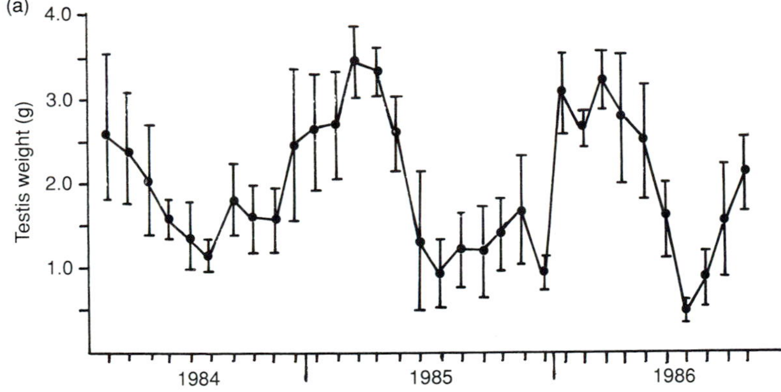

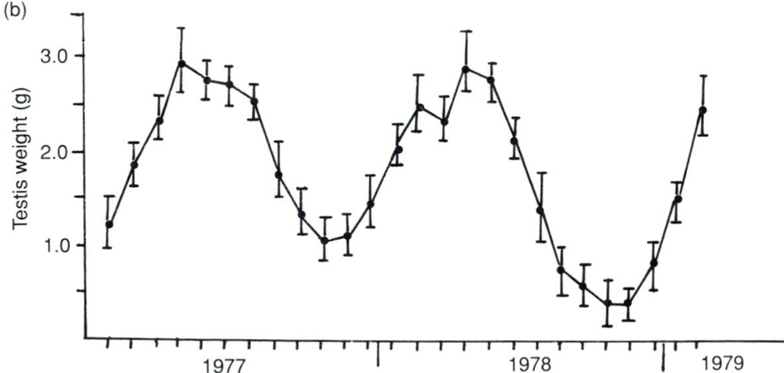

Fig. 3.13 Changes in rabbit testis weights (mean ± standard deviation) in (a) the Camargue and, (b) near Paris (after Arthur 1980 and Vandewalle 1989).

dicted from a combination of climate, food availability, and food quality, although the effect of each may vary between sexes (Table 3.12). Male reproduction is influenced mainly by climatic factors, particularly radiation and temperature, which accounted for 65 to 79 per cent of the variance. Mineral elements in food accounted for only some 5–10 per cent of the variation in males, and the organic components were not relevant. For females, climate was also important, but less so than the quality (organic and inorganic fractions) of food.

In summary, therefore, the rabbit's reproductive activity in Europe is governed by climate within a framework of photoperiod, particularly winter conditions in the north, and summer conditions in the south. Food quantity and quality, also influenced by climate, have a lesser role, albeit an important one, influencing mainly the length of the breeding season. The rabbit's reproductive cycle fits the broad seasonal pattern of the Mediterranean climate. But Mediterranean weather is unpredictable; our experience is that exceptional years seem to be normal! The rabbit is well placed to take advantage of them as they arise.

3.5.2 Survival and mortality

Survival and mortality again show distinct regional differences. The survival rate of the young rabbits in southern Spain is much lower than in France. Conversely, amongst adults the average survival rates are much higher in Spain. At the three sites within France survival rates are similar (Table 3.13).

In Ile-de-France there are two important periods of adult mortality, May–July (15–20 per cent per month) and September–October (20 per cent). From January to April, adult mortality is almost nil (Fig. 3.15). For all ages combined, monthly mortality in the Ile-de-France samples ranged from a high of 34 per cent from May to July, dropping to an average of 20 per cent until December, and then less than 10 per cent from January to April (Arthur unpublished). In

Table 3.9 Proportion of female rabbits pregnant each month across Europe

Month	Spain				Portugal	France		Holland	Sweden
	SW	S1	S2[a]	N	S	S	N	NW	S
January	0.36	0.60	0.29	0.20	0.11	0.0	0.0	0.0	0.0
February	0.55	0.33	0.67	0.78	0.22	0.05	0.28	0.0	0.0
March	0.69	0.33	0.22	1.00	0.75	0.65	0.77	0.12	0.20
April	0.65	0.0	0.0	0.86	0.67	0.68	0.57	0.12	0.71
May	0.38	0.0	0.0	0.67	0.0	0.61	0.80	0.36	0.59
June	0.0	0.0	0.10	0.25	0.0	0.21	0.65	0.07	0.68
July	0.0	0.0	0.0	0.67	0.0	0.0	0.65	0.01	0.56
August	0.0	0.0	0.0	0.27	0.0	0.0	0.44	0.0	0.13
September	0.0	0.0	0.0	0.0	0.11	0.0	0.32	0.0	0.0
October	0.0	0.0	0.14	0.0	0.31	0.0	0.07	0.0	0.0
November	0.30	0.0	0.29	0.0	0.20	0.0	0.0	0.0	0.0
December	0.36	0.0	0.05	0.05	0.67	0.0	0.01	0.0	0.0
Mean	0.28	0.10	0.15	0.40	0.25	0.18	0.38	0.06	0.24

[a] Irrigated plot.
Sources:
France S—Camargue: Rogers (1979). Vandewalle (in preparation).
France N—Ile-de-France: Arthur (1980 and unpublished data).
Holland: Wallage-Drees (1983). Data are frequency of births.
Portugal: Lopez Ribeiro (1981).
Spain (north): Ceballos (unpublished).
(south): Soriguer (1981*a*, 1983*b*); Soriguer and Myers (1986).
Sweden: Andersson *et al*. (1979).

Table 3.10 Some mean annual reproductive parameters of wild rabbits in France and Spain

	Camargue	Vaucluse	Paris region	Andalucía
% pregnant	64 ± 6	58 ± 18	64 ± 9	–
Length of reproductive season (days)	94 ± 30	142 ± 31	174 ± 10	90–150
Date of first conception (week number)	2 (1–7)	2 (1–6)	5 (1–8)	–
No. of pregnancies per female	2.0	2.7	3.4	
No. of embryos per female	4.9 ± 0.4	5.0	4.4–0.3	3.7
No. of live births per female per year	9.8 ± 3.0	13.1 ± 4.2	17.4 ± 2.7	–

Sources: C. P. Arthur (Paris—unpublished); C. P. Arthur and Gaudin (Vaucluse—unpublished); P. M. Rogers (Camargue litters—1979); R. C. Soriguer (Andalucía—unpublished); P. Vandewalle (Camargue other—unpublished).

Table 3.11 Litter sizes of rabbits in Europe and the Mediterranean

Location	Mean litter size	Source
Spain		
NW Andalucía	3.21	Soriguer (1981*a*)
SW Andalucía	3.88	Delibes and Calderon (1979)
Navarra (N. Spain)	4.11	Ceballos (in prep.)
France		
Camargue	5.20	Rogers (1979)
	4.89	Vandewalle (in press)
Vaucluse	5.00	Arthur and Gaudin (unpublished)
Ile-de-France	4.40	Arthur (unpublished)
Holland	5.00	Wallage-Drees (1989)
Sweden (S)	4.70	Andersson *et al.* (1979)
Great Britain		
Wales	4.36	Stephens (1952)
Caernarvonshire (1941)	4.89	Brambell (1944)
Caernarvonshire (1942)	5.64	Brambell (1944)
Morocco	3.7	Soriguer (unpublished)

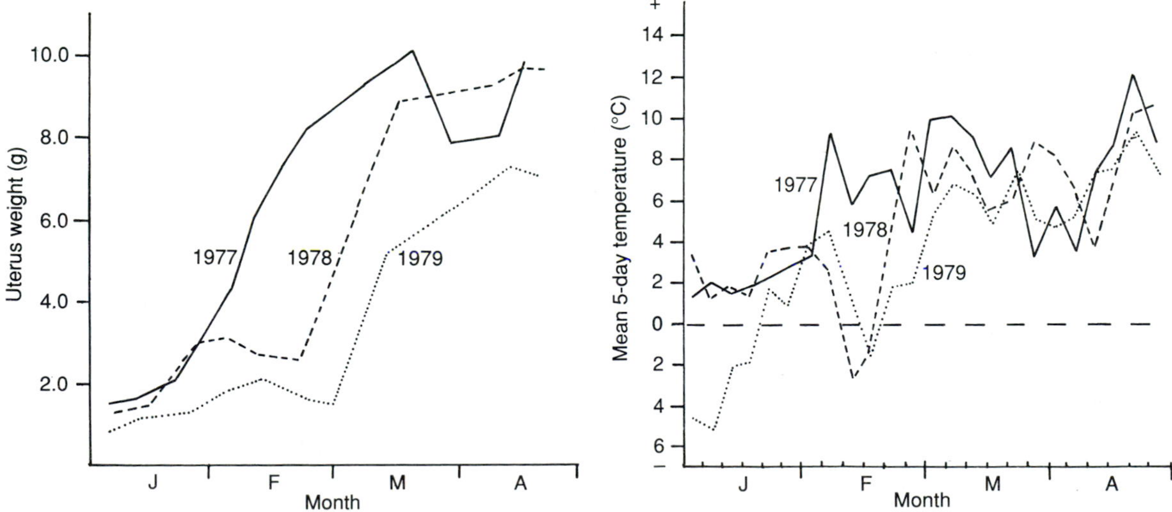

Fig. 3.14 Relationship between uterus weight and ambient temperature in rabbits near Paris, 1977–79. Source: Arthur (unpublished).

Table 3.12 Canonical correlation coefficient (CCR) and probability (in parentheses) for reproductive correlates of rabbits in southern Spain

Canonical variables	CCR—males	CCR—females
A Climate + food availability	0.87 ($<10^{-4}$)	0.84 ($<10^{-4}$)
B Organic fraction	0.65 (0.06)	0.75 (0.03)
C Inorganic fraction	0.71 (0.002)	0.82 (0.001)
A + B + C	0.98 (0.003)	0.99 ($<10^{-4}$)

Source: Soriguer and Myers (1986).

Table 3.13 Survival rates (%) of wild rabbits in France and Spain

Age (months)	Spain	France		
	South	Camargue	Vaucluse	Paris region
0–3[a]	16	32	34	25
4[b]–8	63	58	52	26/49[c]
adult	85	58	42	52/58[c]

[a] 3.7 months for the Spanish data.
[b] 3.8 months for the Spanish data.
[c] 26 and 52 in years with hunting, 49 and 58 in years without hunting.
Source: Arthur (1980), Soriguer (1981, 1983*a*) and unpublished data from C. P. Arthur, J. C. Gaudin, and P. Vandewalle.

the Camargue, monthly mortality (all ages) also peaks in April and July. During periods of reproduction it reaches 20–30 per cent; in autumn and winter it falls to 5–10 per cent (Rogers 1979). We have few data available on causes of mortality in Europe. Possibilities include hunting, predation, myxomatosis, other diseases, flooding, and injuries, for example from farm machinery.

Survival rates of rabbits over 4-months-old in the three French populations were fairly similar (averaging 42–58 per cent) when they were not hunted—that is, in the Camargue and the Vaucluse all the time, and in the Ile-de-France in some years. But in the years with hunting in Ile-de-France, the mortality of 4–8-month-old rabbits was markedly higher, although the adults were little affected.

Arthur collected dead rabbits whenever he found them throughout a four-year study in Ile-de-France. Excluding over 1000 rabbits that were shot, for the other 419 rabbits collected, the most frequent causes of death were predators or myxomatosis (Table 3.14), followed by injuries and various diseases (pasteurella, staphylococcus, coccidia, pseudotuberculosis). Agricultural operations (especially mowing and harvesting) were an important hazard for younger rabbits. The importance of predation and myxomatosis was confirmed by radio-tagging rabbits from the same population in non-hunting years (Table 3.15).

In the north, the most important predators of young rabbits are mustelids, and of older rabbits, foxes. However, foxes take many young as well: they may be responsible for 30–50 per cent of nestling mortality in spring (Arthur 1980; Mulder and Wallage-Drees 1979), when young rabbits may comprise 70–80 per cent of the diet of foxes (Fig. 3.16; Julliot

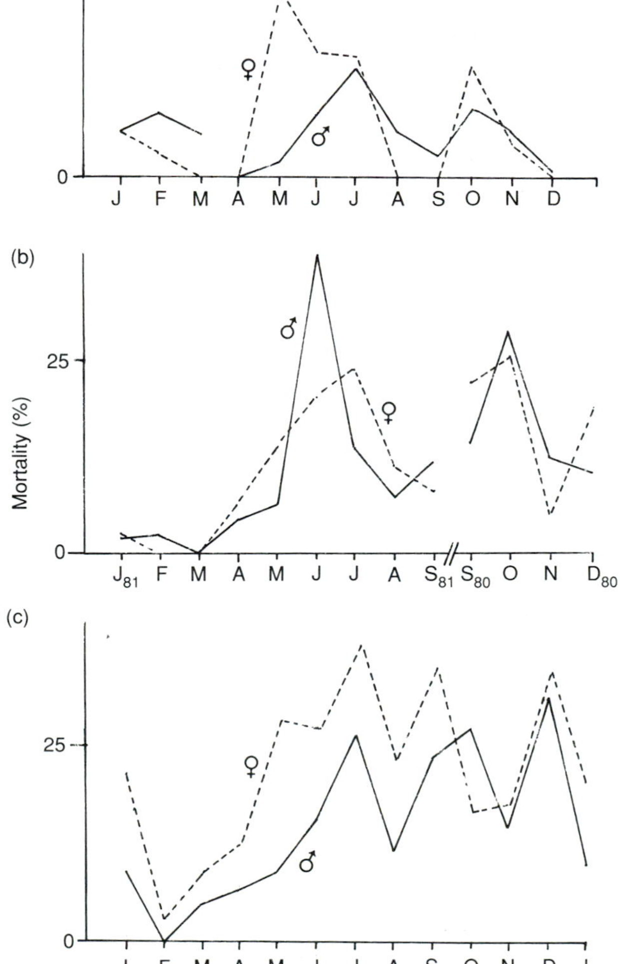

Fig. 3.15 Monthly mortality rates near Paris (a) of rabbits (>4 months) with radio tags, using Trent and Rongstat's (1974) method, 1984–86 years pooled (Arthur unpublished), (b) of marked adult (>9 months) rabbits, 1980–81 (Arthur unpublished), (c) of rabbits known to be alive (all ages >1 month), 1979–80 (Arthur unpublished).

1987). Mortality from myxomatosis may be 10–50 per cent, according to year and season, and evidently influences more immediate causes of death.

In the Camargue, most mortality is due to a combination of predation and myxomatosis (Rogers 1979). Myxomatosis facilitates predation, since the proportion of rabbits in the diet of foxes rises to 70 per cent during myxomatosis epizootics, or during flooding (Reynolds 1979). As rabbits can recover from severe myxomatosis that must affect their susceptibility to predation, the two must to some degree be additive. Over three years myxomatosis was responsible for 13–25 per cent of total mortality annually (Vandewalle 1986; Arthur 1988), while in the Vaucluse it varied from 10 to 50 per cent over three years (Arthur and Gaudin in press).

In the west Mediterranean in general, and in southern Spain in particular, the number of species of predator is exceptionally high (Table 3.16). Furthermore rabbits are an important item of most predators' diet, especially in southern Spain, where, with the exception of wolf, all predators of medium and large size (3 kg for mammals, 1.5 kg for birds) rely to a large extent on rabbits for food (Delibes and Hiraldo 1981; Jaksić and Soriguer 1981; Soriguer 1981*a*, 1981*b*, 1983*b*; Soriguer and Rogers 1981).

Table 3.14 Causes of death of rabbits found dead in Ile-de-France, in four body-weight classes

	<300 g		300–1000 g		Immature/ sub-adult		Adult (>9 months)	
	n	%	n	%	n	%	n	%
Predation	33	34.3	19	22.1	31	29.8	55	41.7
Myxomatosis	0	–	26	30.2	48	46.2	38	28.8
Road kill	19	19.8	11	12.8	7	6.7	7	5.3
Injuries/disease	0	–	15	17.4	11	10.6	20	15.1
Agricultural machinery	8	8.33	7	8.1	5	4.8	2	1.5
Unknown	27	28	8	9.3	2	1.9	10	7.6
Total	87		86		104		132	

Source: Arthur (1980).

Table 3.15 Causes of death of radio-tagged rabbits in the Paris region during a three-year study

	Myxomatosis	Other disease	Predation	Road kill	Miscellaneous	Number tagged
Juveniles (<1000 g)	5	5	6	4	3	45
Adults and sub-adults (>1000 g)	12	5	9	2	4	92

Source: C. P. Arthur—unpublished data.

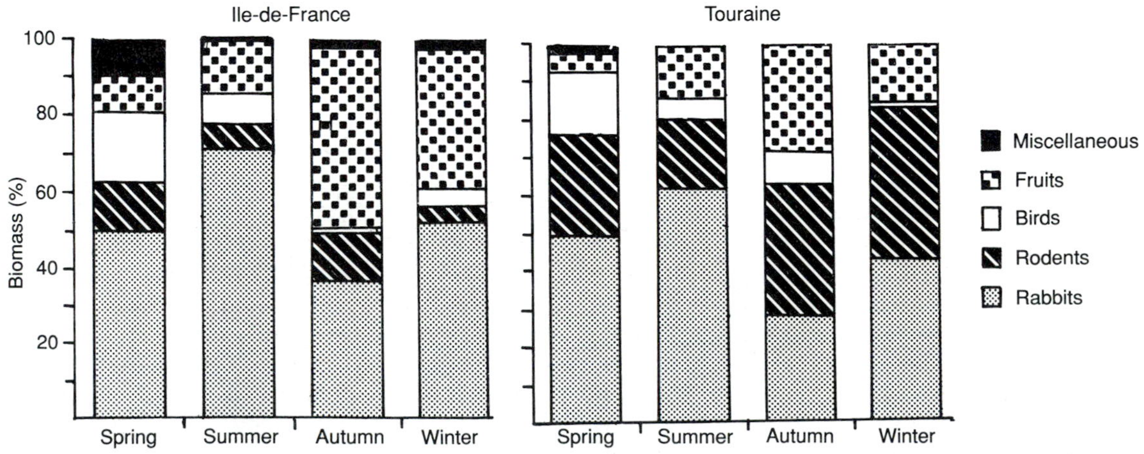

Fig. 3.16 Diets of foxes in two areas of France (Fragner *et al*. 1990). Touraine is 200 km SW of Paris.

Table 3.16 Number of predator species in various parts of the rabbit's range worldwide

Class	Australia	New Zealand	UK	France	Spain
Mammals	3	3	10	10	17
Birds	14	1	7	13	19
Reptiles	6	0	0	7	4
Totals	23	4	17	30	40

Source: Soriguer (1981*a*), Soriguer and Rogers (1981) except Australia (Myers *et al.*, Chapter 5) and UK (Thompson, Chapter 4).

3.6 Myxomatosis (see also Chapter 7)

3.6.1 Incidence

The rapid spread of myxomatosis throughout Europe, starting from Dr Delille's estate near Paris in June 1952, has been well-documented, but substantial study of the disease in wild rabbits in continental Europe did not begin until 20 or so years later. Up to the 1970s we have data only from domestic rabbits. In France, the number of domestic rabbits with the disease climbed to a peak in 1955, and in 1960 stabilized at around 100 000 cases a year (Fig. 3.17). A clear seasonal pattern became established, in which incidence ranged from a maximum in August–September to virtually zero from December to May–June (winter to early summer), although spring epizootic were not unknown (Fig. 3.18; Joubert *et al.* 1972).

Data on myxomatosis in wild rabbits are now available from different parts of France (Table 3.17), including the Camargue, the Vaucluse, and the Ile-de-France. In the Camargue, epizootics recur regularly every summer, and in the Vaucluse, every autumn. In Ile-de-France, however, myxomatosis may break out at any time. In the Isère, near Lyon, the timing of epizootics is also variable, although it is usually in late spring or early summer (Fig. 3.19; Gilot and Joubert 1980). In the north, epizootics last about 3 months, often longer than in the Camargue and significantly longer than in the Vaucluse (2 months).

In southern Spain, July to September/October used to be the usual period for epizootics, but since 1977 they have been most frequent in winter–spring, lasting 6 months in 1976 and 10 months in 1977, although some years are without the disease altogether (Soriguer 1980*a*, 1981*a*). More recently the highest mortality has been from July to October (Soriguer 1981*a*; Soriguer and Rogers 1981). The mean interval between successive epizootics is the same (9–10 months) in the three regions for which we have data, but is most variable in the north, where it may be anything from 4 to 24 months.

Estimated mortality, calculated from the ratio of acute to total cases including recoveries, varies from 39 to 61 per cent with little difference between years. The highest mortality is in summer epizootics (43 per cent over 6 months), lower in autumn (13 per cent over 2.3 months), and lowest in winter epizootics (6 per cent in 4.5 months). There is, nevertheless, great variability in mortality rates between areas, and again the most variable region is Ile-de-France. In both the Camargue and the Vaucluse, total losses per epizootic are similar, 32 and 36 per cent respectively. The short autumn epizootics in the Vaucluse seem to be especially intense, affecting some 83 per cent of the juveniles and sub-adults but total losses are few. Many young rabbits there exhibit high levels of resistance to myxoma virus in the laboratory (no deaths out of 11 young rabbits tested in the Vaucluse compared with 5 deaths out of 7 young rabbits tested in Ile-de-France; Arthur, Gaudin and Guénezan, unpublished data).

The proportion of rabbits affected varies with age-class, region and year (Table 3.18). In southern Spain

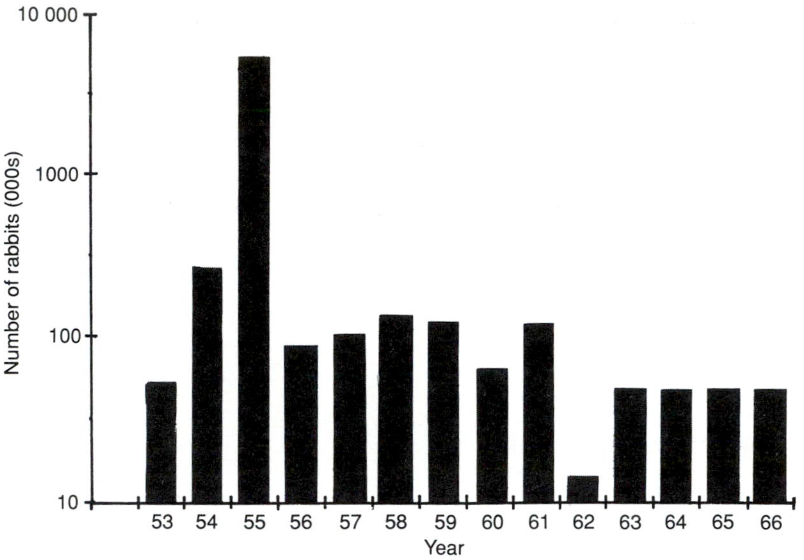

Fig. 3.17 Numbers of domestic rabbits with myxomatosis in France, 1953–56 (after Joubert *et al*. 1972).

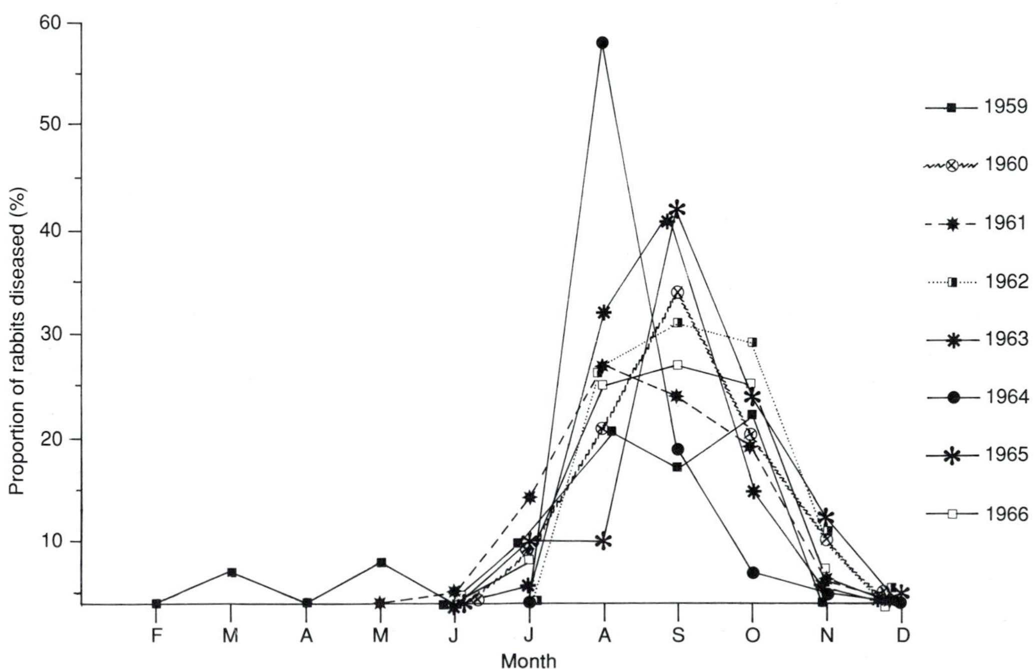

Fig. 3.18 Annual changes in myxomatosis rates in wild and domestic rabbits in France, 1959–66 (after Joubert *et al*. 1972).

Table 3.17 Characteristics of myxomatosis epizootics in three sites in France

	Vaucluse n = 3	Camargue n = 5	Paris region n = 9
1. Duration (months)	2.2 ± 0.6	3 ± 1	3.1 ± 1.9
2. Interval between two epizootics (months)	9.8 ± 0.4	9.7 ± 0.6	9.4 ± 6.5
3. Proportion of animals affected (% cumulative)	83 ± 29[a]	59.9 ± 27[a]	41 ± 31
4. Estimated mortality (%)	39 ± 10	61 ± 10	54 ± 13
5. Population losses (%)	32 ± 8[b]	36 ± 11[b]	22 ± 20
6. No. of epizootics in			
spring	0	0	3
summer	0	5	4 (summer and autumn)
autumn	3	0	
winter	0	0	2

[a] Only juveniles, immatures and sub-adults were affected.
[b] Losses were from the young of the year only.
3. = proportion of infected animals, summed monthly.
4. = proportion of animals with acute and severe clinical sign.
n = number of epizootics observed.
Source: Paris—Arthur (1988); Camargue—Rogers (1979), Vandewalle (1986); Vaucluse—Arthur and Gaudin (in press).

the proportion of juvenile rabbits found with clinical signs is low because, we suspect, they succumb to the disease quickly, often before showing any external sign; the proportion of juveniles with antibodies (13 per cent) is about half of that in adults (Soriguer and Lopez, in press). In the Camargue and the Vaucluse few juveniles are diseased, whereas in Ile-de-France juveniles represent 19 per cent of affected animals. In the north, 33 per cent of adults captured showed clinical symptoms, whereas in the Camargue and the Vaucluse almost no adults are affected.

3.6.2 Epidemiology

Many of the different regional characteristics of the disease reflect variation in transmission routes and in the immunological status of the populations. The relationship between rabbits and myxomatosis has also changed over the years.

In France and Spain, as elsewhere, there are two known vectors, fleas and mosquitoes. Their relative importance and specific roles vary from one region to another. Rabbits in Spain support at least 22 genera of ectoparasites, versus 8 in Australia, all but one of which are also recorded from Spain (Table 3.19).

In the Camargue at least 12 species of mosquito bite rabbits from late March to end-October; they are especially active in May and October, when they may inflict 1200 bites per rabbit per day. Although mosquitoes are relatively uncommon in July–August, when epizootics of myxomatosis are usually at their height, each rabbit is still bitten 60 to 300 times a day during the months that mosquitoes carry the virus. Some 20 to 30 per cent of mosquitoes are carriers, and 50 per cent in *Aedes detritus* and *A. caspius*, the two species most attracted to rabbits (Legrand 1986). The proportion of mosquitoes which are carriers reduces gradually through the season, to nil in October. In contrast, in the Isère some 10 species of mosquito bite rabbits from the beginning of May to the beginning of October, averaging only 40 bites/rabbit/day. In Ile-de-France only 5 species

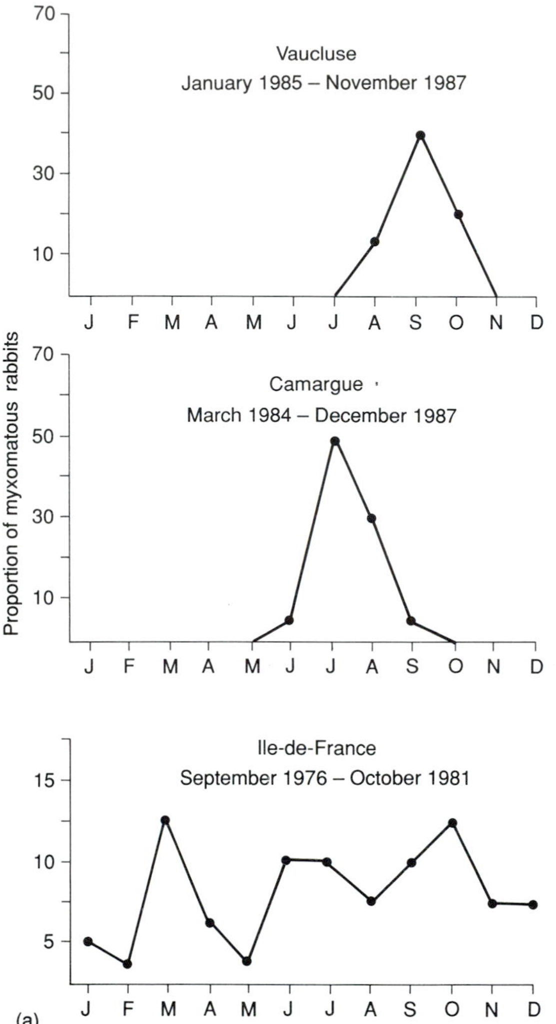

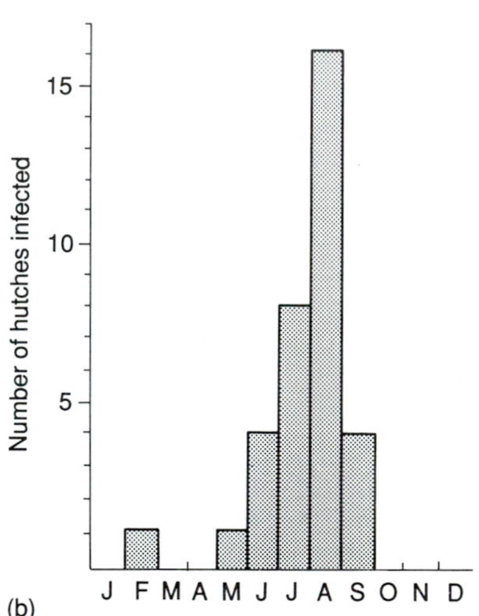

Fig. 3.19 Monthly changes in mean myxomatosis rates in France (a) in wild rabbits in three regions (after Arthur 1988), (b) domestic rabbits in Courtenay, Isère (Gilot and Joubert 1980).

Table 3.18 Percentage of rabbits captured (numbers in parentheses) having clinical sign of myxomatosis during epizootics in France (years pooled) and Spain

	Vaucluse	Camargue	Ile-de-France	Spain	
				1976	1977
Juveniles (<500 g)	1.8 (648)	1.9 (321)	19.1 (833)	0.0 (130)	9.6 (52)
Immatures and sub-adults	47.7 (235)	62.8 (137)	35.4 (768)	8.2 (49)	38.9 (36)
Adults	0.0 (83)	6.0 (73)	32.8 (481)	4.0 (228)	17.6 (131)

Source: Arthur (1988); Soriguer (1980*a*).

Table 3.19 Genera of ectoparasites found on rabbits in Spain and Australia

Spain	Australia
Amblyoma	*Bdellonypsus*
Boophilus	*Cheyletiella*
Caenopsylla	*Echidnophaga*
Cheyletiella	*Haemadipsus*
Chorioptes	*Haemaphysalis*
Ctenocephalides	*Listrophorus*
Dermacentor	*Spilopsyllus*
Desmodex	*Xenopsylla*
Haemadipsus	
Haemaphysalis	
Hyaloma	
Ixodes	
Linguatula	
Listrophorus	
Notoedes	
Odontopsyllus	
Pulex	
Psoroptes	
Rhipicephalus	
Sarcoptes	
Spilopsyllus	
Xenopsylla	

Source: Soriguer (1980*b*).

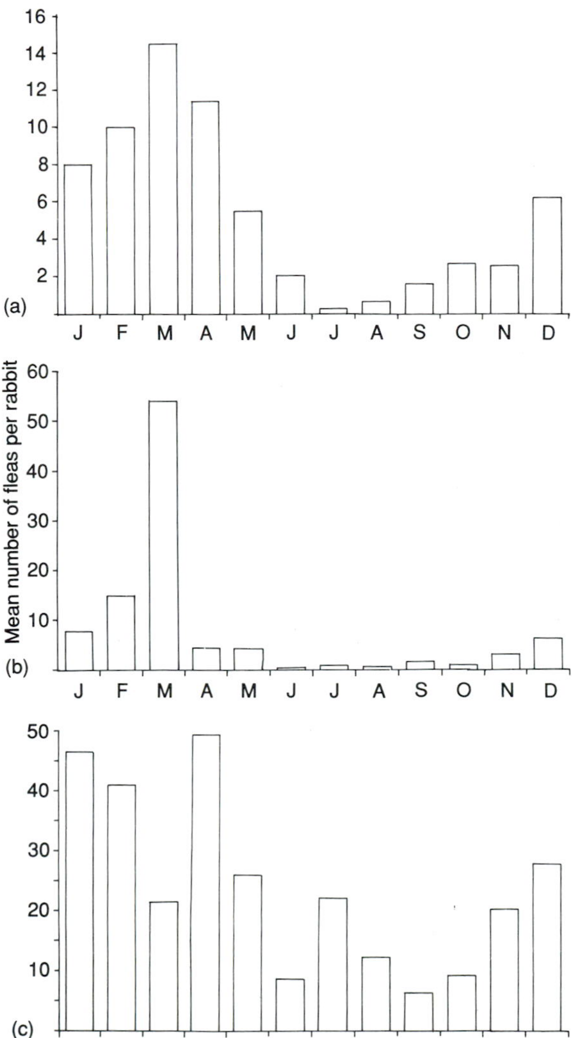

Fig. 3.20 Monthly numbers of fleas on rabbits in France. (a) Camargue (Vandewalle unpublished), (b) Vaucluse (Arthur and Gaudin unpublished), (c) Paris (Launay and Arthur unpublished).

bite rabbits, from May until September, and the maximum number of bites per rabbit per day is 12 in July (Vandewalle 1981).

The abundance of fleas, at least on French rabbits, varies inversely with that of mosquitoes. In both the Camargue and Vaucluse the average number of fleas per rabbit (adults and sub-adults) is greatest from March to April (Camargue, 20–30 fleas per rabbit; Vaucluse, 40–60 fleas per rabbit), and is virtually zero during the season of epizootics. In Ile-de-France rabbits carry the most fleas, 60 to 120 per rabbit, from February to May. In summer the number is still 10 to 20 (Launay 1980; Fig. 3.20).

The virus is present on the mouth-parts of fleas throughout the year in all three areas studied in France. That is, while fleas are proven carriers during epizootics, they also carry the virus even when no disease is evident among the rabbits, also in May even in the absence of any sign of the disease (Gourreau *et al.* in preparation). The presence of the virus on known vectors apparently in the absence of

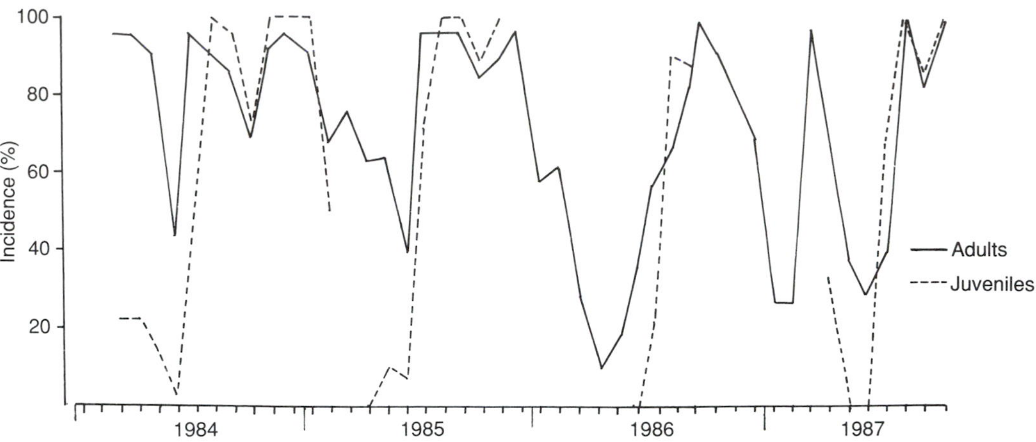

Fig. 3.21 Proportion of juvenile and adult rabbits with myxoma antibodies, Camargue, southern France, 1984–7 (after Vandewalle, unpublished).

any clinical sign of disease is puzzling, and raises important questions about, among other things, the immunological status of the rabbit population.

In the Camargue there is a clear seasonal cycle in the proportion of rabbits with antibodies (Vandewalle 1986). All adult rabbits have antibodies after an epizootic (usually in late August or early September) (see also Chapter 7). They all continue to carry antibodies until January–February, when the proportion gradually decreases to 20–30 per cent in April–May, just before the start of the next epizootic. A variable proportion (0 to 25 per cent) of juveniles carry antibodies until the age of 2 months, but during an epizootic the proportion of juveniles climbs quickly to reach 100 per cent by the end of August (Fig. 3.21). Thereafter they follow the adult pattern.

In the Vaucluse all adults have antibodies from October to June, after which the proportion decreases to about 50 per cent in August. From September, stimulated by the appearance of a new epizootic, the proportion climbs back to towards 100 per cent. Again, some 10 to 50 per cent of young rabbits carry antibodies until the age of 2 months. By June–July usually less than 10 per cent of young still have them, but the proportion rises to 50–60 per cent by the end of October, and 90–100 per cent in January (Fig. 3.22; Arthur and Gaudin, in press).

Results from Spain confirm this pattern. After an outbreak over 90 per cent of rabbits carry antibodies, declining to 50 per cent five months later. Mortality particularly affects juveniles and sub-adults, because 40 per cent of young rabbits (up to 800 g) do not carry antibodies (Soriguer 1980*a*, 1981*a*).

In the Ile-de-France, there are two different cyclical patterns. When there is no autumn or winter epizootic, the 50 to 60 per cent of adults which have detectable antibodies in May declines to 10–20 per cent by January–February. During the summer, 15–20 per cent of young (up to 2 months) have maternal antibodies, but in August–September they lose their antibodies, and from October–November, when the young of the previous season make up three-quarters of the total population, the proportion with detectable antibodies stays at zero until the onset of the next epizootic. After a new epizootic, the proportion of surviving rabbits carrying antibodies increases rapidly the following spring, according to the intensity of the outbreak, to stabilize by August at 60–70 per cent. About 30–40 per cent of the young of the year have antibodies.

The second type of cycle starts after an autumn or winter outbreak, and its effect is to stimulate a higher level of resistance in winter and spring. After an autumn epizootic, 70–90 per cent of adults and young are carriers by November; or 30–40 per cent by February–March after a winter epizootic, the majority of them being young of the previous season (Arthur, unpublished data).

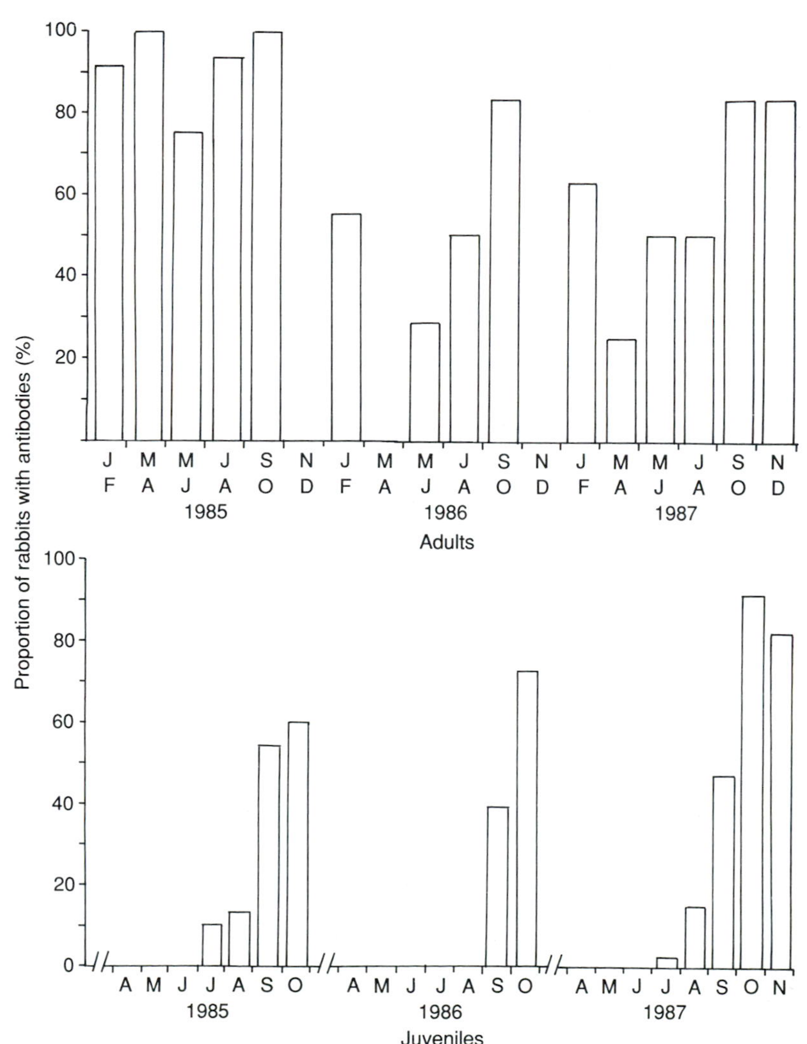

Fig. 3.22 Changes in the proportion of juvenile and adult rabbits with myxoma antibodies, Vaucluse, south-central France, 1985–7 (after Arthur and Gaudin in press).

3.7 Viral haemorrhagic disease

Viral haemorrhagic disease (VHD) was first described in domestic rabbits in China. The cause was a small (28–33 nm diameter) round icosahedral virus, without envelope. The nucleic acid is single-strand RNA (Liu *et al.* 1984). The virus could agglutinate erythrocytes in the blood of sheep, poultry and humans (type 'O'). The incubation period for the disease was 48–72 hours, with death from several hours to one or two days later. The typical symptoms was epistaxis, although often no clinical sign was observed. Characteristic pathological changes included punctate haemorrhage in the respiratory and digestive systems, the spleen, cardiac muscles, and occasionally the kidneys. A rabbit tissue vaccine of virus inactivated with 0.4 per cent formaldehyde was developed with satisfactory results; later modifications improved its efficiency (Arguello 1989; Pages 1988).

A few years after Liu described it in China, the same disease appeared in Europe, under several aliases: virus 'x', hepatosis, rabbit haemorrhagic disease, haemorrhagic septicemia syndrome,viral haemorrhagic disease. It caused high mortality in domestic and wild rabbits, and also hares, in Italy (Cancelloti *et al.* 1988), France (Morisse 1988), Spain (Arguello *etal.* 1988), Germany (Loliger *et al.* 1989), Austria (Nowotny *et al.* 1990). VHD is similar to another disease of hares, European brown hare syndrome, which produces VHD-like lesions. VHD can be transmitted to hares, experimentally at least, and in Czechoslovakia hares have been found with VHD antibodies (Morisse *et al.* 1991).

VHD was first detected in Spain in the spring of 1988, in both domestic and wild rabbits from two widely separated areas, Asturias in northern Spain and Almeria in the south-east (Arguello 1989). A little more than a year after these first two foci were detected, only the south-west parts of the country were VHD-free. The disease was recorded in the adjacent region of Murcia, at Sucina in December 1988, and at Lorca in April 1989 (León Vizcaino, unpublished); one year later the first rabbit with VHD was found in Doñana National Park. In only two years the virus had spread throughout the whole of Spain.

In France, the first outbreak of VHD was recorded in the summer of 1988, in several small rabbitries in the north-east (Haute-Saône, Vosges, Côte d'Or). The deaths were so unusual and unexpected that they were at first attributed to radiation from Chernobyl or some other environmental pollution. Their infectious origin was not discovered until the end of the year (Morisse 1989). Of over 100 dead rabbits taken from all over France in 1989, about half had died from VHD (ONC 1990). By the end of 1991 VHD was established in 56 out of the 95 départements of France.

In parts of France (in the Camargue, Vaucluse, Hérault) several VHD epizootics break out each year. Their effect on wild rabbit populations, although largely unquantified, has often been dramatic. Soriguer and Cooke (unpublished) observed one of the initial outbreaks of VHD in wild rabbits at El Alquian (Almeria, Spain) in June–July 1988. Because the disease progresses so rapidly, some rabbits died while trying to escape them, yet without showing any external symptom. There were many more dead rabbits on the ground than the predators and scavengers could remove (León Vizcaino, Cooke and Soriguer, personal observations). In Doñana National Park, where systematic observations had been made for some years, VHD was associated with an additional mortality of 60 to 70 per cent or more (Villafuerte and Moreno personal communication). On a reserve in the Vaucluse, the reduction in population was estimated at 70–80 per cent (Gaudin, personal communication).

From an ecological point of view in the predator-rich community of the Mediterranean, VHD represents a catastrophic event. Studies now in progress show that rabbits have begun to develop some genetic resistance to the disease, although annual outbreaks are normal. It is still too soon to make any predictions, but we expect that some equilibrium between the VHD virus and the rabbit will develop eventually, as with the myxoma virus.

3.8 Conclusion

The picture of the rabbit in Europe which emerges from this review is that of a biological opportunist bound by ecological constraints. Like any mammal, the rabbit needs high quality food for reproduction and growth. Its reproductive strategy is flexible enough to allow exploitation of favourable periods of plant growth—indeed it has evolved physiological links with photoperiods to enable it to predict them. Its habitat requirements are not too specific, somewhere to dig a hole and food nearby. Given its requirements the rabbit's productivity is proverbial (see, for example, Thompson and Worden 1956; or Sheail 1971).

There must, however, be constraints to hold such potential in check. In southern Spain, which is perhaps the nearest contemporary equivalent of the rabbit's ancestral home and where we have relatively undisturbed study sites, there seem to be three groups of constraints: terrain, climate, and predation. The same three reappear in the south of France as part of the recurring theme of latitudinal trends. Terrain determines the security of the burrows.

While it is not uncommon for some rabbits to live above ground, whole populations thrive only where the soil is suitable for building burrows, or where some substitute is available (some thriving populations in Spain use gaps between rocks, where it is impossible to burrow). Climate limits the reproductive season of the rabbit, via its effects on vegetation, quantity in winter, and quality in summer. But even so the rabbit might be able to outstrip its resources but for the third and final constraint, predation. In many places today this may be imposed by shooting, but an impressive variety of natural predators depends on rabbits, especially in the south. Many translocation/reintroduction efforts fail in Spain because predators kill the rabbits before they can establish stable burrows in their new home. The rabbit has also managed to come to terms with catastrophic events such as the arrival of myxomatosis and VHD. Such ecological adaptability allows a stability in southern rabbit populations that is not to be found further north.

Historical changes in land use allowed the rabbit to extend its range northwards. It was a serious pest of agriculture in the north of France by the sixteenth century. But the damage it did was outweighed by its value as a game species, and from the end of the nineteenth century active management by gamekeepers encouraged the rabbit to attain very high population densities between 1920 and 1940. The introduction of myxomatosis was a spectacular setback, but recovery began quickly and continued throughout Europe until 1970. Since 1980, further changes in land use and the appearance of new epizootics have reduced the numbers of rabbits available for shooting. The efforts of hunters in the south to augment game populations have been copied in the north. Conservationists also work to maintain rabbit populations, in order to protect the many Mediterranean predators that depend largely on rabbits—some exclusively, such as Bonelli's eagle *Hieraetus fasciatus* during its nesting season (Iborra *et al.* 1990). The same is true for the Spanish lynx *Lynx pardina* and the imperial eagle *Aquila adalberti* (*A. heliaca*), two of the most endangered species in Europe.

The biology and ecology of the rabbit in Europe, from Sweden where it is an introduced species to its ancestral home in the south of Spain, show a tremendous range which spans the differences measured in other parts of the world. In its ancestral home the rabbit is an essential part of a complex natural environment. Only when it is removed from the demands and constraints of that environment does the ecological opportunism, which allows it to survive at home, turn it into a pest elsewhere.

3.9 Summary

Oryctolagus, first known from the Miocene in southern Spain (5–7 million years ago) is of uncertain ancestry. Of its three known species (*layensis*, *lacosti*, and *cuniculus*) only one is extant, *O. cuniculus*. It was first found in southern Spain in the mid-Pleistocene and co-existing with the other extant lagomorph, *Lepus*. Over the last 15 000 years *O. cuniculus* has become smaller. Today it also exhibits a latitudinal gradient in body size, rabbits in the north of Europe being some 50 per cent larger than the south.

Genetics and ectoparasites indicate 3 groups of rabbits. In southern Spain, their presumed ancestral home, they have been separate from those of southern France (and of northern Spain) for over 50 000 years. Northern European rabbits represent a third distinct group to which domestic rabbits and those introduced elsewhere are most closely related.

Rabbits spread through Europe in the last 10 000 to 20 000 years from the Mediterranean. Their domestication by monks in France in the Middle Ages, where rabbits have long been managed for food or as game, is well documented. Population numbers gradually increased until the 1920s, much higher in the north than the south, staying high until the introduction of myxomatosis in 1952 near Paris. The disease spread throughout most of Europe in little more than a year, decimating populations. After 4 or 5 years, numbers gradually increased until the early 1970s, and more rapidly thereafter.

Inherent population cycles, disease, and climate may all influence changes in population numbers.

Terrain characteristics and changes in land use are also important. The rabbit has the status of game throughout most of Europe, and open seasons and hunting methods are strictly controlled. In France and elsewhere, rabbits (wild-type and crosses) are bred for restocking wild populations. Releases of alien species, notably *Sylvilagus* spp. have occurred, but with little success, and are illegal. Conflict between farmer and hunter is reduced by compensation systems, but not in Spain, where rabbits are not officially regarded as a pest.

Extensive studies show that rabbit diets in Europe reflect landscape and climate. Grasses and forbs are preferred, and a small number of plant species may make up a large proportion of the diet. Exclosure experiments have emphasized the role of rabbits as primary consumers, and in changing floristic composition. Rabbits may have an important, so far unmeasured, role in seed dispersal especially in the south of Spain.

The length and timing of most reproductive parameters also show strong latitudinal trends. Northern rabbits starting breeding older, later, and for longer than their counterparts in the south and produce more young per year. In Mediterranean areas breeding is extremely variable. In southern Spain male reproductive cycles can be predicted mainly from the climate, particularly radiation and temperature; climate is also important for females, but less so than food quality. These characteristics reflect the opportunist strategy of rabbits, a response to the variability of Mediterranean weather patterns.

Survival rates of young rabbits in southern Spain are much lower than in France. Conversely, survival rates of adults are much higher in Spain. Causes of mortality include hunting, predation, often in combination with disease, and injuries from farm machinery. In the west Mediterranean the number of predator species is high. Rabbits are important in most predators' diets, especially in southern Spain, where all medium and large predators except wolf rely largely on rabbits, and some exclusively during breeding.

Both myxomatosis and viral haemorrhagic disease are endemic in Europe. Genetic resistance to myxomatosis is well-established, and resistance appears to be developing to the latter. Both have an extensive influence on population numbers.

In its ancestral home the rabbit is an essential part of a complex natural environment, valued by hunters and conservationists alike. The demands and constraints of that environment have prevented it becoming the pest that it is elsewhere.

Acknowledgements

The authors are grateful for financial support from the Fondation Tour du Valat and the National Research Council of Canada (to PMR), the Office National de la Chasse (to CPA) and the Consejo Superior de Investigaciones Científicas (to RCS). Much of PMR's input is derived from work done at the Station Biologique de la Tour du Valat from 1975–9. CPA was employed by the Office National de la Chasse whilst collecting the date used in this chapter. Section 3.1 was written with N. Lopez Martinez, Universidad de Madrid and E. Bernaldez, Estación Biológica de Doñana, CSIC, Sevilla, and O. Lopez Ribeiro, Portugal. PMR and RCS also wish to acknowledge a debt of gratitude to Dr Ken Myers for support, guidance and fruitful collaboration over many years—no Ken, no chapter.

References

Andersson, M., Dahlback, M., and Meurling, P. (1979). The biology of the wild rabbit, *Oryctolagus cuniculus*, in southern Sweden. I. Breeding season. *Viltrevy*, **11**, 103–27.

Arguello, J. L. (1989). Ovejero (Spain) develops vaccine for a new disease in rabbit. *Animal Pharmacology*, **172**, 6.

Arguello, J. L., Llanos, A., and Perez-Ordoño, L. L.

(1988). Enfermedad vírica hemorrágica del conejo en España. *Medicina Veterinaria*, **5**, 645–51.

Arthur, C. P. (1980). Démographie du lapin de garenne, *Oryctolagus cuniculus*, L. 1758, en région parisienne. *Bulletin spécial scientifique et technique de l'Office National de la Chasse*, December 1980, 127–61.

Arthur, C. P. (1988). A review of myxomatosis among rabbits in France. *Revue scientifique et technique de l'Office International des Epizootics*, **7**, 937–57.

Arthur, C. P. and Chapuis, J. L. (1985). Le lapin de garenne, *Oryctolagus cuniculus*. In *Atlas des mammifères sauvages de France* (ed. A. Fayard), pp. 204–5. Société Française pour l'Etude et la Protection des Mammifères, Paris.

Arthur, C. P. and Gaudin, J. C. (in press). Influence de la vaccination sur la myxomatose du lapin de garenne, *Oryctolagus cuniculus*, dans le département du Vaucluse. *Revue de Médecine Vétérinaire*.

Arthur, C. P. and Guenezan, M. (1986). Le prélévement cynégétique en lapins de garenne en France, saison 1983–1984. *Bulletin mensuel de l'Office national de la Chasse*, **108**, 23–32.

Arthur, C. P., Chapuis, J. L., Pages, M. V., and Spitz, F. (1980). Enquête sur la situation et la répartition écologique du lapin de garenne en France. *Bulletin spécial scientifique et technique de l'Office National de la Chasse*, December 1980, 37–86.

Bailey, J. A. (1968). A weight–length relationship for evaluating physical condition of cottontails. *Journal of Wildlife Management*, **32**, 835–41.

Bassett, P. (1978). *The effects of grazing in the horse pasture*. Fondation Tour du Valat, eighth meeting of the board, 25–7.

Bassett, P. and Rogers, P. M. (1979). Exclosure experiments. Project R26 of the Fondation Tour du Valat, in Rogers, P. M. (1979), *Ecology of the European wild rabbit,* Oryctolagus cuniculus *(L.), in the Camargue, southern France*, 164–71. Ph.D. thesis, University of Guelph. Microfiche, National Library of Canada, Ottawa.

Beaucournu, J. C. (1980*a*). Les puces du lapin de garenne. In *First International Conference on Fleas*, Ashton, UK (eds R. Traub and H. Starcke), pp. 383–9. Balkema, Rotterdam.

Beaucournu, J. C. (1980*b*). Les ectoparasites du lapin de garenne: apports à son histoire. *Bulletin spécial scientifique et technique de l'Office National de la Chasse*, December 1980, 23–36.

Bernaldez, E. and Soriguer, R. C. (1989). Morphometric evolution in *Oryctolagus cuniculus* during the Holocene in SW of the Iberian Peninsula. *Abstracts of the 5th International Theriological Congress*. Rome.

Bhadresa, R. (1977). Food preferences of rabbits *Oryctolagus cuniculus* (L.) at Holkman sand dunes, Norfolk. *Journal of Applied Ecology*, **14**, 287–91.

Biju-Duval, C., Ennefaa, H., Dennebouy, N., Monnerot, M., Mignotte, F., Soriguer, R. C., El Gaaïed, A., El Hili, A., and Mounolou, J. C. (1991). Mitochondrial DNA evolution in lagomorphs: origin of systematic heteroplasmy, organisation of diversity in European rabbits. *Journal of Molecular Evolution*, **33**, 92–102.

Birkan, M. G. and Pépin, D. (1984). Tableaux de chasse et de piégeage d'un même territoire entre 1950 et 1971: fluctuations numériques des espèces et facteurs de l'environnement. *Gibier Faune Sauvage*, **2**, 97–112.

Bodson, L. (1978). Données antiques de zoogéographie. L'expansion des léporidés dans la Méditerrannée classique. *Naturalistes belges*, **59**, 66–81.

Brambell, F. W. R. (1944). The reproduction of the wild rabbit *Oryctolagus cuniculus* (L.). *Proceedings of the Zoological Society of London*, **114**, 1–45.

Cancelloti, F. M., Villeri, C., Renzi, M., and Monfredini, R. (1988). Le insidie della malattia X del coniglio. *Rivista di Coniglicoltura*, **9**, 41–6.

Ceballos, O. (in prep.). Biología y ecología del conejo en Navarra, N. España.

Chaline, J. (1976). Les lagomorphes. In *La Préhistoire française I*. pp. 419. Centre National de la Recherche Scientifique, Paris.

Chapuis, J. L. (1979). Le régime alimentaire du lapin de garenne, *Oryctolagus cuniculus* (L. 1758) dans deux habitats contrastés: une lande bretonne et un domaine de l'Ile de France. Unpublished doctoral thesis, Université de Rennes.

Delibes, M. and Calderon, J. (1979). Datos sobre la reprodución del conejo *Oryctolagus cuniculus* en Doñana s.o. España, durante un año seco. *Doñana Acta Vertebrata*, **6**, 91–9.

Delibes, M. and Hiraldo, F. (1981). The rabbit as prey in the Iberian Mediterranean ecosystem. In *Proceedings of the world lagomorph conference* (1979) (ed. K. Myers and C. D. MacInnes), pp. 614–22. University of Guelph, Ontario.

Delort, R. (1984). *Les animaux ont une histoire*. Seuil, Paris.

Donard, E. (1982). Recherches sur les léporinés quartenaires (Pléistocéne moyen et supérieur, Holocène). Unpublished doctoral thesis. Université de Bordeaux.

Duncan, P. (1992). *Horses and grasses. The nutritional ecology of equids and their impact on the Camargue*. Springer-Verlag, New York.

Fenner, F. and Ratcliffe, F. N. (1965). *Myxomatosis*. Cambridge University Press.

Fragner, J., Stahl, P., and Arthur, C. P. (1990). Relation entre régime alimentaire du renard (*Vulpes vulpes* L.) et l'abondance des proies en melieux semi-bocager. *XXIIème colloque francophone de mammalogie*, SFEPM Paris, pp. 64–86.

Gayot, P. (1865). *Lièvres, lapins et léporidés*. Librairie de la Maison rustique, Paris.

Giban, J. (1956). Répercussion de la myxomatose sur les populations de lapins de garenne en France. *Terre et Vie*, **3–4**, 179–88.

Gilot, B. and Joubert, L. (1980). Le rôle vectoriel des Culcidés dans l'épidémiologie de la myxomatose: bilan critique des études effectuées dans la région Rhône-Alpes de 1975 à 1980. *Bulletin spécial scientifique et technique de l'Office National de la Chasse*, December 1980, 243–64.

Gourreau, J. M., Arthur, C. P., Gaudin, J. C., and Vandewalle, P. (in preparation). Présence et persistence du virus myxomateux sur la puce du lapin, *Spilopsyllus cuniculi*, dans trois régions de France. Relations avec l'épidémiologie de la myxomatose. *Revue Médecine Vétérinaire*.

Hart, J. S., Pokl, M., and Tener, J. S. (1965). Seasonal acclimatization in varying hare (*Lepus americanus*). *Canadian Journal of Zoology*, **43**, 731–44.

Homolska, M. (1985). Die Nahrung einer Population des Wildkaninschen (*Oryctolagus cuniculus*) auf dem böemischemärischen Höhenzug. *Folia Zoologica*, **34**, 303–14.

Iborra, O., Arthur, C. P., and Bayle, P. (1990). Importance trophique du lapin de garenne pour les grands rapaces provençaux. *Vie et Milieu*, **40**, 2–3, 177–88.

Jacotot, H., Vallée, A., and Virat, B. (1956). Apparition en France d'un mutant naturellement atténué du virus de Sanarelli. *Annales Institut Pasteur*, **89**, 231–6.

Jaksić, F. and Soriguer, R. C. (1981). Predation upon the European rabbit (*Oryctolagus cuniculus*) in the Mediterranean habitats of Chile and Spain: a comparative analysis. *Journal of Animal Ecology*, **50**, 269–85.

Joubert, L. (1979). *Compte rendu de l'état d'avancement des travaux sur la myxomatose*. Convention no. 78–21, Ecole Nationale, Vétérinaire, Lyon. Office National de la Chasse, March 1979.

Joubert, L., Leftheriotis, M., and Mouchet, P. (1972). *La myxomatose*. 2 volumes. L'expansion scientifique, Paris.

Launay, H. (1980). Approche d'une prophylaxie de la myxomatose: écologie des puces du lapin de garenne. *Bulletin spécial scientifique et technique de l'Office National de la Chasse*, December 1980, 213–42.

Legrand, S. (1986). *Contribution à l'épidémiologie de la myxomatose en Camargue. Rôle et importance vectorielle des culicidés*. Rapport BEPA Cynégetique, LEPA Vendôme. Office National de la Chasse, Paris.

Liu, S. J., Xue, H. I., Pu, B. Q., and Quin, N. M. (1984). A new viral disease in rabbits. Rabbit viral haemorrhagic disease. *Animal Husbandry and Veterinary Medicine*, **16**, 253–5.

Loliger, H. C., Matthes, H. S., and Liess, E. B. (1989). Uber das Austeten einer Infectioner hemorrhagischen Erkrangunen bei Hauskanischen in der Bundesrepublik Deutschland. *Tierärztliche Umschau*, **44**, 22–5.

Lopez Martinez, N. (1977*a*). Nuevos lagomorfos (Mammalia) del Neógeno y Cuaternario Español. *Trabajos N/O, Madrid CSIC*, **8**, 7–45.

Lopez Martinez, N. (1977*b*). *Revisión sistemática y bioestratigráfica de los lagomorfos (Mammalia) del Neógeno y Cuaternario de España*. Unpublished doctoral thesis, Universidad de Madrid.

Lopez Martinez, N., Michaux, J., and Villalta, J. F. (eds) (1976). Rongeurs et Lagomorphes de Bagur 2 (Province de Gerone, Espagne). Nouveau remplissage des fissures du début de Pleistocène moyen. *Acta géologica hispaniensis*, **11**, 2, 46–54.

Lopez Ribeiro, O. (1981). Quelques données sur la biologie du lapin de garenne (*Oryctolagus cuniculus*, L.) au Portugal (Contenda Sud-est de Portugal). *XV Congresso International de Fauna Cinegética y Silvestre*, Trujillo, Spain, 607–13.

Lumley-Woodyear, H. de (1969–1971). *Le paléolithique inférieur et moyen du Midi méditerranéen dans son cadre géologique*. Supplément à Gallia-Préhistoire, 2 volumes. Centre National de la Recherche Scientifique, Paris.

Middleton, A. D. (1934). Periodic fluctuations in British game populations. *Journal of Animal Ecology*, **6**, 231–49.

Morisse, J. P. (1988). Haemorrhagic septicemia syndrome in rabbits. First observations in France. *Le Point Vétérinaire*, **20**, 117, 835–9.

Morisse, J. P. (1989). La maladie hémorragique virale du lapin (VHD). Etat des recherches et évolution de la maladie un an après son apparition en France. *L'éleveur de lapins*, **26**, 28–34.

Morisse, J. P., Le Gall, G., and Boilletot, E. (1991). Hépatite d'origine virale des Léporides: introduction et hypothèse étiologique. *Revue scientifique et technique de l'Office International des Epizootics*, **19**, 269–82.

Mulder, J. L. and Wallage-Drees, J. M. (1979). Red fox predation on young rabbits in breeding burrows. *Netherlands Journal of Zoology*, **29**, 144–9.

Muñoz-Goyanes, G. (1960). *Anverso y reverso de la mixomatosis*. Dirección General de Montes, Caza y Pesca Fluvial, Madrid.

Nowotny, N., Fuchs, A., Schilcher, F., and Loupal, G. (1990). Zum Auftreten der Rabbit Haemorrhagic Disease (RHD) in Österreich: I. Pathomorphologische and virologische Untersuchungen. *Wiener Tierärztliche Monatsschrift*, **7**, 19–23.

ONC (1976). Enquête statistique nationale sur les tableaux de chasse et tir pour la saison 1974–1975. Premiers résultats. Le lapin de garenne. *Bulletin spécial scientifique et technique de l'Office National de la Chasse*, **4**, 38–40.

ONC (1990). Bilan 1989 de la surveillance sanitaire de la faune sauvage, SAGIR. *Office National de la Chasse, Service de Protection de la Faune*, Paris.

Pages, M. V. (1980). *Statut du lapin de garenne*, Oryctolagus cuniculus *L. 1758, dans certains milieux du Languedoc*. Unpublished doctoral thesis. Université de Montpellier I.

Pages, A. (1988). Consideraciones técnicas de la sueroterapia y de la profilaxis vacunal en la enfermedad hemorrágica vírica del conejo (RHDV). *Medicina Veterinaria*, **6**, 285–91.

Panaget, C. (1983). *Etude comparative des régimes-alimentaires du lapin de garenne et du lièvre d'Europe sur un domaine de l'Ile de France*. Rapport de stage, Diplome Universitaire de Technologie, Tours. Muséum National d'Histoire Naturelle, Paris.

Reynolds, P. (1979). Preliminary observations on the food of the fox (*Vulpes vulpes*) in the Camargue, with special reference to rabbit (*Oryctolagus cuniculus*) predation. *Mammalia*, **43**, 3, 295–307.

Richardson, B. J., Rogers, P. M., and Hewitt, G. M. (1980). Ecological genetics of the wild rabbit in Australia. II. Protein variation in British, French and Australian rabbits and the geographical distribution of the variation in Australia. *Australian Journal of Biological Science*, **27**, 671–5.

Rogers, P. M. (1979). *Ecology of the European wild rabbit,* Oryctolagus cuniculus *(L.), in the Camargue, southern France*. Ph.D. thesis. University of Guelph. Microfiche, National Library of Canada, Ottawa.

Rogers, P. M. (1981). Ecology of the European wild rabbit *Oryctolagus cuniculus* (L.) in Mediterranean habitats. II. Distribution in the landscape of the Camargue, S. France. *Journal of Applied Ecology*, **18**, 355–71.

Rogers, P. M. and Myers, K. (1979). Ecology of the European wild rabbit *Oryctolagus cuniculus* (L.), in Mediterranean habitats. I. Distribution in the landscape of the Coto Doñana, S. Spain. *Journal of Applied Ecology*, **16**, 691–703.

Ross, J. (1982). Myxomatosis: the natural evolution of the disease. *Symposia of the Zoological Society of London*, **50**, 77–95.

Servan, J. (1972). Rapport de stage sur le lapin de garenne, *Oryctolagus cuniculus*. Unpublished D.E.A. d'écologie, Université Paris 6.

Sheail, J. (1971). *Rabbits and their history*. David & Charles, Newton Abbot.

Skulyatvev, A. A. (1987). Wild rabbit (*Oryctolagus cuniculus*) in USSR hunting grounds. *Transactions of XVIII UIGB Congress*, Stockholm, p. 184.

Siriez, H. (1957). *La myxomatose, moyen de lutte biologique contre le lapin, rongeur nuisible*. Société des Editions Pharmaceutiques, Paris.

Soriguer, R. C. (1980*a*). Mixomatosis en una población de conejos de Andalucía Occidental. Evolución temporal. Epidemia invernal. Resistencia genetica. *Actas I. Reunión Iberoamericana de Zoología y Conservación de Vertebrados*, 241–50.

Soriguer, R. C. (1980*b*). Ciclo anual de parasitismo por pulgas y garrapatas en el conejo de campo

(*Oryctolagus cuniculus* L.) en Andalucía Occidental, España. *Revista Iberica de Parasitología*, **40**, 4, 539–50.

Soriguer, R. C. (1980*c*). El conejo (*Oryctolagus cuniculus*) en Andalucía Occidental: parámetros corporales y curva de crecimiento. *Doñana Acta Vertebrata*, **7**, 83–90.

Soriguer, R. C. (1981*a*). Biología y dinámica de una población de conejos (*Oryctolagus cuniculus*, L.) en Andalucía Occidental. *Doñana Acta Vertebrata (Vol Especial)*, **8**, 1–379.

Soriguer, R. C. (1981*b*). El conejo y las comunidades de plantas y vertebrados terrestres de Doñana: un caso particular de estudio. *Seminario Sobre Reservas de la Biosfera en España*. Universidad de la Rabida, Huelva.

Soriguer, R. C. (1983*a*). Consideraciones sobre el efecto de los conejos y los grandes herbívoros en los pastizales de la Vera de Doñana. *Doñana Acta Vertebrata*, **10**, 155–68.

Soriguer, R. C. (1983*b*). El conejo: papel ecológico y estrategia de vida en los ecosistemas mediterráneos. *Actas XV. Congresso International de Fauna Cynegética y Sylvestre*, 517–42.

Soriguer, R. C. (1988). Alimentación del conejo (*Oryctolagus cuniculus* L. 1758) en Doñana. SO España. *Doñana Acta Vertebrata*, **15**, 141–50.

Soriguer, R. C. and Herrera, C. M. (1984). Impacto de los grandes herbívoros en el matorral del Parque Nacional de Doñana. *Resúmenes I Jornadas Sobre la Investigatión en el Parque National de Doñana*.

Soriguer, R. C. and Lopez, M. (in press). Mixomatosis en el conejo de campo: antigenos y anticuerpos circulantes en sangre. *Medicina Veterinaria*.

Soriguer, R. C. and Myers, K. (1986). Morphological, physiological and reproductive features of a wild rabbit population in Mediterranean Spain under different habitat management. *Mammal Review*, **16**, 197.

Soriguer, R. C. and Rogers, P. M. (1981). The European wild rabbit in Mediterranean Spain. In *Proceedings of the world Lagomorph conference* (1979) (ed. K. Myers and C. D. MacInnes), pp. 600–13. University of Guelph, Ontario.

Staniforth, R. J. and P. B. Cavers (1977). The importance of cottontail rabbits in the dispersal of *Polygonum* spp. *Journal of Applied Ecology*, **14**, 261–7.

Stephens, M. N. (1952). Seasonal observations on the wild rabbit *Oryctolagus cuniculus* (L.) in western Wales. *Proceedings of the Zoological Society of London*, **122**, 417–34.

Tapper, S. (1985). Rabbit numbers 1961–1983. *Game Conservancy Annual Review*, for 1984, 69–73.

Terroine, E. R. and Trautmann, S. (1927). Influence de la température extérieure sur la production calorifique des homéothermes et loi des surfaces. *Annales Physiologiques Physiochemiques et Biologiques*, **3**, 422–57.

Thompson, H. V. and Worden, A. N. (1956). *The rabbit*. Collins, London.

Trent, T. T. and Rongstat, O. J. (1974). Home range and survival of cottontail rabbits in southwestern Wisconsin. *Journal of Wildlife Management*, **38**, 459–72.

Van der Loo, W. and Arthur, C. P. (in preparation). Gene diversity and population differentiation of the immunoglobulin chain in French rabbit populations.

Van der Loo, W., Ferrán, N., and Soriguer, R. C. (1991). Estimation of gene diversity at the B-locus of the constant region of immunoglobulin light chain in natural populations of European rabbit (*Oryctolagus cuniculus* L.) in Portugal, Andalusia and on the Azorean Islands. *Genetics, 127,* 789–99.

Vandewalle, P. (1981). *Contribution à l'épidémiologie de la myxomatose en Région Parisienne. Rôle et importance vectorielle des Culicidés*. Rapport BEPA Cynégétique, Lycée Agricole Vendôme. Office National de la Chasse, Paris.

Vandewalle, P. (1986). *Etat d'avancement des travaux sur l'écologie du lapin et la myxomatose en Camargue*. Convention Office National de la Chasse–Fondation Sansouire no. 86–26, état d'avancement des travaux mai, 1986.

Vandewalle, P. (1989). Le cycle reproducteur du lapin de garenne (*Oryctolagus cuniculus*) en Camargue. Influence des facteurs environnementaux. *Gibier Faune Sauvage*, **6**, 1–25.

Vandewalle, P. (in preparation). Fertilité et productivité d'une population de lapins de garenne en Camargue, France. Variations interannuelles et influence de l'âge.

Wallage-Drees, J. M. (1983). Effects of food on the onset of breeding in rabbits, *Oryctolagus cuniculus* (L.), in a sand dune habitat. *Acta Zoologica Fennica*, **174**, 57–9.

Wallage-Drees, J. M. (1989). The influence of food supply on the population dynamics of rabbits,

Oryctolagus cuniculus (L.) in a Dutch dune area. *Zeitschrift für Saügetierkunde*, **54**, 304–23.

Williams, O. B., Wells, T. C. E., and Wells, D. A. (1974). Grazing management of Woodwalten Fen: seasonal changes in the diet of cattle and rabbits. *Journal of Applied Ecology*, **11**, 499–516.

Zeuner, F. E. (1963). *A history of domesticated animals*. Hutchinson, London.

4

The rabbit in Britain

Harry V. Thompson

4.1 Introduction and establishment

The fossil record indicates that the European rabbit originated in Iberia (see Chapter 1), died out over much of Europe during the Pleistocene glaciations, and only reappeared in Britain when introduced in the twelfth and thirteenth centuries AD (Veale 1957; Sheail 1971, 1978, 1991). Island or coastal sites were favoured and rabbits flourished on Drakes Island, Devon, in 1135 (Hurrell 1979), on the Isles of Scilly in 1176, on Lundy Island in 1274 and Skokholm Island in 1324. They fetched high prices and were rated as highly as sucking pig (Barrett-Hamilton 1912). In 1389, references to conynges, warrens and ferrets first appeared in the statute book and, in the fourteenth and fifteenth centuries, rabbits were on the menus at great feasts, such as the coronation of Henry IV in 1399, the installation of the Archbishop of Canterbury in 1443 and of the Archbishop of York in 1465.

By the sixteenth century rabbits were abundant; the German naturalist Gesner describes them as living in rocky places in Spain, but in open country in England. Rabbits and conies are mentioned several times by Shakespeare. By the early seventeenth century many estates had rabbit warrens and the cony was much valued for providing food and fur; this obtained until late in the nineteenth century, when the import of carcases and skins from Australia reduced the value of the native product, and more efficient agriculture made wild rabbit production uneconomic.

Although apparently abundant near Edinburgh since the sixteenth century, rabbits were sparse in the rest of Scotland before the early nineteenth century, except on some islands, highlands and coastal areas (Fraser Darling 1947).

Rabbits are recorded from Wales in the thirteenth century, but spread mainly in coastal areas up to the late eighteenth century. The spread of rabbits in Scotland in the nineteenth century has been attributed to improvements in agriculture, but in Wales agriculture was in decline and rabbits were probably introduced and preserved in areas they had not previously colonized. The Normans probably introduced rabbits to Ireland in the thirteenth century and skins were exported by the fourteenth century (Flux and Fullagar 1992).

4.1.1 The importance of warrens

The Phoenicians, sailing to the Iberian peninsula in 1100 BC, noted large populations of rabbits (an animal unfamiliar to them) and named the country after them. In translation the word is Hispania, or Spain. Rabbits continued to be associated with Spain, and the Emperor Hadrian (AD 117–38) struck a coin which figured a rabbit on the reverse side. The Greek historian Polybius, writing of Corsica in the second century BC, refers to the rabbit, which he called *Kunikloi*, and Varro (116–27 BC) in a book on the history of agriculture *De re rustica* suggested keeping rabbits with other game in *Leporaria*; these walled enclosures were the models for the warrens constructed in France and Germany in the Middle Ages. The geographer Strabo, about 30 BC wrote of a plague of rabbits in the Balearic Islands of Majorca and Minorca, which was so severe that the settlers petitioned the Emperor Augustus for help (Schwenk 1986).

While lacking the speed and agility of the hare, the rabbit can run at some 40 kph for short distances until it finds shelter, usually in a burrow. Although it can live above-ground in dense cover, the burrow is the rabbit's natural refuge and the digging of natural warrens enables its colonies to spread into a wide range of habitats. This habit is reflected in the vernacular names for the rabbit in central European languages; all are derived from the Latin *cuniculus*, which also signifies an underground passage. The Italian is *coniglio*; Spanish *conejo*; Portuguese *coelho*; Belgian *konin*; Danish and Swedish *kaning*; German *kaninchen*; Old French *connin*; Welsh *cwningen* and old English *conyng* and *coney*. The term *rabbit*, or rather *rabbet*, *rabytt*, or *rabette*, was used in the fifteenth century for young rabbits only and is apparently derived from the Walloon *rabett*. A warren is defined as 'a piece of uncultivated ground on which rabbits breed wild in burrows', while a burrow is 'a hole or excavation in the ground for a dwelling place by rabbits, foxes and the like' (Oxford English Dictionary 1989).

For seven centuries after their introduction to Britain, rabbits were bred in warrens for domestic use and for sale. The warrens were either unenclosed, as on the sand-dunes of Pembrokeshire and Carmarthenshire, or protected by stone walls or rabbit-proof fences (Figs 4.1, 4.2). Mounds of earth might be provided for burrowing, and holes of up to a metre in length bored in them to encourage rabbit occupation (Sheail 1971). A warren might be farmed or used for sport; in the latter case coverts were planted, and the rabbits shot as they crossed the rides. In the nineteenth century battues were held and under good management a yield of 125 to 250 rabbits per hectare could be expected (Simpson 1908). Crops such as sow thistles, dandelions, and parsley were grown within the warren and the rabbits fed with turnips, swedes, gorse, or hay to boost the yield and discourage them from escaping. The word warren is derived from the French 'garenne' and their abundance is testified by the frequency on our maps of such names as conygarth, coneygarth, conigree, conigre, and warren. The keeper of a warren was responsible for its husbandry, the control of predators such as foxes, cats, mustelids, and birds of prey, including eagles and kites, and the discouragement of poaching.

Warren management was carefully defined in leases, so that both landlord and warren manager shared the profit from the rabbit crop. Rabbit was sometimes favoured above sheep or cattle, which often shared the warren, as in the Breckland of Norfolk. A warren on shallow, poor, sandy, chalky, or moorland soil could provide a useful income when other farming enterprises failed. Public records and agricultural reviews testify to the increasing importance of rabbits in the rural economy from the thirteenth century onwards.

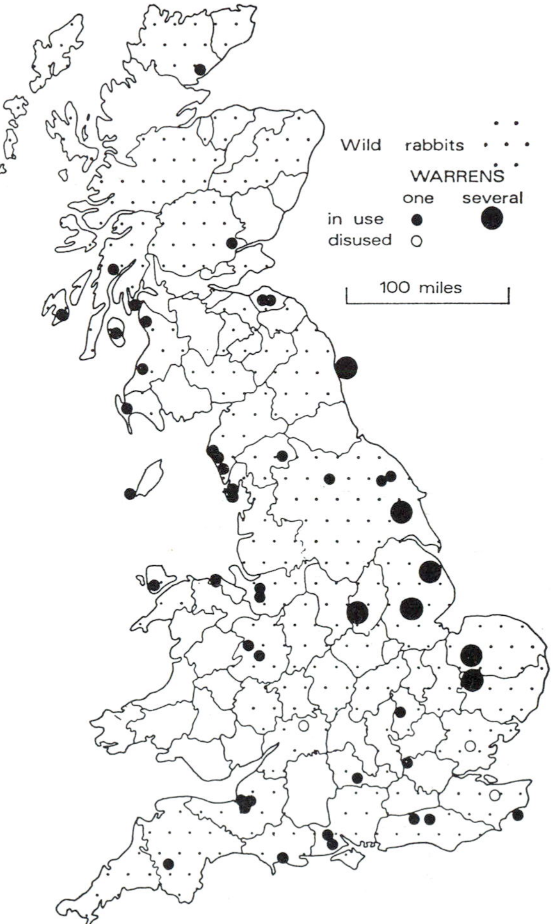

Fig. 4.1 Distribution of wild rabbits and warrens in Britain in the 1790s (based on Board of Agriculture reports) (Sheail 1971).

Rabbits were introduced to many islands off the coast of England, Wales, and Scotland, and often flourished to the extent of being the main source of local income (Ritchie 1920; Matheson 1941; Sheail 1971; Flux and Fullagar 1992).

Although inferior to the skins of beaver, mink, and

(a)

(b)

Fig. 4.2 Ditsworthy Warren, near Plymouth. (a) An artificial mound, and (b) part of the wall round the warren. (Copyright A. S. Thomas.)

many other mammals used in the trade, there was a great demand for rabbit fur in Europe and Asia from medieval times onwards, which encouraged the spread of warrens in Britain. By 1880, skins were being imported from Germany and from Australia and New Zealand; while under The *Ground Game Act* of 1880 tenant farmers could legally kill rabbits on their land and, indeed, sell them more cheaply than the warreners. Wild rabbit meat, with its gamey flavour, maintained its popularity and as the warrens flourished it ceased to be a luxury food by the mid-nineteenth century. Imports from Europe and, when refrigeration was developed, from Australia and New Zealand, were cheaper than from the warrens, which thereafter declined.

4.1.2 Rabbits as game

Rabbits were kept as an amenity on estates in the seventeenth century and increasingly in the eighteenth and nineteenth centuries, often overrunning parkland and causing damage to plantations and adjacent farmland as they spread over the countryside. Away from protected warrens and parkland, rabbits were regarded as a nuisance and treated as such until the increasing interest in small game (pheasants, partridges, and hares) towards the end of the eighteenth century favoured rabbits also. To preserve the game, landlords and sporting tenants employed keepers whose duties included the killing of mammalian and avian predators. Encouraged on game estates, rabbits increased and became an important part of the bag. Concurrently fox-hunting grew in popularity and coverts planted for fox denning and protection incidentally gave cover for rabbits, which also augmented the foxes' diet (Sheail 1971).

4.1.3 Legislation

Legislation about rabbits dates from 1389 and was enacted to protect property rights in rabbits as game, reaching its apogee in the early and mid-nineteenth century. The owner of land, or the person to whom he grants his rights, has a right of property in the wild rabbits on that land and an exclusive right to take alive or to kill them. Warrens were protected by the *Larceny Act* of 1861.

A considerable threat to warren keeping was posed by poachers and drastic measures were used against them, including mantraps and barbed-wire entanglements. The nineteenth century was notable for its legislation against trespassers and poachers. It is an act of trespass to go on to, or send a dog on to, another's land without permission to kill or take rabbits. The provisions are in the *Game Act* 1831, the *Night Poaching Acts*, 1828 and 1844, and the *Poaching Prevention Act*, 1862. It is an offence to be found on land unlawfully in search of, or in pursuit of, rabbits during the daytime, that is from the beginning of the last hour before sunrise to the end of the first hour after sunset. During the night it is an offence unlawfully to take or destroy rabbits in any land, on public roads, highways, and paths and openings from land thereon. Special penalties are provided for second and third offences: originally for a third offence the accused might be sentenced to transportation for seven years, but this was altered to imprisonment (Worrall 1956).

Farmers' legitimate complaints of damage by rabbits gradually gathered force over the centuries and received legal recognition in the *Ground Game Act* of 1880, which gave the occupier of land a right to destroy rabbits on such land. Common law, which is case law based on usage and custom, entitled a tenant to kill rabbits to protect his crops but, prior to the 1880 Act, the rights could be limited or excluded by the terms of the lease. The *Ground Game Act* was stated to be 'in the interests of good husbandry, and for the better security for the capital and labour invested by the occupiers of land in the cultivation of the soil, that further provision should be made to enable such occupiers to protect their crops from injury and loss by ground game', and gave the occupier of land a limited right to kill and take rabbits and hares concurrently with the right of any other person entitled to do so on the same land (Worrall 1956). Since 1880, the rabbit's potential for becoming a pest has been well-recognized and resulted in the *Prevention of Damage by Rabbits Act* 1939, the *Agriculture Act* 1947, and the *Pests Act* 1954.

4.2 Changing attitudes to the rabbit

Views of the economic value of the rabbit have varied greatly over the centuries as most of the chapters in this book indicate. It is greatly prized in domesticity and as ground game, and has been introduced into over 800 islands or groups of islands throughout the world, mostly for food or as game (Chapter 2; Flux and Fullager 1992); at the same time, it has caused untold damage to agriculture.

4.2.1 Impact on the British countryside and agriculture

As rabbits spread into the wild, conflict with traditional agriculture seemed inevitable. Besides grazing closer to the ground than sheep, rabbits can convert heathland *Calluna vulgaris* to grassland dominated by common bent *Agrostis tenuis* and sheep's fescue *Festuca ovina* as shown by Farrow (1917, 1925) in the Breckland of Norfolk. By the eighteenth and nineteenth centuries, farmers realized that heathland, if cultivated, would support grass, corn, and commercial woodland, so warrens were ploughed and sown, rents were raised, crop rotation practised and the landscape greatly changed. As the warrens were destroyed in chalkland, wold, and heath the populations of rabbits fell. But rabbits living wild were still worth catching and some naturalists thought that, without the protection of warrens, the species would not long survive (Bewick 1814). This turned out not to be so, as there was still sufficient scrub and other cover for rabbits to find shelter.

Jealously preserved on estates and protected by the Game Laws, rabbits consumed the crops of adjoining tenant farmers which led to much ill-will, poaching, and a black market in game; this was remedied by legalizing the sale of game (Game Act 1831). Some landlords, appreciating the extent of damage by rabbits, either reduced their numbers, compensated the tenants in cash or by reducing rents, or allowed them to kill rabbits, other than by shooting. By the mid-nineteenth century the Game Laws provoked additional criticism on the grounds that an abundance of rabbits and hares reduced the yield and increased the cost of home-grown corn, while the Corn Laws limited the import of foreign corn. Parliamentary committees considered the issues, the Corn Laws were repealed and in 1880 the *Ground Game Act* made rabbits and hares the joint property of landlord and tenants, while winged game remained solely the property of the landowner.

Over the last 150 years the fortunes of the rabbit have waxed and waned inversely with the prosperity of farming; when cereal prices fell, less land was ploughed and rabbits colonized the resulting scrub. In 1881, a Royal Commission on Agriculture even considered that parts of Norfolk were more profitable under rabbits and game than if cultivated. In 1871, over 5.5 million hectares of land in England and Wales were cultivated, thereafter the area was reduced and even under First World War wartime shortages did not exceed 5 million hectares. Between 1918 and 1939 the cultivated area fell to some 3.5 million hectares (Sheail 1971).

From the Boer War period (1899–1902) the trade in rabbit skins and meat expanded, facilitated by rail and later road freight, the breaking up of estates and, particularly, the increased use of the gin trap. Catching wild animals for food, for their skins, and in some cases to avoid being killed by them, is fundamental to the life of primitive man, who has used cages, snares, pitfall, and deadfall traps. Bronze Age portable wooden traps with treadle plate and bow spring have been found in peat bogs in Ireland and Wales (Lloyd 1962). Iron traps are recorded from the fifteenth century, and a trap identical to the modern gin was in use by the mid-seventeenth century. The word 'gin' has a host of meanings, including 'a product of ingenuity', various 'engines' and 'a contrivance for catching game' (Oxford English Dictionary 1989). By the eighteenth century large double-springed gin traps (mantraps, some 1.3 metres long and with jaws 0.5 metres in diameter) were in use on estates, to discourage poachers. These were essentially the same type of trap as used for bears in North America, where the traps of various sizes were all referred to as 'steel' or 'leg-hold' traps and where there is a history of trap development for companies such as the Hudson Bay Company and the American Fur Company (Bateman 1971). The gin trap most used in Britain, for mammals varying in size from weasels to foxes, was that with 10-centimetre jaws. It had a flat spring, set under tension, so that when an animal stepped on a treadle plate and released the spring, the two hinged jaws were clamped round the animal's leg. Until the outbreak of myxomatosis in 1953–4, over three million gin traps were in regular use in Britain (Lloyd 1962). It is an efficient, if inhumane, trap and in expert hands can be used to eradicate rabbits locally, but may also become the tool of 'rabbit farming' i.e. supplying the market regularly with rabbits, by trapping 100 or more a night and moving the traps to other fields or farms as the catch falls off.

The spread of rabbits across West Wales and the growth of commercial trapping is well-documented (Buckley 1958). As a schoolboy in Carmarthenshire before 1914, Lt Col Buckley noted the spread of gin trapping from Pembrokeshire to Carmarthenshire and its increase during the food shortage of the war years 1914–18. One 80-hectare farm initially yielded 12 foxes, 30–40 weasels and stoats and only 250 rabbits; in subsequent years the annual catch of rabbits was over 3000. A wild rabbit industry became established, not only in Wales but in south-west England, East Anglia, parts of northern England and in Scotland. The Board of Agriculture, no doubt with the food supply in mind, looked benignly upon the increase in rabbits (Sharpe 1918). Returning to farm his estate in 1930, Buckley was appalled to find it overrun with rabbits, while many sheepdogs, cats, and some foxes had three legs, cattle sometimes were found with gin traps on lips or tongues and there was a virtual absence of pheasants, partridges, hares, and woodcock. To restore agricultural production, he renovated the hillsides, using crawler tractors, re-seeded grassland and wrote to the Press and

Parliament urging the abolition of rabbit farming based on open-trapping with the gin.

Meanwhile he dealt with his own rabbit problem. Acting on advice from Australia, and with help from Lord Melchett of Imperial Chemical Industries, Buckley fumigated his rabbit warrens with cyanide powder and killed thousands of rabbits (Buckley 1935). Within eight years his farm was restored to high productivity, and his success elicited great interest and support from farmers and landowners throughout the country and, initially, threats of legal action by trappers and dealers who had bought the rabbiting rights on land adjoining his own. Buckley's articles in the Press brought him into contact with Captain (later Major) C. W. Hume, who founded the University of London Animal Welfare Society (now the Universities Federation for Animal Welfare), and this led to a most fruitful collaboration both in the battle against the gin trap and the encouragement of research on the biology of the rabbit.

At a meeting in 1929, organized by ULAWS, the writer Sir William Beach Thomas put forward the view that instead of reducing the numbers of rabbits, trappers maintained them, somewhat as butchers handling a supply of sheep. Hume, who was already campaigning against gin traps on grounds of cruelty, was impressed by this argument, and also a few years afterwards by reports of a similar attitude in Australia, where cyanide gassing was being used successfully. ULAWS were busy drafting a *Gin Trap (Prohibition) Bill* when Buckley's rabbit gassing activities came to their attention, and they immediately joined forces. The Bill came before the House of Lords in May, 1935, as a Private Member's Bill and, although the cruelty of the trap was accepted, the Bill was narrowly defeated by those favouring the monetary value of wild rabbits. But the extent of rabbit damage to crops was increasingly realized by landowners and farmers, some of whom had forbidden the use of gin traps on their land. The campaign against the gin trap was supported by the National Federation of Women's Institutes, the County Councils Association, the Royal Society for the Prevention of Cruelty to Animals, and the Scottish SPCA. In December 1935 the issue was raised again in the House of Lords and a Select Committee on Damage by Rabbits was appointed. It reported in 1937, accepted that rural opinion was justifiably concerned about damage by rabbits, and recommended that local authorities be empowered to issue orders requiring the destruction of rabbits wherever excessive damage was being caused to adjoining properties. In respect of the gin trap, they proposed that its use in the open should be banned but that it should continue to be used in the mouths of burrows; also that the Ministry of Agriculture should seek for a less cruel trap to replace the gin. Hume next addressed the Parliamentary Science Committee on the subject of the rabbit problem, and it was decided to introduce a Bill based on the recommendations of the Select Committee. The Ministry of Agriculture, which had introduced grassland improvement grants and sought to protect the increased production, supported the Bill on both agricultural and welfare grounds and the *Prevention of Damage by Rabbits Act* was passed in 1939 (Hume 1958, 1962; Worrall 1956).

As Hume pointed out:

> It may seem odd that a humane society should have pioneered the view that wild rabbits should be treated solely as a pest. The explanation is that wild rabbits were being farmed in order that every year between thirty and forty million of them might be tortured to death in traps and thus put money into the pockets of the rabbit-trapping industry. This industry prospered because farmers said 'the rabbits pay my rent', and because gamekeepers said 'rabbits are a valuable source of food', and because hatters used some 20 million rabbits' skins per annum for making felt hats. If once we could persuade the public that, from a national point of view, rabbits do not pay, all justification for the rabbit-trapping industry would disappear.

4.2.2 Research stimulated by Animal Welfare interests and Research Councils

In 1937, at a meeting of the British Association for the Advancement of Science, Hume appealed for systematic research to be done on wild rabbits (Hume 1939) and this was taken up by Charles Elton, the pioneering animal ecologist who founded the Bureau of Animal Population at Oxford University in 1932 and was its Director for 35 years. The result was a study of rabbit ecology and behaviour which broke much new ground (Southern 1940*a*, 1940*b*, 1942*a*, 1948*a*), partly financed by the Universities Federation for Animal Welfare. At the outbreak of war in 1939 this research was subsumed by the Bureau devoting its efforts to research on the control of the brown rat, the black rat, and the house mouse, to conserve crops and stocks of food (Chitty and Southern 1954). Some work was done on the control of rabbits, advocating trapping for food in the winter and gassing burrows in the summer (Middleton 1942). As a complement to the work of Southern and Middleton, studies began in 1941 at the University College of North Wales on the reproduction of the wild rabbit (Brambell 1942, 1944, 1948 and many other papers listed in Thompson and Worden 1956). This work was supported by the Agricultural Research Council, and it demonstrated the remarkable extent to which pre-natal mortality occurs in rabbit populations of high density.

In May 1945, Charles Elton invited members of UFAW to meet his staff at the Bureau of Animal Population to discuss research needed for the humane control of mammals. One result of this was a study in West Wales, based at University College, Aberystwyth, partly financed by UFAW and with support from the Ministry of Agriculture, Fisheries and Food and the BAP. This work showed that rabbits damaged the quality as well as the quantity of pasture, that professional rabbit trapping removes only 30 to 40 per cent of the population, and that a farm could be cleared of rabbits by a combination of gassing and ferreting (Phillips *et al.* 1952; Phillips 1953, 1955*a*, 1955*b*). Based on this work a larger clearance scheme was carried out by MAFF on 1076 hectares of coastal farms in the Mathry district of North Pembrokeshire from 1948–50. The scheme of rabbit clearance on a block-by-block basis was most successful, and became the pattern for subsequent Rabbit Clearance Society work. UFAW again supported a scientific observer, at University College, Aberystwyth, who recorded seasonal data on body weight, breeding, coprophagy, and the incidence of disease and parasites (Stephens 1952).

4.2.3 National control of rabbits

With the outbreak of war, the need for maximum agricultural production made it necessary for the Ministry of Agriculture and Fisheries to invest County War Agricultural Executive Committees with powers under the Defence Regulations for the control of harmful mammals and birds on agricultural land; one of the Regulations again permitted the trapping of rabbits in the open. The CWAECs exerted considerable authority in local farming matters, with executive, scientific, and technical staff, under the guidance of local people of repute. There was much conscientious rabbit control, but without solving the usual problem of reconciling some damage to crops with the food value of the carcases. With the ending of the war there was renewed recognition that the rabbit was an agricultural pest of major importance, that had been well-controlled by some landholders and neglected by others. For example, Colonel Buckley returned to his estate after five years' absence to find the rabbit damage as bad as or worse than in 1930. In the process of remedying this situation he provided generous facilities for the research work based at Aberystwyth (Phillips *et al.* 1952).

The Parliamentary review of agriculture in the mid-1940s resulted in the *Agriculture Act*, 1947 which, *inter alia*, transferred the power to require the destruction of rabbits from local authorities to the Minister of Agriculture, Fisheries and Food, but the Minister delegated his functions to the County Agricultural Executive Committees; similar powers were exercised in Scotland by the Secretary of State. In 1948, concern over reducing imports of food and timber led a senior civil servant, Max Nicholson (who later became Director-General of the Nature Conservancy), to argue a case for virtually eradicating rabbits, with a consequent saving in foreign exchange. This imaginative initiative did not receive sufficient support to be implemented; but the need for drastic action on a voluntary basis was accepted, and endorsed by the farming and landowning organizations. But in the absence of any financial inducement the response was meagre, despite the provisions of the 1947 Act (Sheail 1991).

Humanitarian concern about the gin trap had increased, both within and outside Parliament, and the subject of trapping became part of the remit of a *Select Committee on Cruelty to Wild Animals*, which reported in 1951. It supported a policy of exterminating the rabbit under compulsory clearance orders, stigmatized the gin trap as 'a diabolical instrument' and recommended banning the use of all non-approved traps. Meanwhile, rabbit numbers increased and by October 1953 further legislation was being considered. While alternative traps to the gin were devised and tested and gassing experiments carried out with compressors and impellers, the scene was transformed by the first outbreak of myxomatosis, on 13 October 1953 (see below, Section 4.3; Chapter 7). During 1954 it became clear that myxomatosis would devastate the rabbit population as it had done already in Australia and parts of Europe, so landholders were urged to take the opportunity of eliminating the rabbits which survived the disease.

Concurrently, Parliament debated the issues and by November had passed the *Pests Act*, 1954. This extended the Minister's powers, to designate by order 'rabbit clearance areas', to be freed as far as practicable of wild rabbits. An order could also regulate the manner of clearance, and the destruction of harbourage. During the passage of the Bill there was much discussion both of the gin trap and of myxomatosis. There was agreement about the abolition of the gin (the arguments put forward on animal welfare grounds having been accepted) but hesitation about when to do so, in the absence of an efficient and humane alternative. It was finally agreed that the use of spring traps would be banned in England and Wales after 31 July 1958, except where their design had been approved; furthermore, any approved trap set to catch wild animals must be inspected at least once every day between sunrise and sunset. After the passing of the 1954 Act, a *Humane Traps Advisory Committee* was appointed to stimulate inventors and to approve humane traps, while the Treasury authorized the sum of £5000 for the payment of awards for approved designs. Net traps, anaesthetics, poisonous, explosive, and electrical devices were not considered. Traps were required to be of similar efficiency to the gin, but to be humane. Most effort was concentrated on mechanical spring traps, designed to be set in burrow entrances and to kill effectively and quickly. Between 1954 and 1958, the Committee received 260 patterns and designs and the most promising were field-tested by the MAFF. A basic pattern of successful trap was found to have two arms striking the neck or head of the rabbit in a near vertical plane and holding it. Development work was done by the National Institute of Agricultural Engineering, and by 1957 two effective rabbit traps, the Imbra and the Juby, were developed and approved. Four other spring traps, for use against smaller mammals, were also approved; the Sawyer, the Fenn, the Lloyd, and the Fuller squirrel trap. Subsequently, larger models of the Fenn were approved as rabbit traps (Lloyd 1962, 1963*a*; Bateman 1971) and also the Springer No 6 Multi-purpose trap (and see section 4.7.3).

The most characteristic symptom of myxomatosis is a great swelling of the eyelids and the base of the ears, so that affected rabbits are unable to see and become insensitive to noise. They blunder on to roads and are killed by passing traffic. Public opinion was revolted by the sight of sick and dying rabbits and strongly opposed to the deliberate spreading of the disease; this had never been done officially but it was well-known that individuals had privately moved infected rabbits from one place to another. With support from the Prime Minister Winston Churchill, an amendment was added to the Pests Bill, making it an offence to use a rabbit infected with myxomatosis to spread the disease among uninfected rabbits.

Another legislative measure (under the *Destructive Imported Animals Act*, 1932) was the making of the *Non-Indigenous Rabbit (Prohibition of Importation and Keeping) Order*, 1954. This arose from the fear that non-European rabbits, immune to myxomatosis, might be misguidedly introduced into Britain, to fill the niche left vacant by the European rabbit—as was attempted in France (Chapter 3). The Order prohibits the importation into Britain, and the keeping within Britain, of all species of non-indigenous rabbits.

During the initial spread of myxomatosis, some 99 per cent of affected rabbits died (Thompson and Worden 1956). But pockets of susceptible rabbits missed by the disease, and the few immune rabbits, continued to breed. Landholders co-operated in the setting up of 'rabbit clearance areas' and formed Rabbit Clearance Societies, taking as their model the Rabbit Boards of New Zealand (see Chapter 6). Some societies began on a voluntary basis as early as 1954, and a particularly successful one was the Easter Ross Society in Scotland, which covered 20 000 hectares and owed much to the enthusiasm of one farmer, Mr G. C. D. Budge. In 1958, the Government introduced a 50 per cent grant for co-ordinated rabbit control and the number of societies in Britain increased to 655 by 1960 (Fig. 4.3). By 1964 46 per cent of the farmed land in England and Wales, (6 000 000 hectares) was included within 750 societies. In some areas the rabbit control was most effective, success depending as usual on the enthusiasm and drive of

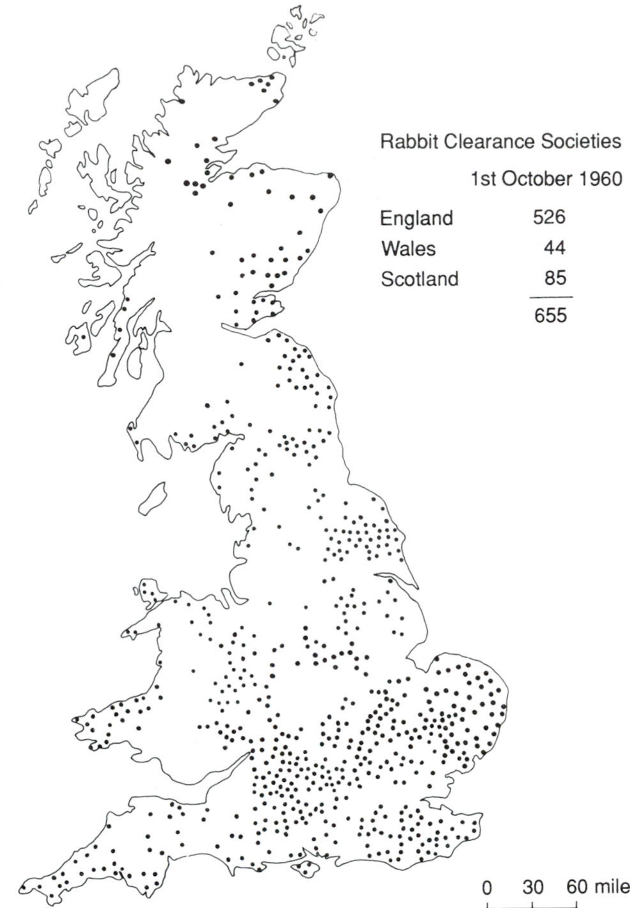

Fig. 4.3 Location of Rabbit Clearance Societies on 1 October 1960 (Thompson 1961).

local individuals, such as Mr Frank Ward, then chairman of the Shropshire Agricultural Executive Committee. With myxomatosis still taking its toll of rabbits, concern over damage subsequently declined, and the number of societies in England and Wales fell to 550 in 1971 and, when grant was withdrawn that year, to 280. A MAFF survey in 1987 found that the number of societies and rabbit groups had fallen to 75 and, despite enthusiasm, there were often financial problems, which resulted in emphasis on the sale of rabbits and a greater use of ferrets than of fumigation (Thompson 1961; McKillop 1988).

4.3 Myxomatosis (see also Chapter 7)

A synoptic account of this virus disease is given in Chapter 7, but some aspects particular to Britain are dealt with here. The disease had been of interest in the 1930s, when some experiments were done at Cambridge in collaboration with Australian scientists. These showed that the disease spread rapidly through rabbit populations kept in large outdoor enclosures (Martin 1936). With Martin's help, three attempts were made to establish myxomatosis in the dense rabbit populations of the island of Skokholm, Pembrokeshire, but they were all unsuccessful (Lockley 1940). It was subsequently found that fleas

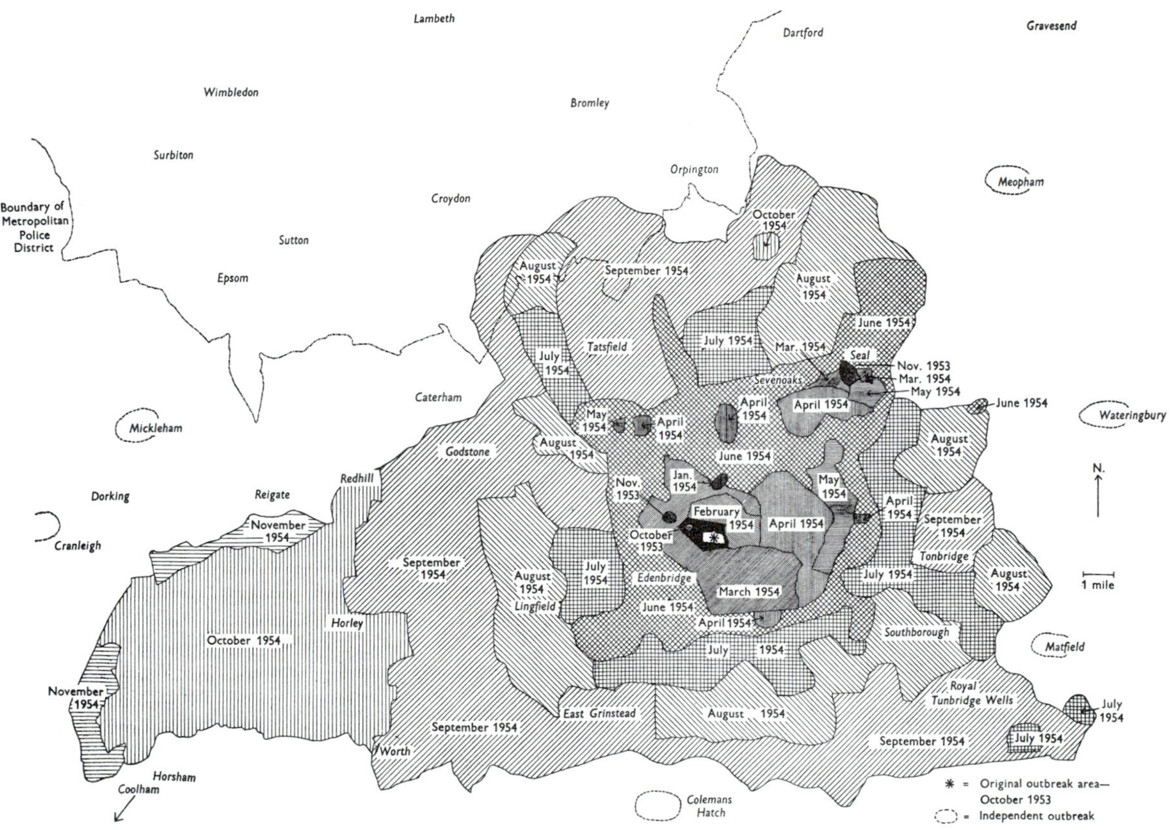

Fig. 4.4 Initial spread of myxomatosis in southern England in 1954 (Armour and Thompson 1955).

appeared to be absent from Skokholm rabbits, and this is the likely reason for the failure of the disease to spread (Lockley 1955). During the Second World War, the use of myxomatosis for rabbit control was discussed officially, but dismissed as unpromising. It is worth repeating that, although it had frequently been suggested that myxomatosis should be introduced to Britain, no official attempt has ever been made to introduce the disease on to the mainland. This remained the official view after the spread of myxomatosis in Australia from 1950, and its introduction to France in 1952. The reasons for this decision were: conditions here are very different from those in Australia; it was not known whether suitable vectors occurred here; it was doubtful if the disease would provide a permanent form of control; there was the possibility that domestic rabbit stocks would be endangered, and there was a very strong bias against using a disease as a method of controlling a mammalian pest.

As the disease spread across Europe it seemed inevitable that it would reach Britain, naturally or by human agency. A local man brought a myxomatous rabbit from France and the first outbreak was confirmed near Edenbridge, Kent, on 13 October 1953, and a second outbreak on 27 October, in East Sussex (Thompson 1954; Ritchie *et al.* 1954).

In the hope of wiping out, or at least confining, the disease, the affected areas were surrounded by rabbit-proof netting and efforts were made to kill all the rabbits within the enclosures. It seemed possible that the disease might die out over winter, but nine further small outbreaks appeared in Kent, Sussex, Essex, and East Suffolk by February 1954. In early

Table 4.1 Rate of spread of myxomatosis in southern England in 1954 (Armour and Thompson (1955))

	Monthly increase in area infected (square km)				Maximum linear spread (km/month)
	From Edenbridge	From Seal	Isolated pockets	Total	
1953					
October	1.88	—	—	1.88	1.4
November	—	0.73	0.52	1.25	—
December	—	—	—	—	—
1954					
January	—	—	1.55	1.55	—
February	4.17	—	—	4.17	1.8
March	21.42	1.63	—	23.05	3.8
April	22.46	6.84	2.56	31.86	3.8
May	8.05	2.12	0.39	10.56	2.9
June	130.23	*	0.36	130.59	8.2
July	93.06	—	3.60	96.66	5.8
August	192.93	—	**	192.93	4.3
September	327.48	—	—	327.48	8.5
October	128.63	—	—	128.63	14.6
November	20.62	—	—	20.62	1.8
Total	950.93	11.32	8.98	971.23	

* On 12 June 1954, the Seal outbreak merged with the Edenbridge.
** There were nine small pockets of infection which became incorporated in the spread from Edenbridge.

November 1953, an *Advisory Committee on Myxomatosis* was appointed with Lord Carrington as chairman; they concluded that further efforts to prevent the spread of the disease would be fruitless, but that no attempt should be made to assist its spread or to introduce it into unaffected areas of the country (Advisory Committee on Myxomatosis 1954). Mosquitoes were known to be the prime vectors in Australia and, with this experience in mind, the modest spread of myxomatosis in our winter was attributed to the seasonal absence of adult woodland mosquitoes and an epizootic was anticipated in the spring. Although cold weather delayed the emergence of adults in 1954, the woods were alive with mosquitoes by June, but they showed little interest in rabbits. This was confirmed experimentally by exposing six domestic rabbits in separate hutches near the first outbreak of myxomatosis in Kent, in areas where wild rabbits were visibly affected by the disease. Each hutch was attached to a tree, some 1.5 metres from the ground, with one side of the hutch covered only by 25 mm mesh wire-netting to allow insects to enter freely. The domestic rabbits were fed and watered normally and remained in good health throughout the rest of the year. Studies of mosquito–rabbit ecology showed that, although numerous *Aedes cantans* and *Aedes annulipes* were present, they were not attracted to the hutched rabbits (Muirhead-Thomson 1956*a*, 1956*b*). During the spread of myxomatosis in 1954 it was generally expected that domestic rabbitries would be affected, since many breeders neither vaccinated their rabbits with Shope's fibroma vaccine nor screened them against mosquitoes. The few cases of myxoma infection among domestic rabbits were associated with the presence of the coastal mosquito, *Anopheles maculipennis atroparvus*, but since the virus was not recovered from mosquitoes caught outside the

Table 4.2 Rabbits surviving myxomatosis in the Edenbridge area (Armour and Thompson (1955))

1954	Rabbits seen*	Blood samples tested for immune antibodies	
		Negative	Positive
April	4	—	—
May	2	1	—
June	6	6	—
July	6	6	—
August	—	—	—
September	—	—	—
October	1	—	—
November	26	10	—
December	14	3	1
Total	59	26	1

* These were all individuals, except for one colony of nine, caught in November.

As well as making a continual survey of places where myxomatosis was active, periodic visits were paid to areas behind the disease 'front' to look for survivors. As shown in Table 4.2, fifty-nine survivors were found, fifty of these being single individuals and nine forming a colony. Blood samples from twenty-seven of these rabbits were sent to the Ministry's veterinary laboratories at Weybridge for complement–fixation tests, and only one was found to possess immune antibodies to myxomatosis; the remaining rabbits were susceptible to the disease.

hutches these insects are unlikely to be important vectors (Muirhead-Thomson 1956*c*).

The spread of the disease in the Edenbridge area was plotted weekly, and by November 1954 had covered 97 500 hectares and reached its limits, bounded in the north by Greater London and to the south, east, and west by independent outbreaks. Although there were reports of diseased rabbits being taken from the area and released elsewhere, this had little effect on the local spread (Fig. 4.4; Tables 4.1 and 4.2; Armour and Thompson 1955).

Assuming that this spread was natural, it moved an average 5.6 km a month from February to November 1954; similar rates of spread were noted in other parts of the country. In the study area, myxomatosis took about six weeks to pass through a group of rabbit warrens. The disease was not reported outside the south-east of England until May 1954, when isolated outbreaks occurred in Cornwall, Norfolk, and Radnor. It reached, or was taken to, Scotland in July and by the end of 1954 there were 498 outbreaks in Britain. Throughout 1955 myxomatosis continued to spread, mostly maintaining its high virulence and causing a mortality of about 99 per cent. Some attenuation of the virus was to be expected, and was first noted in April 1955 (Hudson and Mansi 1955; and see Chapter 7). By the end of the year the disease was present throughout England, Wales, and Scotland, although there were some areas, with few rabbits, that were unaffected.

At an early meeting of the Advisory Committee on Myxomatosis, one of its members, Miriam Rothschild, suggested that the European rabbit flea (*Spilopsyllus cuniculi*) could be a very important vector in Britain (Rothschild 1953). This flea was absent from Australia at that time, but it was subsequently introduced there, as an additional vector (Sobey and Menzies 1969). Work in Australia had shown the native stickfast flea (*Echidnophaga myrmecobii*) to be an effective carrier of myxomatosis (Bull and Mules 1944), and the similar capacity of the rabbit flea was demonstrated (Lockley 1954; Shanks *et al.* 1955;

Table 4.3 Seasonal reproductive activity of fleas on wild rabbits (Mead-Briggs (1977))

Month	Fleas from female rabbit hosts						
	No. ♀♀ examined (No. ♀ hosts involved)	Ovarian development % in Stage:			% ♀♀ impreg.	% ♀♀ with 'corpora lutea'	% ♀♀ impreg. and with 'corpora lutea'
		1	2	3			
Sept	35 (5)	88	9	3	0	3	0
Oct	50 (6)	100	0	0	6	6	6
Nov	20 (2)	100	0	0	0	0	0
Dec	15 (2)	100	0	0	0	0	0
Jan	55 (7)	96	4	0	0	18	0
Feb	47 (6)	74	13	13	2	2	2
March	45 (5)	4	25	71	4	11	11
April	50 (6)	10	48	42	58	64	54
May	79 (8)	32	53	15	42	56	41
June	70 (7)	47	14	39	14	21	15
July	70 (7)	58	36	6	7	15	8
Aug	20 (2)	70	30	0	5	5	5

Month	Fleas from male rabbit hosts						
	No. ♀♀ examined (No. ♂ hosts involved)	Ovarian development % in Stage:			% ♀♀ impreg.	% ♀♀ with 'corpora lutea'	% ♀♀ impreg. and with 'corpora lutea'
		1	2	3			
Sept	45 (7)	84	16	0	2	4	2
Oct	15 (2)	100	0	0	0	7	0
Nov	40 (5)	97	3	0	0	0	0
Dec	12 (2)	91	9	0	0	0	0
Jan	15 (2)	100	0	0	0	0	0
Feb	5 (1)	100	0	0	0	0	0
March	20 (2)	0	0	100	5	5	5
April	20 (2)	10	85	5	65	70	60
May	8 (1)	62	38	0	13	25	13
June	20 (2)	65	35	0	10	30	0
July	0 (0)	—	—	—	—	—	—
Aug	10 (1)	90	0	10	0	0	0

Seasonal nature of the reproductive activity of female rabbit fleas taken from healthy wild rabbits shot in Kent, September 1958–August 1960. Ovarian development is classed as Stage 1 (immature, largest oocyte follicles <100 μm in length), Stage 2 (initial stage of maturation or final stage of regression, follicles 100–150 μm), or Stage 3 (stage of rapid yolk deposition and maturation, or of rapid yolk resorption and regression, follicles >150 μm); the percentages of fleas impregnated (spermatozoa present in spermatheca), with 'corpora lutea' (evidence of past ovarian maturation) and with both features (strong evidence that flea has laid fertile eggs) are detailed.

Table 4.4 States of the reproductive organs of 88 female rabbits collected in Kent, September 1958–August 1960 (Mead-Briggs (1977))

Month	Number of ♀ rabbits examined	Number of ♀ rabbits			
		In anoestrus (1)	Pregnant not lactating (2)	Pregnant and lactating (3)	Lactating not pregnant (4)
Sept	6	6	0	0	0
Oct	6	5	0	0	1
Nov	3	3	0	0	0
Dec	3	3	0	0	0
Jan	10	9	1	0	0
Feb	7	3	4	0	0
March	5	0	1	4	0
April	7	0	1	5	1
May	8	0	2	6	0
June	11	1	0	7	3
July	14	1	0	7	6
Aug	8	1	0	5	2
Total	88	32	9	34	13

(1) Non-breeding; (2) probably first pregnancy of season; (3) probably second or later pregnancy of season; (4) probably entering anoestrus and suckling final litter of season.

Muirhead-Thomson 1956*a*). Some 20 years of research on the European rabbit flea was carried out by staff of MAFF, and the subject was comprehensively reviewed by Mead-Briggs (1977). *S. cuniculi* has been recorded from every county in Britain (Mead-Briggs 1964*a*) and careful work on its reproductive system showed it to be completely dependent on the breeding of the rabbit. The flea's ovarian activity, from February to August, coincides with the rabbit's breeding season, and both female and male fleas develop to maturity only on pregnant rabbits, and only in the last seven days of the rabbit's 30 day pregnancy (Tables 4.3 and 4.4; Mead-Briggs 1964*b*; Mead-Briggs and Rudge 1960).

Fleas were found not to mate on the pregnant doe, but mostly to leave her at parturition, feed avidly on the newly born young and mate (Mead-Briggs and Vaughan 1969). The flea's eggs are laid in the fur lining the rabbit's nest. It was later found that the breeding of the flea was controlled by the reproductive hormones of the host rabbit (Rothschild and Ford 1964*a*, 1964*b*, 1966, 1969, 1973). Research on the flea is still in progress (Rothschild 1991).

Rabbit numbers remained very low until the 1960s, although some damage to crops continued (Andrews *et al.* 1959). Regular outbreaks of myxomatosis recurred where rabbit numbers increased (Lloyd 1970*b*). Surveys showed a gradual rise in rabbit numbers to some 20 per cent of those before myxomatosis by 1984 (Rees *et al.* 1985); my personal view is that the figure may now be 40 per cent. Studies of the epidemiology of myxomatosis are detailed in Chapter 7. Strains of virus of lesser virulence arose and killed a smaller percentage of rabbits infected, resulting in higher proportions of immune rabbits. While the selective increase in the number of animals with genetic resistance favoured rabbit survival, this was offset to some extent by a subsequent increase in the virulence of the disease. For a variety of reasons the interactions between the rabbit and myxomatosis in Australia and Europe have varied but, as postulated in Chapter 7, progressive selection for genetic resistance is improbable. It seems likely that myxomatosis will remain a fairly severe disease of the European rabbit but, since the rabbit can withstand a very high juvenile mortality, its population will oscillate around a level somewhat lower than before the introduction of the virus, though still sufficiently substantial to cause agricultural damage unless control measures are taken.

4.4 Ecological effects of rabbit reduction in Britain

A census is seldom an exact procedure, even when most of the subjects assist by providing written returns, and burrowing mammals are more difficult to enumerate than many other species. Pre-myxomatosis estimates of abundance were made, mainly based on weights of rabbits sent by rail and road (Thompson and Worden 1956). For example, in 1945 35 kg of rabbit meat were produced in West Wales, for every 50 kg of carcass meat (beef, veal, mutton, and pork). The total population of rabbits in Britain, in the summer season, was estimated to be around 100 million; this figure was substantiated by the research which showed that professional rabbit trapping removes 30–40 per cent of the population at risk, and the figures supplied to the *Advisory Committee on Myxomatosis* by the rabbit meat and fur trade (Advisory Committee on Myxomatosis 1955), whose intake of home-produced wild rabbits during 1950–3 averaged 40 million annually. Populations were most dense in the west of England and Wales (a situation which has changed today, see Section 4.5.5); some 75–100 per hectare on the island of Skokholm, in mainland West Wales 40/ha, elsewhere many moderate numbers of 7–25/ha and an average of 2–5/ha (Fig. 4.5). Comparable figures for Scotland were unavailable, but many areas were over-run by rabbits (Thompson 1956).

Although the role of the rabbit as an agricultural pest was well-recognized by the 1950s, the extent to which its presence affected the flora and fauna of Britain was realized only when myxomatosis reduced its population by 99 per cent. The effects of this 'involuntary experiment', the virtual removal of the major wild herbivore, were indeed dramatic, persistent and often surprising (Fig. 4.6). In order to study the ecological consequences of the disease, there was liaison from the outset between MAFF and the Nature Conservancy (Moore 1987). On the botanical side, the changes on sample nature reserves were studied by recording plants on a number of transects, totalling over seven kilometres, and taking photographs from the same places in 1954 and 1957 (Fig. 4.7; Thomas 1956*a*, 1956*b*, 1960, 1963). In the absence of rabbit grazing, herbs and grasses grew taller, flowering was more prolific and more seedlings survived, encouraging plant successions. On a transect at Lullington Heath, Sussex, the turf height increased from 1.3 cm to 9 cm and on another from 5 cm to 23 cm between March 1954 and September 1955; on another, at Horn Heath, Cambridgeshire, ling *Calluna vulgaris* increased from 2.5 cm to 15 cm in a year. Immediately after myxomatosis, a greater variety of plants was found, but the subsequent dominance of some species led to the decline of others. On chalk grassland relatively rare

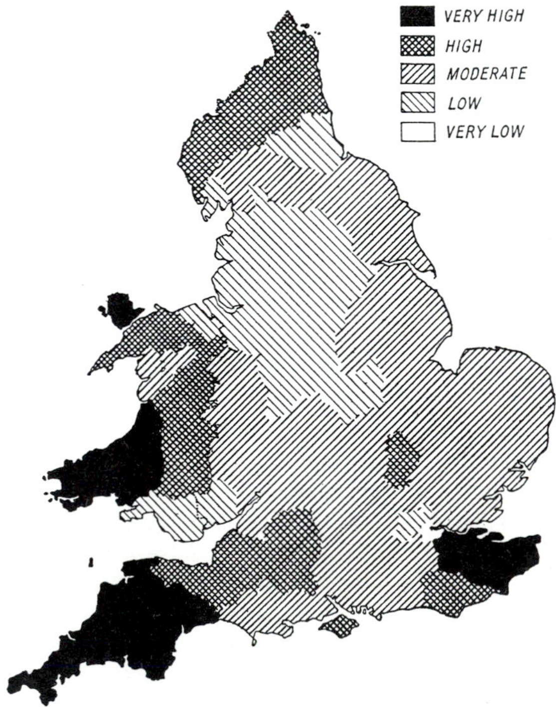

Fig. 4.5 Distribution of the rabbit in England and Wales, before myxomatosis (Thompson and Worden 1956).

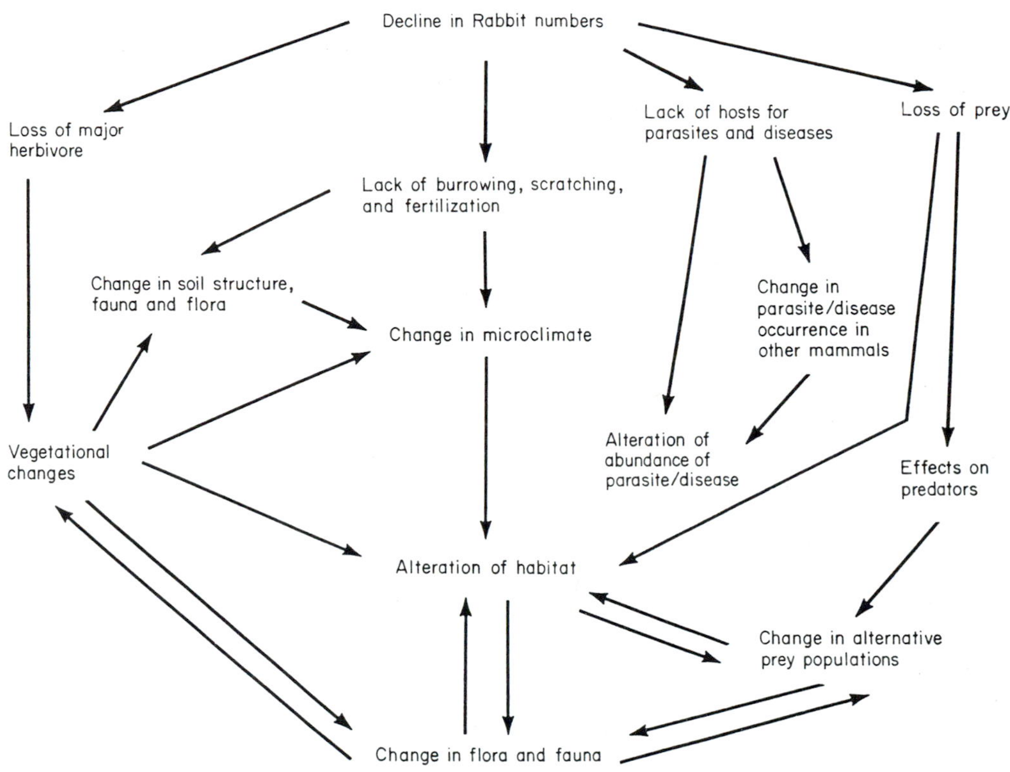

Fig. 4.6 Ecological effects of the decline in rabbit numbers after myxomatosis (Sumption and Flowerdew 1985).

species such as the Pasque-flower *Anemone pulsatilla*, rockrose *Helianthemum chamaecistus* and cowslip *Primula veris* flourished in 1955–6 (Thomas 1960), until these species were overgrown or grazed. Shrubs and trees, such as gorse *Ulex europaeus*, bramble *Rubus* spp, hawthorn *Crataegus monogyna*, juniper *Juniperus communis*, and oak *Quercus robur* regenerated and colonized grassland, unless this was grazed by livestock. Many botanists recorded changes in vegetation (Smith 1980; Watt 1981), well reviewed by Sumption and Flowerdew (1985).

Animals of many species were greatly affected by the eclipse of the rabbit, both directly by its absence as prey, or host for parasites, and indirectly by the changes in vegetation following the cessation of rabbit grazing. Dozens of studies recorded the effects of the rabbit's removal on vertebrates and invertebrates, indicating the extent to which the rabbit and its activities influenced their survival. The publications are, again, most effectively discussed by Sumption and Flowerdew (1985) and only a few particularly pertinent species will be considered here. The Nature Conservancy in collaboration with the British Trust for Ornithology studied the effects of myxomatosis on the buzzard *Buteo buteo* (Moore 1956, 1957, 1987). Buzzards are found in the west and north of the country, preferring moorland and also because their distribution is negatively correlated with the density of gamekeepers. In 1954, before myxomatosis reached them, the rabbit was found to be the buzzard's most commonly recorded food, and most pairs bred successfully. But in 1955, with most of the rabbits gone, fewer than one seventh of the pairs bred, and there was a reduction in clutch size. In subsequent years buzzard numbers recovered, presumably because of their ability to find alternative food. The tawny owl *Strix aluco* and its usual prey, bank voles *Clethrionomys glareolus* and wood mice *Apodemus sylvaticus* were studied in Wytham Wood, Oxford, for many years (Southern 1956*a*, 1956*b*, 1970). Some 20 pairs of birds produced about 20 young annually, but in 1955 only four chicks were

(a)

(b)

Fig. 4.7 Recovery of vegetation on the South Downs, Lullington Down, Sussex, after myxomatosis. (a) Shows erosion caused by rabbit burrowing—April 1954. Myxomatosis removed the rabbits later that year, and in (b), taken at the same place in June 1957, the vegetation is seen to have recovered. (Copyright A. S. Thomas).

successfully fledged. This was indirectly due to myxomatosis, since, in the absence of rabbits, foxes, stoats and weasels turned to eating greater numbers of bank voles and wood mice, reducing their numbers by about 90 per cent. In a year or two the owls regained their normal breeding levels as numbers of voles and mice increased, helped by a reduction in numbers of stoats.

Under normal conditions in Britain the rabbit has, probably for several centuries, been a major item in the diet of the stoat *Mustela erminea*, which also eats many small rodents and birds. The extent to which stoats rely on rabbits was not fully realized until myxomatosis spread through the country and stoats suffered a food shortage (Southern 1956*b*). Stoats were obliged to compete with weasels for small rodents and with foxes, feral cats, and raptors for larger prey, in most cases to the stoat's disadvantage. There were no population studies of stoats in Britain at the time, but the records of predators killed on game estates provided useful information. For about 10 years after the spread of myxomatosis stoats appeared infrequently in gamekeepers records and were relatively scarce for the next 10 years. From the early 1970s rabbit numbers began to increase, followed by stoats, but with every new outbreak of myxomatosis the original pattern was repeated (Fig. 4.8; King 1980, 1989). The indirect effects of myxomatosis on another mustelid, the weasel *Mustela nivalis*, was totally different. Weasels specialize in feeding on small rodents, and the unprecedented growth of herbage, after the removal of the rabbit-

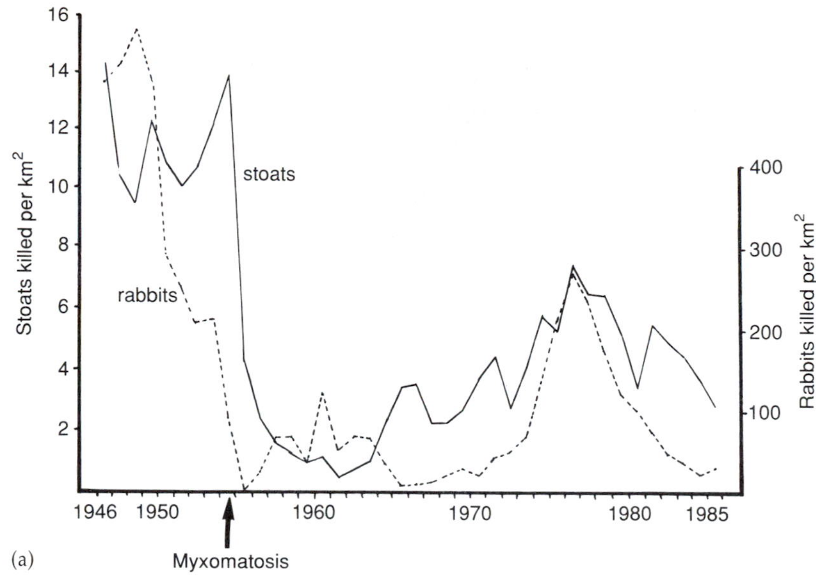

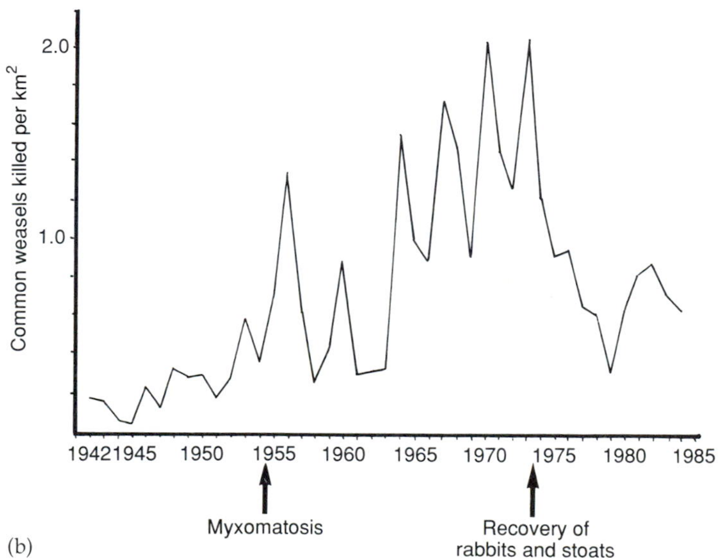

Fig. 4.8 Linked changes in the numbers of rabbits and of stoats and weasels. (a) The vermin bag records of a Suffolk estate illustrate well the close relationship between stoats and rabbits in much of England (data from the Game Conservancy, courtesy of S. Tapper). (b) The number of common weasels killed by gamekeepers increased suddenly after the arrival of myxomatosis, and on average remained much higher than previously until the mid-1970s, when the rabbits and stoats began to return (data from the Game Conservancy, courtesy S. Tapper) (King 1989).

grazing pressure, resulted in a glut of bank voles, field voles *Microtus agrestis* and wood mice. While stoats were short of food, weasels had a surplus and, since weasels, unlike stoats, can greatly increase their fertility under favourable conditions, their population soared; gamekeepers' records again indicated this (Fig. 4.8; Jefferies and Pendlebury 1968; King 1989). From the early 1970s rabbits partially recovered their former numbers and there was a consequent decline in the numbers of weasels and an increase in those of stoats (Tapper 1982). The records kept on game estates indicate that the numbers of rabbits killed annually are continuing to increase (Fig. 4.9; Tapper 1992).

There was much speculation about the repercussion on the 'balance of nature' that might follow

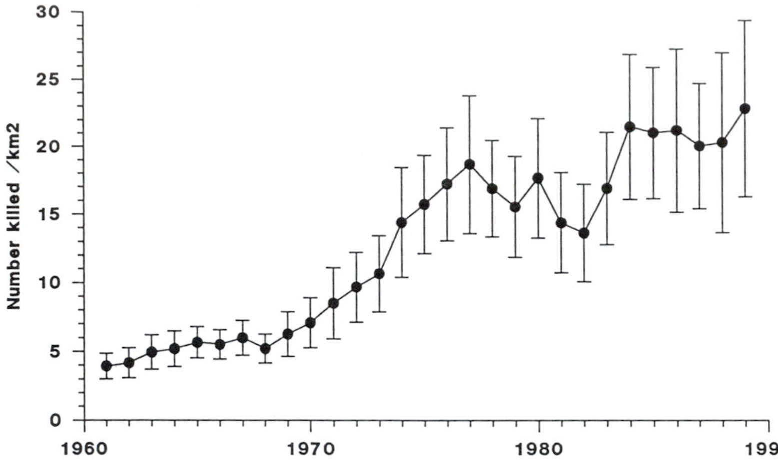

Fig. 4.9 Trend in the average bag of rabbits on game estates since 1961; mean ± standard deviation (Tapper 1992).

Table 4.5 Food items from fox stomachs taken in the pre-myxomatosis period (Lever (1959))

	1955	Numbers of stomachs	Rabbit and hare	Rodents	Sheep	Deer	Birds	Insects
Westmorland	March–May	8	4	1	5	2	2	4
Cheshire	March	1	1	–	–	–	–	1
Derbyshire	March–April	2	2	1	1	–	1	–
Argyll	May	6	1	1	5	–	2	3
Totals		17	8	3	11	2	5	8

from the great reduction in the rabbit population. The best that could be said was that the balance was always in a state of fluctuation, that nature is remarkably resilient and that some new state of equilibrium would be reached before long. There was much concern that the fox *Vulpes vulpes*, in particular, would take a much greater toll of poultry, lambs, sheep, and game. In the early 1950s there was little published information on the feeding habits of the fox, apart from a preliminary study of its summer food (Southern and Watson 1941), which showed rabbits to be a very substantial item in stomachs and faeces. MAFF expressed their concern in a three-year study of fox diet (Lever *et al.* 1957; Lever 1959) which showed the main food items to be small mammals, predominantly the field vole *Microtus agrestis*, in both lowland and hill areas. Brown rats *Rattus norvegicus* were an important item in some areas; worms and vegetable matter were eaten; while insects, especially in samples from hill areas, were more frequent than in the pre-myxomatosis study (Tables 4.5 and 4.6). Findings in Scotland were generally similar (Lockie 1956) and also in Northern Ireland (Fairley 1966). Lamb remains were found in stomachs but could have been carrion. There were complaints of excessive fox damage, particularly in the Welsh uplands and south-west England, and more foxes than usual were killed in 1954–5 (Thompson and Worden 1956), but over the whole country fox predation of stock did not greatly increase. The number of foxes killed, in connection with various bounty schemes, rose in England and Wales (Lloyd 1981) and, in Scotland, very sharply in the winters of 1954–5 and 1955–6 (Hewson and Kolb 1973), suggesting that abundant diseased rabbits and rabbit-carrion favoured fox-breeding and survival. The

Table 4.6 Post-myxomatosis record of food (% frequency of occurrence) of 478 adult foxes (Lever (1959))

Area		Period	Specimens: Stomachs	Scats	Total	Rabbit and hare	Small rodents: Voles†	Rat	Mice‡	Total	Grey squirrel	Sheep	Carrion§ & deer	Hedge-hog	Game and poultry	Total birds	Insects¶
LOWLANDS																	
Beds. (2), Bucks. (2), Cheshire (1), Derby (10), Essex (23), Herts. (1)	Lincs. (1), Middx. (3), N'hants. (3), Notts. (8), Oxon (2)	Jan. 1955–Aug. 1957	54	2	56	30	46.5	9	3.5	59	0	3.5	5	0	39	64	28.5
Berkshire		Nov. 1957–Apr. 1958	0	48	48	14.5	43.5	23	2	68.5	4	2	8	2	14.5	52	18.5
Western Kent:																	
Chevening Park Estate		Mar. 1955–Nov. 1957	55	1	56	27	41	12.5	12.5	66	3.5	1.5	5	0	32	70	18
Penshurst Park Estate*		Feb. 1955–Nov. 1957	55	3	58	31	38	8.5	12	58.5	8.5	3.5	1.5	1.5	38	62	13.5
Hants (32), Mid-Kent (8)	Surrey (5), Sussex (7)	Mar. 1955–Jan. 1958	32	20	52	32.5	38.5	13.5	9.5	61.5	7.5	0	2	0	23	63.5	25
Cornwall (25)*, Glos. (4), Herefs. (7)	Pemb. (3), S'set. (2)	Mar. 1955–Apr. 1958	23	18	41	39	53.5	5	5	63.5	0	22	5	2.5	22	50	63.5
Denbigh (2)	Flint (53)	Nov. 1956–Dec. 1957	52	3	55	23.5	34.5	5.5	5.5	45.5	0	23.5	5.5	2	36	60	29
Fife (4)	Kinross (2)	Apr. 1955–Aug. 1956	6	0	6	0	33.5	16.5	0	50	0	16.5	0	0	16.5	83	16.5
Totals and averages			277	95	372	24.5	41.5	11.5	6	59	3	9	4	1	27.5	63	26.5
HILLS																	
Westmorland		July 1955–Feb. 1958	52	14	66	21	51.5	1.5	4.5	57.5	0	45	4.5	0	10.5	33.5	12
Brecon (5)	M'gomery (5)	May 1955–Nov. 1957	9	1	10	0	70	10	0	80	0	20	0	10	0	20	10
Aberdeenshire		Apr. 1955–July 1957	12	0	12	58	33	8.5	0	41.5	0	33	0	0	25	66	25
Argyll		Feb. 1956–June 1957	5	0	5	0	80	0	0	80	0	60	0	0	20	40	60
Banff (8)	Inverness (8)	Apr. 1956–Apr. 1958	0	13	13	92	23	7.5	0	30.5	0	7.5	7.5	0	7.5	38	23
Totals and averages			78	28	106	34	51.5	5.5	1	58	0	33	2.5	2	12.5	40	26

* *Mus musculus*, Penhurst Park and Cornwall. † Chiefly *Microtus*. ‡ *Apodemus* unless otherwise recorded. § Non-sheep carrion. ¶ Excluding larval Calliphoridae.
This table excludes seventeen pre-myxomatosis records given in Table 4.5 and thirteen empty stomachs.

banning of the gin trap, under the *Pests Act*, 1954, was fully effective in England and Wales in 1958 (but not so for foxes in Scotland until 1973); this gave further encouragement to an increase in fox numbers.

In nature, the brown hare *Lepus europaeus* and rabbit do not much associate, although they have several parasites in common (Thompson and Worden 1956, Dunn 1969, Mead-Briggs and Page 1975). Brown hares are found to a varying extent throughout most of Britain, but there is some evidence that they do not thrive where rabbits are abundant; their decrease in Wales is correlated with the increase of rabbits, which spread from the coastal areas as agriculture declined at the end of the nineteenth century (Matheson 1941). After the removal of rabbits by myxomatosis, a collaborative effort was made by the Nature Conservancy, the Mammal Society, and MAFF to measure any effect on hares (Moore 1956). It was found that hares were more numerous in 1955 than in previous years, but there was little evidence of re-colonization in south-west Wales or the peninsula of Cornwall and Devon. On the other hand, some shooting records indicated large increases in hare numbers in the years after myxomatosis (Rothschild and Marsh, 1956; Rothschild 1958, 1961). Rabbit fleas have been recorded from hares as occasional stragglers, but after myxomatosis an increase in the numbers of rabbit fleas on hares was reported from Ashton, near Peterborough (Rothschild 1963). Subsequently, gravid fleas were found on hares, at Ashton and in Berkshire, but whether the fleas on these hares were reproducing, or their population was dependent on local rabbits is uncertain (Rothschild and Ford 1965; Mead-Briggs 1977). A comparison, based on night counts and shooting bag records, found no evidence of an inverse correlation between numbers of rabbits and hares pre- or post-myxomatosis, nor of competition between these species (Barnes and Tapper 1986).

Changes in the vegetation after myxomatosis altered many habitats, and the numbers of some species of insects increased, while others decreased. One in particular, the large blue butterfly *Maculinea arion* must be mentioned, since it has become extinct. The larva of this species feeds on thyme until its last instar. It is then taken by an ant *Myrmica subuleti* to its nest where, until pupation and emergence, the butterfly larva feeds on the ants' offspring and produces a secretion on which the ants feed. The ant is found only in turf up to 1 cm in height and, in the absence of rabbit grazing, the thyme, the ant, and the large blue all disappeared (Thomas 1980*a*, 1980*b*).

4.5 General biology

The interactions between the rabbit and its environment embrace aspects of geography, phytogeography, physiology, intra-specific, and inter-specific behaviour which are discussed in various contexts in this book. The rabbit can adapt to a variety of habitats but in Britain, as elsewhere, prefers short grass, on machairs, dry heaths, and agricultural grassland up to about 600 metres. It needs refuge areas near its feeding grounds and readily finds cover in natural vegetation, among rocks or man-made structures such as hedgerows and the earth and stone banks of Wales and south-west England. Burrows, once established, provide good protection from predators other than the smaller mustelids. The excavations may be very extensive, have been found to a depth of over 2.7 metres, and some have been carefully surveyed and mapped (Barrett-Hamilton 1912; Thompson and Worden 1956; Kolb 1985). In many warrens there were found to be no connections between contiguous sets of holes, suggesting the haphazard accumulation of small burrows.

Rain, unless it is very heavy, does not deter rabbits from normal feeding, and although preferring dry land they are found in marshy areas. They have been seen swimming across rivers and in the sea (Harting 1898; Thompson and Worden 1956; Marchington 1978; Swan and Thompson 1981). The rabbit's normal gait is hopping, with the forefeet placed in front of the hind; as its speed increases and its suspension becomes longer, the hindfeet move further forwards and land in front of the forefeet. It gives the appearance of running quite rapidly and in a short sprint the speed has been variously estimated as 40 km/h and up to 56 km/h (Thompson and Worden 1956; Cowan and Bell 1986). Rabbits are not adapted for climbing, although they have been observed attempting to climb wire-netting paw over paw. They will mount an inclined surface such as a sloping branch or fence support. While a well-constructed netting fence 0.75 m in height will usually keep them out of a plantation or field (McKillop *et al.* 1988) rabbits have occasionally been seen to jump such a fence when hotly pursued (Thompson and Worden 1956).

4.5.1 Effects on vegetation and plant succession

The reverse effects of the removal of rabbit pressure on vegetation, by myxomatosis, have already been noted. The influence of rabbit-grazing on natural vegetation was first clearly demonstrated in the Brecklands of Norfolk where ling on Cavenham Heath was converted into grassland, dominated by common bent and sheep's fescue (Wallis 1904; Farrow 1917, 1925). In the absence of grazing the natural climax vegetation is considered to be woodland, of birch and Scots pine. Tansley (1949) stated that 'grassland of some sort is the inevitable fate in our climate of most of the land which is regularly grazed', and pre-myxomatosis rabbits were the primary grazers among wild animals. They are mainly crepuscular and nocturnal but graze close to their burrow during the day and, with constant passage, keep these areas bare of vegetation. The turf beyond is grazed to within some 2 cm of the soil and, apart from species avoided by rabbits (Table 4.7), few plants other than hemicryptophytes, that put out vegetative shoots on or in the soil, can survive. Nature conservationists, managing fen, grass heath, or chalk downland welcome some rabbit-grazing (in the absence of sheep or cattle) since it will arrest plant succession and prevent the establishment of tree seedlings and aggressive grasses (Williams *et al.* 1974). Where rabbits are very numerous, a hundred or more hectares may be affected by a single extended warren, no woody plants can become established, and under extreme rabbit pressure the grasses are replaced by mosses or lichens, as in some hill grazings in Scotland (Fenton 1940). There is also the likelihood of erosion, as on rabbit-infested coastal sand-dunes. As so frequently, a study of islands provided instructive results. A comparison was made between the rabbit-infested Skokholm (105 ha), which is dominated by thrift *Armeria maritima*, Yorkshire fog *Holcus lanatus*, bracken *Pteridium aquilinum*, and ling *Calluna vulgaris*, and rabbit-free Grassholm (9 ha), which is covered with red fescue *Festuca rubra*. The flora is suppressed by strong winds on both islands but also, more importantly, by rabbit-grazing on Skokholm. When rabbits were excluded from some areas of Skokholm, the vegetation changed to the Grassholm type (Gillham 1955). The effect of grazing on the growth form of the sea plantain *Plantago maritima* was striking (Fig. 4.10).

Table 4.7 Plant species disliked or avoided by rabbits (Smith 1980; Thompson and Worden 1956)

Urtica urens	Small Nettle
Urtica dioica	Nettle
Arctium spp.	Burdock
Senecio jacobaea	Ragwort
Verbascum thapsus	Common Mullein
Solanum dulcamara	Woody Nightshade
Solanum nigrum	Black Nightshade
Sedum acre	Stonecrop
Thymus spp.	Thyme
Arenaria spp.	Sandwort
Myosotis spp.	Forget-me-not
Poa pratensis	Meadow-grass
Stellaria media	Chickweed
Glechoma hederacea	Ground Ivy
Bryonia dioica	White Bryony
Campanula glomerata	Clustered Bellflower
Cerastium vulgatum	Common Mouse-ear Chickweed
Cirsium arvense	Creeping Thistle
C. vulgare	Spear Thistle
C. palustre	Marsh Thistle
Centaurium erythraea	Common Centaury
Cynoglossum officinale	Hound's Tongue
Helianthemum chamaecistus	Common Rockrose
Teucrium scorodonia	Wood Sage
Conium maculatum	Hemlock
Aphanes arvensis	Parsley Piert
Sambucus nigra	Elder
Rubus spp.	Bramble
Sarothamnus scoparius	Broom

Fig. 4.10 Effects of rabbit grazing on the growth form of plants. 1 and 2, *Plantago maritima*; 3 and 4, *Festuca rubra*. 1 and 4 are ungrazed, 2 and 3 are grazed. (Magnification ×0.22) (Gillham 1955).

4.5.2 Behaviour and social organization

Pioneering work was done at Oxford University in the late 1930s (Section 4.2.2; Southern 1940*b*, 1948*a*) on a warren in Berkshire, maintaining about 150 rabbits on a hectare of land. Observations were made from tree cover, using a pancratic telescope, and many rabbits were earmarked with numbered, coloured, celluloid discs. Observations of sexual behaviour included courtship chasing, tail-flagging, and enurination; aggressive chasing and leaping were also studied. Many of these observations were confirmed subsequently (Thompson and Worden 1956) and greatly extended by research in Australia (Chapter 5). Further work in Britain paralleled that done overseas and extended our knowledge of stable social groups, linear hierarchies among males within groups, dominance among females, the establishment of territories, and the importance of chin and anal glands' secretions and urine in communication (Bell 1980, 1986; Cowan 1987; Cowan and Garson 1985; Cowan and Bell 1986). Behavioural studies require clear identification of individual animals and Southern's discs were succeeded by aluminium chicken wing-tabs and tattooing (Thompson and Armour 1954). They indicated home ranges of up to three hectares and, exceptionally, foraging of two kilometres. More recently radio-telemetry has been used (Kolb 1986; Trout and Sunderland 1988).

4.5.3 Physiology

The European rabbit is certainly a successful species. It is capable of thriving on a varied diet, has good food-conversion, is generally resistant to diseases and parasites, is adaptable and resilient under adverse conditions, and has a high reproductive potential. Its digestive system, and especially the phenomenon of re-ingestion, refection or coprophagy and its discovery and 'rediscovery' over the centuries are well-known (Eden 1940*a*, 1940*b*; Taylor 1940*a*, 1940*b*; Thompson and Worden 1956). By analogy with rumination, there are nutritional advantages in the production, during the day, of soft green faeces which are eaten direct from the anus and pass through the gut a second time, to emerge at night as the typical hard pellets of rabbit dung. Rabbits living on chalk, compared to those on sand or clay, have been found to have larger digestive organs, excepting the appendix (Sibly *et al.* 1990). This is attributed to the vegetation on chalkland being more fibrous, and of poor quality. Monk (1989) showed that rabbits are inefficient at digesting fibre. Recent studies have sought to relate the social status of dominant and subordinate rabbits to their underlying physiology and to explore the action of pheromones which influence the rate of sexual maturation, the timing of oestrus cycles or induction of spontaneous abortion (Bell 1981, 1986; Bell and Mitchell

1984). An interesting field of work, following from Australian studies, is being developed.

4.5.4 Reproduction

The first thorough study of the reproduction of the wild rabbit was done at Bangor, Gwynedd, by F. W. Rogers Brambell and his colleagues (Section 4.2.2; Brambell 1942, 1944, 1948). They found the main breeding season to be from January to the end of June, with sporadic breeding in other months, and pregnancy 28–30 days. The rabbit is an induced ovulator and, in wild rabbits, gestation and lactation were found to be concurrent. Does were pregnant for an average of 16 weeks; the mean number of corpora lutea, indicating the number of ova shed, was 5.36 and the mean litter size at birth 4.87. It was calculated that the mean number of young born per adult doe per season was 11.52; that this number was not greater was accounted for by the death of all the embryos of many litters in mid-pregnancy. The embryos, embryonic membranes, and maternal placental tissues were resorbed at the site of attachment in one or two days, around the 13th day of pregnancy, resulting in an average pre-natal mortality of 60 per cent. This work (summarized, with later studies in Thompson and Worden 1956) was done on a rabbit population of high density, and no explanation for the intra-uterine mortality was apparent at the time.

As already indicated, myxomatosis provided remarkable opportunities for biological studies and it seemed to MAFF most desirable to re-examine the rabbit's reproduction, now it was living at low density. The specimens for the work were obtained in 1957, as far as possible from the county of Gwynedd from which Brambell's material came. As an indication of the changed conditions, the number of rabbits caught was only 348 (202 females), for an expenditure of effort which would have yielded some 10 000 rabbits before myxomatosis (Lloyd 1963*b*). The breeding season was found to extend to mid-July and to be at least 22 weeks, compared with 15–17 weeks in 1941–2; litter size at parturition was 5.95, and the mean number of young born alive to one doe per season was estimated as 29.5. No intra-uterine mortality of entire litters was found at mid-term (Table 4.8). Other observations confirmed that a long and fertile breeding season is typical of an expanding rabbit population. Boyd and Myhill (1987), working in Cambridgeshire, found that on average a doe had 23.9 conceptions and suckled 17.2 young annually. Subsequently, it has been shown that in years where the climate is mild the length and productivity of the breeding season increases, and a high mortality between birth and first emergence of young has been demonstrated (Bell and Webb 1991).

It seems likely that the distinct patterns of reproduction in rabbit populations of different densities are determined largely by territorial aggressiveness and the resultant stress (Mykytowycz 1958, 1959, 1960, 1961) although shortage of food may be an overriding factor in shortening the breeding season and lowering fertility. Later work showed an increase in fecundity and fertility and a greater proportion of juveniles breeding in the year of their birth (Lloyd 1970*a*; Boyd 1985) but many of these young failed to survive to adulthood. Samples from most counties of England and Wales, taken in November/December, over the period 1965–75,

Table 4.8 Rabbit reproduction before and after myxomatosis

Year	Rabbit density	Average breeding period of doe (weeks)	Prenatal mortality (%)	Mean number of ova ovulated	Mean litter size at birth	Annual production per doe	Reference
1942	Heavy	16	60	5.36	4.87	11.52	Brambell (1944)
1957	Very light	22	virtually nil	6.78	5.95	29.5	Lloyd (1963*b*)
1975	Light–medium	20	8	—	—	22	MAFF (1981)

showed a mean recruitment rate (adults:juveniles) of 1:1.3. Since the reproductive output was over 20 live young per doe, the mortality of juveniles was approaching 90 per cent (Lloyd 1981). On chalk downland the mortality in the first year of life was found to be about 95 per cent (Cowan and Roman 1985). The loss of juveniles could not be attributed solely to myxomatosis and will be considered in the next section.

4.5.5 Predators

It is improbable that high populations of rabbits can be reduced by predation alone, but there is evidence that predators can inhibit the expansion of rabbit numbers at low densities. Rabbits were very abundant in Britain before myxomatosis, providing a staple food for the fox *Vulpes vulpes*, feral cat *Felis catus*, stoat *Mustela erminea*, and buzzard *Buteo buteo*; a dietary item for the badger *Meles meles*, weasel *Mustela nivalis*, brown rat *Rattus norvegicus*, carrion crow *Corvus corone*, and great black-back gull *Larus marinus* and, in some parts of the country, the polecat *Mustela putorius*, wild cat *Felis silvestris*, pine marten *Martes martes*, mink *Mustela vison*, red kite *Milvus milvus*, golden eagle *Aquila chrysaetos*, raven *Corvus corax*, and the reintroduced goshawk *Accipiter gentilis*. This considerable 'combined predator force' certainly did not control rabbits then, partly because none of the predators breed as fast as rabbits; the fox produces four to five young annually, the stoat up to nine, and the rabbit between 10 and 30+ (Corbet and Harris 1991). Both prey and predators also had to contend with human activities, commercial trapping of rabbits, control of cats, stoats, and foxes on game estates, agricultural and forestry work. By contrast after myxomatosis had reduced the rabbit population by some 99 per cent (Thompson and Worden 1956) recovery was slow and uncertain (MAFF 1978), perhaps because it was inhibited by predation.

During the 1980s rabbit numbers and distribution increased (MAFF 1981; Trout *et al.* 1986), but in general only slowly and patchily. It is notable that rabbits were found to be more abundant in the eastern counties (Fig. 4.11), whereas before myxomatosis the highest densities were in the west of England and Wales (Fig. 4.5). Much mortality, especially of young rabbits, could be attributed to

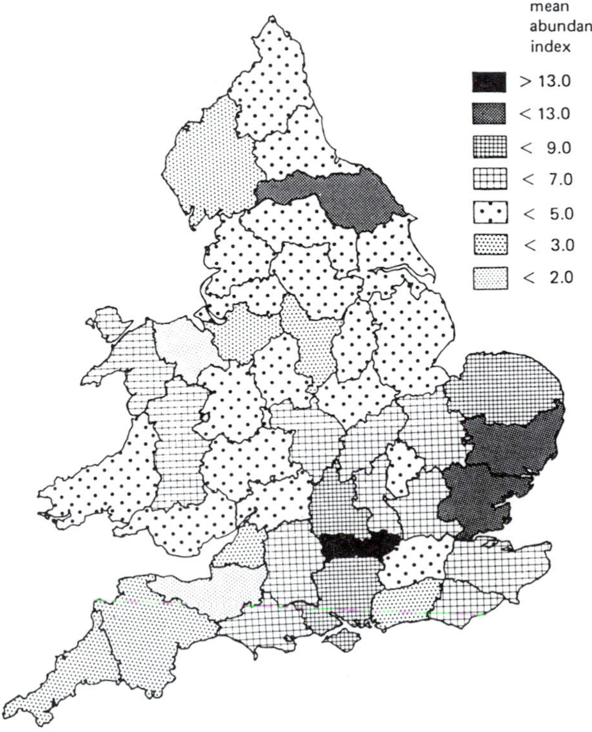

Fig. 4.11 Distribution of the rabbit in England and Wales, 25 years after myxomatosis. (The abundance in Berkshire is anomalous and is attributed to a sampling error.) (MAFF 1981, Crown copyright.)

myxomatosis (Vaughan and Vaughan 1968) but the shortage of cover, caused in part by more intensive agriculture and also the collapse of disused burrows, made rabbits more vulnerable to predators (Lloyd 1981). An extensive review of the effect of predators on rabbit population density (Trout and Tittensor 1989) commented on observations at many sites in Britain, mostly based on twilight counts of rabbits, and confirmed that high predator numbers were associated with low rabbit numbers and vice versa. Records from 450 sites in England and Wales, in the spring of 1984, showed rabbits to be twice as abundant and 1.5 times as widespread on sites with no predator control, compared to those with such control. It was concluded that: when rabbit numbers are reduced by some means, such as myxomatosis or a drought, predators can hold rabbits below their previous density, for a variable time; that when a rabbit population is expanding into a new area or into marginally favourable habitat, predators can slow the rate of spread; that if predators crash in numbers or are controlled, for example by gamekeepers on shooting estates (Tapper 1977), rabbits become abundant and widespread. In Australia, the rabbit is a substantial item in the diet of dingoes and diurnal birds of prey, but it is the main food of the introduced cat and fox; a crash in a local rabbit population is followed by a major reduction in numbers of foxes and cats (Chapter 5). The thousands of cats, stoats, ferrets, and weasels brought to New Zealand in the nineteenth and early twentieth centuries, bred and released to control rabbits, failed to reduce their numbers from the high densities typical of most of the country until 1948. But since then rabbit populations have been reduced by man's control measures, and predators are today regarded as allies (Chapter 6). The once common belief that predators regulated the numbers of their prey, thus keeping herbivores in balance with their food supply, was undermined in the 1930s by the finding that the defence of territory could result in a predator limiting its numbers to the food available in a particular area. Work on muskrats and mink (Errington 1946, 1963) and on grouse and raptors (Jenkins *et al.* 1964) had indicated that the surplus prey animals, often young individuals dispersing, were most at risk of being caught. It follows that the availability of prey regulated the numbers of predators rather than the reverse.

This concept, that predators did not control the numbers of their prey, cast doubt on the long-established practice of predator control on game estates, to increase the stocks of birds for shooting. But very thorough work on partridges by the Game Conservancy has indicated that hen and nest survival were much reduced on estates where predators were not controlled (Potts 1986). This led the Game Conservancy to carry out a six-year experiment on two comparable partridge-producing areas on the Salisbury Plain in Hampshire. On one area (A) the predators (crows, magpies, foxes, stoats, and brown rats) were controlled for three consecutive years, while on the other area (B) there was concurrently no such control. For the following three years the management procedures were reversed, with no predator control on (A) and control on (B). The results showed unequivocally that lack of predator control reduced the autumn stock of partridges to less than half of that in the controlled area, while the partridge bag in the controlled area was over three times that in the uncontrolled (Tapper *et al.* 1991). However, although rabbits flourish on game estates, they are not partridges; and when rabbit numbers are high, they are not regulated either by natural predators or the artificial predators on game estates, the guns.

Since our rabbit originated in Spain (Chapter 3), its relationship with predators there is a matter of interest. They are very numerous: 27 nocturnal and diurnal birds of prey, 11 carnivores and 2 snakes (Delibes and Hiraldo 1981). With the exception of the wolf *Canis lupus*, the larger predators, notably the lynx *Lynx pardina* and the eagle-owl *Bubo bubo*, are very dependent on the rabbit, while the smaller predators have much more alternative prey. There is heavy predation of young in the spring breeding season, and rabbit numbers are also affected by outbreaks of myxomatosis and the availability of food (Soriguer and Rogers 1981). The reliance of so many predators on the rabbit is a tribute to its abundance and resilience, while conversely the effect of predators on the rabbit seems more important than elsewhere in Europe. Nevertheless the result, as previously, is that predators can limit a rabbit population when its numbers are low, but when rabbits reach high numbers they can escape the control of predators. As Fenner and Ratcliffe (1965) wrote, 'Predation on the rabbit in Australia may be likened to a poor handbrake on a car, which will hold the vehicle on a gentle slope, but becomes less and less effective after the car starts to move and as it gathers momentum.'

4.5.6 Diseases and Parasites

The most common diseases of domestic rabbits are intestinal and respiratory and there is a very extensive literature (Blount 1945; Weisbroth *et al.* 1974; Adams 1986). Apart from myxomatosis (see Section 4.3; Chapter 7), much less information is available on diseases of wild rabbits, although coccidiosis *Eimeria steidae*, pseudo-tuberculosis *Pasteurella pseudotuberculosis*, and rabbit-syphilis *Treponema cuniculi* are well-known (Stephens 1952; Thompson and Worden 1956; Cowan 1985; Boag 1989). Coccidiosis can cause high mortality from hepatitis among juveniles between two- and four-months old, and is sometimes found in adults. Both coccidiosis and pseudo-tuberculosis are transmissible from the rabbit to the hare; in Hampshire, immediately after myxomatosis there was no winter epizootic of pseudo-tuberculosis in hares as there had been during the two previous years (Thompson and Worden 1956).

Among the rabbit's parasites, much work has been done on the rabbit flea (Section 4.3; Chapter 7) and also on trematodes (liver-flukes), nematodes (round worms), and cestodes (tape worms); (Evans 1942*a*, 1942*b*; Stephens 1952). Where there are no cattle or sheep as final hosts, the trematode *Fasciola hepatica* can be maintained by rabbits in wet areas suitable for the intermediate host *Limnaea truncatula*. A severe liver-fluke infection is associated with cirrhosis, necrosis, and sometimes peritonitis of the surrounding tissues. The nematodes *Graphidium strigosum*, *Passalurus ambiguus*, and *Trichostrongylus retortaeformis* have been found in wild rabbits (Evans 1942*a*) and may cause haemorrhagic gastritis, anaemia, emaciation, and considerable mortality. These round worms are also found in sheep and goats, and they may acquire the worms on pasture contaminated by rabbits (Leiper 1937). It has been suggested that the drastic reduction in rabbit numbers, after 1954–5 led to the disappearance of some parasitic nematodes (Boag 1972, 1985, 1988).

In the Aberystwyth area of Wales, Evans (1942*b*) found six species of cestodes in rabbits: *Cittotaenia denticulata*, *Cittotaenia pectinata*, *Coenurus serialis* (the larval form of *Taenia serialis*), *Cysticercus pisiformis* (the larval form of *Taenia pisiformis*), and two species of *Hymenolepis*. While in West Wales, adult tape worms of two species *Cittotaenia ctenoides* and *Cittotaenia denticulata* were present in 30 per cent of 4100 rabbits examined. In the cystic form *Cysticercus pisiformis* and *Coenurus seialis* were also found in 42 per cent and 1.5 per cent respectively of 4450 carcases (Stephens 1952).

The incidence of cestodes in rabbits after myxomatosis varied; Boag (1972) found fewer parasites, while other workers found numbers broadly comparable with those noted before myxomatosis (Mead-Briggs and Vaughan 1973; Mead-Briggs and Page 1975).

Rabbit syphilis had been identified in various parts of Britain earlier in this century (Middleton 1932). Leptospirosis *Leptospira icterohaemorrhagiae*, has been isolated in rabbits from Skokholm (Twigg *et al.* 1969); and a mycotic infection *Emmonsia crescens* identified (McDiarmid 1962).

Viral haemorrhagic disease, which is causing mortality in wild and domestic rabbits in continental Europe, has appeared in Britain, causing an apparently isolated outbreak involving two domestic rabbitries (Meldrum 1992). Viral haemorrhagic disease seems unlikely to be as devastating as myxomatosis (see Chapter 3).

4.6 Economic aspects

Farmers' complaints of rabbit damage go back many centuries (see Sections 4.1.1, 4.2.1) and became increasingly clamant during the nineteenth century. They were of concern to parliament during the agricultural depression of the 1930s, and contributed to the passing of the *Prevention of Damage by Rabbits Act*, 1939. With the outbreak of war, the need for crop protection became imperative, and it remained so after 1945. The continuation of food rationing sharpened the countryman's appreciation of the toll taken by the rabbit, despite the continued attraction of hunting rabbits for the pot.

(a)

(b)

Fig. 4.12 (a) Rabbit damage to barley in Kent, 1954, before myxomatosis; one-third of the crop failed in this six-hectare field. (b) a good crop of wheat in the same field, in 1955, after myxomatosis. (Crown copyright.)

4.6.1 Cereals

Until the late 1940s bared areas in fields of wheat, oats, and barley, often next to woodland and most obvious at harvest-time, were variously attributed to unfavourable sowing conditions, slugs, cutworms, leather-jackets, birds, wood-mice, or other small mammals. By fencing off parts of affected fields, staff at MAFF demonstrated that most of the crop damage was caused by rabbits (Gough and Dunnett 1950). Visual estimates suggested a 5 per cent loss of cereal tonnage and, after pilot experiments, led to a more precise national survey of damage to autumn-sown wheat in England and Wales during 1951–2. This was a statistically valid survey, based on harvesting crop samples from pairs of randomly located plots, one fenced and the other unfenced against rabbits, on randomly selected farms. The average loss of winter wheat to rabbits was found to be 204 kg of grain per hectare or 6.5 per cent of the total yield (Church *et al.* 1953). Another survey two years later gave broadly similar results, i.e. an 8 per cent loss (Church *et al.* 1956). The spread of myxomatosis in 1954–5 virtually eliminated rabbit damage until the 1970s, when it again became noticeable. The brilliant summer of 1955 was accompanied by bumper harvest of many crops, a sizeable proportion of which could be attributed to the absence of rabbits. Some farmers cut twice their normal crop of hay, and cereal crops grew right up to the hedges, where usually the headland had been eaten by rabbits; other farmers were able to enlarge their beef and dairy herds. On 13 October 1955, the second anniversary of the notification of myxomatosis in England, the Ministry of Agriculture, Fisheries and Food held a Press Conference and emphasized that although the good crops of cereals could be partly accounted for by the excellent weather, they owed a great deal to the land's freedom from rabbits (Fig. 4.12). It was difficult to translate into figures the increased yield attributable to the absence of rabbits, but it was estimated to be some £15 million (equivalent to £210 million in 1992). Subsequent studies on winter wheat indicated that loss of yield was directly correlated with the duration of rabbit grazing and confirmed previous work that had stressed the importance of early defoliation (Crawley 1989; McKillop *et al.* 1991).

4.6.2 Grassland

Rabbits graze grassland throughout the year, most severely during the winter. They are well-known to be selective grazers (Thompson 1953*a*) and can alter the relative abundance of plant species (Section 4.5.1; Crawley 1990), even degrading heath or hill-grazing to moss or lichen areas. Two experimental studies were done prior to myxomatosis. In the first, in Cardiganshire on siliceous pasture, two fields were ploughed in December 1947 and reseeded the following May with perennial ryegrass, Italian ryegrass, clovers, and rape. In each field a 0.40 ha experimental plot was fenced with pig-netting, through which rabbits could pass, and an adjacent 0.40 ha control plot was fenced with rabbit-proof netting. Over twice as many lambs were kept on the control plots and their increase in weight was 237 kg, compared with 120 kg on the experimental plots.

The second study was on chalk grassland in Kent. Two-and-a-half hectares were surrounded by a rabbit-proof guard fence. Within it an existing rabbit warren of 0.30 ha was cut off by rabbit-proof netting from six plots of 0.40 ha each. The rabbits in the warren had access to three of the plots, chosen at random, through nine tunnels let into the netting. The main object of the work was to assess the number of sheep displaced from a given area by a known rabbit population (Thompson and Worden 1956). The experiment lasted from March 1950 to October 1951. During the spring, summer, and autumn each year, the plots were grazed by sheep for two or three weeks at a time and the pasture rested between grazing periods. The sheep were weighed before and after grazing, and the stocking varied on any plot according to the feed available.

During the first year, similar numbers of sheep were maintained on all the plots, but the increase in weight of sheep after being on the rabbit-grazed plots averaged 20 per cent less than on the rabbit-free plots. In the second year the effect of the rabbits was more noticeable. The plots protected from rabbit-grazing maintained sheep for a fortnight in April and a fortnight in May, but the plots which had been grazed by rabbits did not support sheep until June. Owing to the reduced herbage on these plots, the increase in weight of the sheep was 64 per cent less than on the rabbit-free plots. Quantitative measurements of the available herbage were made by cutting one metre square samples inside and outside rabbit-proof cages in each of the plots before and after each sheep-grazing period; they supported the figures for sheep maintenance and weight increase. The numbers of rabbits grazing on the unprotected plots were counted at different times of day and those feeding at night were periodically checked by attaching box traps to the tunnels between the plots and the warren. The rabbit-density (hence grazing pressure) was greatest in June and July, but over the year averaged 47 per hectare: a high density, but by no means uncommon at that time (Section 4.4).

4.6.3 Other crops

Rabbits are a worry to foresters, preventing natural regeneration by eating seedlings, distorting the growth of trees by damaging leading shoots (Forestry Commissioners 1943; Tee *et al.* 1984) and killing trees by ring-barking. The rabbit had been stigmatized as the most destructive forest pest in Britain, pre-myxomatosis, and the annual cost of protecting 300 000 ha of state-owned plantations was £500 000, plus privately owned woodland,

£1 500 000. It is estimated that, in 1992, 30–40% of the cost of establishing farm woodland, under the New Farms Woodland and Farm Premium Woodland schemes, is related to protection from rabbits (R. C. Trout, personal communication). Damage to fruit trees and the expense of netting, orchards and market-garden crops can be considerable, as is damage to field crops such as kale, sugar-beet, turnips, and swedes. Grazing by rabbits of oil-seed rape and fodder rape significantly reduced the yield of seed at harvest (Boag *et al.* 1990) and poses a considerable threat of economic loss to a relatively new major crop. The total cost of rabbit damage before myxomatosis is likely to have been some £50 million annually (£700 million at 1992 prices). By 1984 rabbit numbers were officially estimated to be about 20 per cent of pre-myxomatosis levels and to cause agricultural damage of between £95 and £120 million annually, with the potential to rise to between £240 and £400 million (Rees *et al.* 1985).

4.7 Management

It is natural that man, in common with all other species, should look after his own interests, in his case extended from personal and familial to national and international dimensions. To people, human interests must be paramount in the same sense that the next meal is the most important consideration to a hungry predator, with the proviso that human interests are much wider. They include an appreciation of the variety of living things, an acceptance of responsibility for the conservation of the total environment and of the obligations of a dominant species to other forms of life.

Every problem of damage attributable to wildlife must be considered on its own merits, in the first place to see if it is in fact a real problem. Changes in husbandry and sometimes the use of repellents, including adequate proofing or mechanical exclusion of the pest species, may be more rewarding than destructive measures. Control is not a euphemism for killing, and as Walter Howard wrote (in 1965): 'The principle underlying all control efforts should be to accomplish the desired effect with a maximum of humaneness and with safety to man and to forms of life useful or of neutral value to him.'

Although welcomed and encouraged initially, rabbits have been responsible for such devastation and economic loss in Australia and New Zealand that drastic and wholehearted control measures have been readily adopted. British attitudes, over many centuries, have often been equivocal, because it has often been necessary to reconcile the conflicting objectives of sport, carcase value, conservation interests, and damage to agriculture and forestry. Making the habitat less attractive to rabbits is one form of management, used for example with marked success in Perthshire (Boag 1987). During a nine-year study of a 500 ha farm, fields were enlarged, by removal of walls and the use of rabbit-proof fencing, old quarries filled in, gorse and broom removed and reseeded with grass, heather improved and reseeded. The result was a greatly increased stocking rate of ewes and cattle and a striking reduction of rabbit numbers. There are limits, however, to the extent to which rabbit harbourage can be removed. Interest in wildlife conservation has burgeoned in the past 20 years, leading to a realization of the extent to which wildlife, in the broadest terms, depends upon farmers and landowners, both private and public, providing adequate and varied habitats—including nature reserves, open spaces, woodland, game coverts, rides, fox coverts, and other shelter belts. Furthermore, intensive agricultural practices have produced food surpluses in the European Common Market, resulting in the notorious 'butter mountains', 'wine lakes', surplus grain, and carcase meat. To reduce the surpluses, curb the subsidies, and devise a more rational common agricultural policy, MAFF is encouraging farmers to diversify their activities and to 'set aside' part of their holdings, i.e. take them out of production. While this policy will increase semi-natural areas beneficial to valued wildlife, it will also provide more rabbit harbourage.

4.7.1 Barriers and repellents

Wire-mesh fencing has been used in rabbit control for some two centuries, reaching its most spectacular expression in the Great Barrier Fences, running for thousands of kilometres across New South Wales, Queensland, and Western Australia, in unsuccessful attempts to stop the spread of the rabbit. But netting is used very effectively for fencing holdings, within which the rabbits are destroyed, and crop damage prevented by thorough boundary maintenance (Stead 1935; Rolls 1969). In Britain, netting fences are used extensively for the protection of tree plantations, field headlands, and market-garden crops and only in the years immediately after the spread of myxomatosis has this been unnecessary. Specifications for rabbit-proof fencing abound (Thompson and Worden 1956) and have been revised by MAFF and the Forestry Commission (McKillop *et al.* 1986, 1988). The use of electrified fencing has long been a standard practice for containing livestock and strip-grazing, but its use as a barrier to wildlife has been successful only from the 1960s, with the development of low-impedance, high-voltage units (McCutchan 1980). Early tests against rabbits were ineffective (Southern 1942*b*) but recent trials were sufficiently convincing for farmers to report increased yields (McKillop and Wilson 1987; McKillop *et al.* 1992). Spraying the vegetation along the fence-lines with herbicides avoids short circuits. Some small numbers of animals were found dead near to the electrified fence—for example rabbits, hedgehogs *Erinaceus europaeus*, frogs *Rana temporaria*, and toads *Bufo bufo*. There is now a considerable literature on the use of these fences and the behaviour of animals towards them, well-reviewed by McKillop and Sibly (1988).

Electrified fences can be regarded as 'repellent' in the sense that they are offensive to an animal, rather than acting merely as a physical barrier. Many chemical repellents have been used against animals, mainly for the protection of food, clothing, or other stores from damage by rodents or insects. Such repellents have also been used to protect trees and seed. Rabbit damage has been prevented by painting rosin dissolved in ethanol on trees (Thompson and Armour 1952) and forestry repellents have been reviewed (Thompson 1953*b*; Armour 1963). Research on pheromones suggests firstly, that compounds in the urine of male rabbits can discourage rabbits from feeding on previously attractive food; and secondly, that components of adult female rabbit urine can suppress the development of young males (Bell 1986).

4.7.2 Fumigation

The use of cyanide powder for gassing rabbits in their burrows has already been described (Sections 4.2.1, 4.2.2). It is authorized by the *Prevention of Damage by Rabbits Act* 1939 and is strongly recommended as being humane and efficient. The fumigant is normally used in the form of powdered cyanide compounds which, in contact with moist air, generate hyrocyanic acid gas. It is introduced to the burrows either by placing a spoonful (about 30 g) of powder 15 cm down each active hole and blocking it with a turf, or forcing it throughout the burrow system using a stirrup-pump or a high-volume fan impeller, and blocking all holes from which powder emerges. This technique has been fully tested (Southern 1948*b*; Thompson and Armour 1953; Thompson and Worden 1956; Thompson and Thompson 1966) and the evidence indicates that gassed rabbits either rapidly lose consciousness and die or, at sub-lethal concentrations rapidly recover (Barcroft 1931; Weedon *et al.* 1940; Southern 1948*b*; McNamara 1976).

More recently, trials have been done with aluminium phosphide, which generates phosphine using the spoon-gassing technique (Rees *et al.* 1985; Ross 1986). The results showed phosphine to be as effective as hydrogen cyanide, more convenient and slightly cheaper to use, but possibly less humane. Whenever fumigating, it is essential to remove all surface-living rabbits, as far as possible, either by shooting or driving them into burrows, preferably using trained dogs.

4.7.3 Trapping

Spring traps, especially the history of commercial trapping with the gin, or leg-hold trap, have already been discussed (Sections 4.2.1, 4.2.2, 4.2.3). The traps now approved for rabbits are the Imbra, Juby, Fenn Rabbit, Fenn Mark VI, and Springer No. 6 Multipurpose trap (*Spring Traps Approval Orders* 1975, 1982, 1988). Approved spring traps must catch and kill humanely in at least 85 per cent of cases, be set wholly within the overhang of rabbit holes to reduce the risk of catching other species, and be inspected at least once every day between sunrise and sunset (*Pests Act* 1954; *Protection of Animals Acts* 1911, 1912, and amendments 1927).

It is gratifying that, from 1st January 1995, the European Community will ban the import of furs from countries using gin traps (European Communities 1991). While this ban will severely affect the fur industry, and animal damage control, in continental Europe, Asia, the USA, Canada, Australia, and New Zealand, it will also stimulate the development and use of alternative, more humane traps.

Until fairly recently, cage-traps for catching rabbits alive have shown little promise, but some success has been achieved with wire-mesh cage-traps developed by MAFF for research purposes. Two types have been used; small unbaited traps, set on well-used rabbit runs in thick vegetation, and larger traps baited with fresh, chopped carrot, set a few metres from rabbit harbourage. All cage-traps should be inspected at least twice daily, in early morning and late evening.

4.7.4 Shooting

There is a mainly sporting interest in shooting, but the efficient use of dog and gun can be a highly effective method of dealing with surface-living rabbits, which are often abundant in woodland and scrub. It is generally believed that bucks predominate among these rabbits, a view confirmed by a field study. In a fenced area where every rabbit was killed, over half were does, but three-quarters of the surface-living rabbits were bucks (Thompson and Armour 1951). Shooting can be a humane control method in skilled hands (Blank 1969) but otherwise is often unsystematic and ineffective (Rees *et al.* 1985).

4.7.5 Ferreting

The ferret is closely related to, and possibly just a variety of, the polecat *Mustela putorius*, with which it can be crossed, and has been domesticated for at least 2000 years. The care and training of ferrets figures in many sporting publications and is a matter of absorbing interest to many countrymen (Marchington 1978). Ferrets are used either to bolt rabbits into purse-nets placed over rabbit holes, or to bolt them for shooting. It is seldom possible to bolt all the rabbits present, and those which the ferret has killed or driven into a corner must be dug out. This is laborious but thorough, and has the advantage of destroying part of the burrow system. It is virtually essential to have a trained dog to mark holes likely to house rabbits and to run down any which escape from a purse-net. Some studies have been made of ferreting as a method of control, and it can be a most valuable technique for dealing with an intransigent residue of rabbits (Thompson and Armour 1951; Stephens 1952; Phillips 1955*a*; Cowan 1984).

4.7.6 Snaring

The ordinary wire-snare catches rabbits quickly and efficiently if they are numerous, and the weather conditions are suitable; but results are poor during dry weather or frost.

The *Wildlife and Countryside Act* 1981 prohibits the use of the self-locking snare except under licence, and requires that snares be visited daily. There is much concern about the humaneness of snaring and, since most rabbits are caught at night, dusk and dawn inspections are preferable.

4.7.7 Long-netting

If rabbits are abundant, large numbers may be taken in a short time in long nets. These nets are usually 45 to 135 m long, 1 m wide, and 5 cm mesh. They are run out at night, downwind from the rabbits, about 13 m from a hedge, wood, or warren, when the rabbits are out feeding. There is a line along the top of each net which is supported by 1 m sticks at 4 m intervals, the slack of the net being allowed to lie loosely on the ground. Once the net is erected a circuit is made and the rabbits are driven into the net at speed by dogs and beaters, and their necks dislocated before removal from the net. It is best to use dogs which can hunt mute, and work the rabbits like a sheepdog. In place of dogs a long line, called a 'dead dog', is sometimes trailed across the field by two men. Silence, speed, and a dark night are essential for success.

Long-netting was once greatly favoured by poachers, who frequently used very light nets made of silk which could easily be carried in the pocket. To prevent poaching with long nets, gamekeepers regularly went round their coverts at night and scared the grazing rabbits, thus making them feed in the early morning and evening when they are less easily netted. Keepers also 'bushed' their field during the day; that is, they strewed short pieces of thorn, bramble, or gorse about the fields near to the coverts, so that the long nets would become tangled. Ingenious long nets have been invented which can be left in position for some time and raised and lowered mechanically or even electrically, but they offer no advantages over the simple net.

4.7.8 Poison baiting

In Australia and New Zealand there is great reliance on controlling rabbits by poisoning them (see Chapters 5 and 6), using, since the 1950s, sodium monofluoroacetate (Compound 1080), which is also very toxic to felids and canids. In Britain, as far as mammals are concerned, the use of poisons is governed by the following legislation; the *Protection of Animals Act, 1911*; the *Protection of Animals (Scotland) Act, 1912*; the *Agriculture Act, 1947*; the *Agriculture (Scotland) Act, 1948*, and the *Animals (Cruel Poisons) Act, 1962*.

Briefly, these Acts restrict the use of poisons against mammals to rats, mice, or other small ground vermin (in the Scottish Act of 1912 'vermin' is not qualified by adjectives); permit the use of poisonous gases in holes or burrows to kill rabbits, foxes, or moles; and enable certain poisons to be prohibited by regulations.

It is illegal to lay poisoned baits for rabbits, but it was urged in 1978, as rabbit numbers increased, that it was time to re-examine this position and to carry out field studies to find if wild rabbits, in British conditions, could be attracted to unpoisoned bait (Thompson 1981; Rees *et al.* 1985). If a bait or baiting technique that is specific to rabbits could be developed, the risk of poisoning other species would be minimized. Besides examining poisons used elsewhere in the world, and against other species, it would be advantageous to develop a poison which was not persistent and which did not leave dangerous residues in the rabbit, thereby eliminating the risk of secondary poisoning. No-one is suggesting that techniques of rabbit poisoning appropriate to countries such as Australia and New Zealand can be applied without modification to closely settled areas of Britain, but search is necessary for a method of poisoning—or indeed other means of population limitation—for use at least in areas of sparse human population where labour-intensive methods are unsatisfactory.

Research was begun in 1979, concentrating on the need for a safe, selective, and effective method of baiting. Work on baits of sliced carrot showed that rhodamine B (persistent but only qualitative) and floss fibres (quantitative but not persistent) were acceptable markers (Cowan *et al.* 1984). Sliced-carrot baits marked in this way were freely eaten by wild rabbits, bucks eating more than does and young more than adults (Cowan *et al.* 1987). Other studies showed that rabbits' consumption of carrots was much reduced by treating the carrots with the urine of male rabbits (Bell *et al.* 1983; Bell 1986).

4.7.9 Other techniques

When focused in the beam of a spotlight, a rabbit is dazzled and can readily be caught by a trained dog; this was a favourite method of poachers, on dark and windy nights. Rabbits can be shot at night, using a spotlight, but this is unpopular with game estates and farmers, because of disturbance to stock. It is also an offence to shoot game or deer at night (which is legally from one hour after sunset to one hour before sunrise). Night shooting disturbs other rabbits and is not regarded as an effective means of control (see also Chapter 6).

The inhibition of reproduction, using a long-lasting, single dose contraceptive, has been considered, but deemed unpromising for rabbits (Rees *et al.* 1985). But the research now in progress in Australia (see Chapter 5), which aims to disrupt rabbit-breeding by inducing an auto-immune reaction to antigens, spread by means of a recombinant myxoma virus, is a fascinating new departure.

An acoustic scaring device has proved ineffective against rabbits (Wilson and McKillop 1986). There are many examples of the abuse, and ineffectiveness, of the 'bounty' system, i.e. paying a reward for the physical evidence (usually a tail) of having killed a pest animal (Thompson and Peace 1962). But faith in the discredited bounty persists, and the Isle of Man Department of Agriculture brought in just such a scheme in 1985, setting aside £20 000 to pay for it, and offering 35p per rabbit tail. In the first two-and-a-half years, over 100 000 tails were paid for, and the number submitted is now averaging 25 000 tails a year; amounting to a total of over 200 000 by the summer of 1992 (personal communication Isle of Man Department of Agriculture). Either the rabbit crop has stabilized or rabbit tails are being imported.

Habitat management has already been mentioned and will be considered in Section 4.8.

4.8 Discussion

The rabbit is an adaptable opportunist which can tolerate a wide range of climates. It has catholic tastes in food and high efficiency of protein digestibility (Monk 1989). The extent to which it has been introduced to other parts of the world is detailed in Chapter 2; but the assisted colonist has not always been successful. For example, there is a thriving colony on San Juan Island, Washington, USA (Stevens and Weisbrod 1981). Nearly 40 years ago, wildlife managers were concerned about the translocation of some of the San Juan rabbits to various mainland states of the USA, for sporting purposes (Thompson 1955) but, fortunately, very few colonies have become established.

Much of the rabbit's successful spread in continental Europe, Britain, Australia, New Zealand, and South America can be attributed to human activities which alter the natural environment in favour of rabbits, such as the replacement of woodland by arable crops and grass. Over four-fifths of Britain is occupied by landowners and farmers (19.6 million out of 23.6 million hectares) and the countryside is greatly modified by agricultural and forestry practices. Until recently much of this beautiful but man-made landscape has been relatively unchanged for some 200 years, but modern farming methods, including the use of large machines which operate more effectively in larger fields, has led to the grubbing-up of many thousands of kilometres of hedges, which previously provided pathways and cover for wildlife (Thompson 1973). As emphasized by Elton (1966) habitat interspersion has a stabilizing effect, and over-simplified communities are vulnerable to invaders, particularly exotic species, which may become pests. In a general sense, farmland is the largest nature reserve in the country and, over the past 20 years, landholders have become increasingly aware of their responsibilities as conservationists. There is consequently much co-operation between those who own and occupy land, those who use it for countryside sports, and the many organizations and individuals concerned with wildlife and animal welfare (Thompson 1985, 1990; Cobham 1992).

Habitat modifications have a profound effect on rabbits, and recent changes in agricultural practices and the politics of land use, must lead to an increase in rabbit numbers. Because of overproduction, and the vast expense of storing and disposing of the surplus, the European Community introduced in 1988 a scheme for reducing the area under arable crops, mainly cereals, known as 'set aside'. This entails leaving fields fallow, apart from occasional mowing, and in Britain this scheme now affects about 155 000 hectares, or some three per cent of the area previously under arable cultivation; farmers are compensated for loss of production at a current rate of over £20 million a year. The scheme is at present voluntary, but is likely soon to be virtually compulsory and to cover 15 per cent of eligible land, i.e. 750 000 hectares (The *Times* 1992). Farmers will naturally 'set aside' the less productive land, and it will afford increased harbourage for rabbits.

It is relatively easy to generate enthusiasm for the control of rabbits when their numbers are high and damage to crops is obvious; it is even possible, as in New Zealand for some decades (see Chapter 6), to espouse a policy of extermination. But as Charles Elton (1958) pointed out, invasions of a species from one part of the world to another can be prevented either by 'quarantine' at the point of entry, or by 'eradication' of the first small population of invaders. Once the invaders are well-established, eradication is usually impracticable and numbers can only be kept in check by control. To attain this end, much more fundamental knowledge of the processes governing balanced populations is needed. There are some examples of rabbits being eradicated from limited areas, usually by fencing them and ruthlessly killing the enclosed rabbits. But the most striking case was on Round Island, Mauritius (Merton 1987). This ecologically important island, a basaltic, volcanic cone, 151 hectares in area and 280 metres high, was completely cleared of 1000 or more rabbits using an anticoagulant poison during a three-month exercise. Certainly only persistent, precisely targeted and thorough campaigns, backed up by research, will succeed in controlling rabbits. As their numbers in Britain increase, the considerable body of legislation enforcing control should be used rigorously.

4.9 Summary

Introduced from mainland Europe to Britain in the twelfth century, and cosseted in warrens, the rabbit was much appreciated as a luxury food for over 300 years. As it became more abundant and spread into the countryside, the rabbit was increasingly valued for its carcase and skin, and its potential as game. But as the damage caused to agricultural crops by rabbits became evident, conflicting opinions about them developed and have continued in the twentieth century.

Some sixty years ago another conflict arose over the gin trap. This miniature mantrap, used from the seventeenth century onwards, catches the rabbit by the leg. The widespread use of the trap allowed, by the nineteenth century, the development of a wild rabbit trapping industry in competition with agriculture. The trap was opposed by animal welfare groups, on the grounds of cruelty, and a combination of humane and farming interests sponsored restrictive legislation, and finally a total ban on the gin trap in 1954.

The same combination of interests stimulated and funded both technical and administrative measures of rabbit control, as well as biological research on the rabbit, culminating in a long-term study of myxomatosis, after its unofficial introduction from France in 1953. This devastating disease caused the temporary eclipse of the rabbit by 1955, which sparked research studies of the consequent ecological effects on the flora and fauna of Britain.

As an assisted colonist, the rabbit has been a striking success in Britain. Although almost eliminated by myxomatosis, it has made a substantial recovery. The disease will continue to act as a brake on rabbit numbers, but these are already sufficiently large to need vigilant control, if damage to agriculture is to be effectively limited.

Acknowledgements

It is a pleasure to thank my colleagues for collaboration in research over many years, Dr J. Ross for reading a draft of this Chapter, and the Universities Federation for Animal Welfare for their constant support. I am grateful to Mrs M. E. Mansell-Moullin and Mrs S. Griffin for transmuting my drafts into readable typescript.

I also thank the late Dr A. S. Thomas and the editors and publishers of the following journals for their permission to use the photographs and figures ascribed to them in the individual captions and references: Academic Press, Agricultural Central Co-operative Association, *Annals of Applied Biology*, *Applied Biology*, Blackwells Scientific Publications, Cambridge University Press, Harper Collins, David and Charles, Game Conservancy, *Journal of Animal Ecology*, *Journal of Ecology*, *Mammal Review*.

References

Adams, C. E. (1986). The laboratory rabbit. In *The care and management of laboratory animals* (ed. T. Poole). Longman, London.

Advisory Committee on Myxomatosis (1954). *Myxomatosis*. HMSO, London.

Advisory Committee on Myxomatosis (1955). *Second report on myxomatosis*. HMSO, London.

Agriculture Act (1947) 10 and 11, Geo. 6, Ch. 48.

Agriculture (Scotland) Act (1948) 11 and 12, Geo 6, Ch. 45.

Andrews, C. H., Thompson, H. V., and Mansi, W. (1959). Myxomatosis: present position and future prospects in Great Britain. *Nature*, **184**, 1179–80.

Animals (Cruel Poisons) Act 10 and 11, Eliz. 2, Ch. 26.

Armour, C. J. (1963). The use of repellents for preventing mammal and bird damage to trees and seed: a revision. *Forestry Abstracts*, **24**, 27–38.

Armour, C. J. and Thompson, H. V. (1955). Spread of myxomatosis in the first outbreak in Great Britain. *Annals of Applied Biology*, **43**, 511–18.

Barcroft, J. (1931). The toxicity of atmospheres containing hydrocyanic acid gas. *Journal of Hygiene*, **31**, 1–34.

Barnes, R. F. W. and Tapper, S. C. (1986). Consequences of the myxomatosis epidemic in Britain's rabbit (*Oryctolagus cuniculus*) population on the numbers of brown hares (*Lepus europaeus*). *Mammal Review*, **16**, 111–16.

Barrett-Hamilton, G. E. H. (1912). *A history of British mammals*, Vol. 2. Gurney and Jackson, London.

Bateman, J. (1971). *Animal traps and trapping*, David and Charles, Newton Abbot.

Bell, D. J. (1980). Social olfaction in lagomorphs. *Symposia of the Zoological Society of London*, **45**, 141–64.

Bell, D. J. (1981). Chemical communication in the European rabbit: urine and social status. In *Proceedings of the World Lagomorph Conference* (1979) (Ed. K. Myers and C. D. McInnes), pp. 271–9.

Bell, D. J. (1986). Pheromones and mammalian pests with particular reference to the European rabbit. In *Control of mammal pests* (ed. C. G. Richards and T. Y. Ku), pp. 113–25. Taylor and Francis, London.

Bell, D. J. and Mitchell, S. (1984). Effects of female urine on growth and sexual maturation in male rabbits. *Journal of Reproduction and Fertility*, **711**, 155–60.

Bell, D. J. and Webb, N. J. (1991). Effects of climate on reproduction in the European wild rabbit (*Oryctolagus cuniculus*). *Journal of Zoology*, **244**, 639–48.

Bell, D. J., Moore, S., and Cowan, D. (1983). Effects of urine on the response to carrot-bait in the European wild rabbit. *Chemical signals in vertebrates 3*. (ed. D. Muller-Schwarze and R. M. Silverstein), pp. 333–8. Plenum Press, New York.

Bewick, T. (1814). *A natural history of British quadrupeds*. Newcastle upon Tyne.

Blank, T. H. (1969). Shooting as a humane method of control of animal populations. In *The humane control of animals living in the wild*, pp. 25–6. UFAW, Potter's Bar.

Blount, W. P. (1945). *Rabbits' ailments*. Fur and Feather, Idle, Bradford.

Boag, B. (1972). Helminth parasites of the wild rabbit *Oryctolagus cuniculus* in North East England. *Journal of Helminthology*, **46**, 73–9.

Boag, B. (1985). The incidence of helminth parasites from the wild rabbit *Oryctolagus cuniculus* in eastern Scotland. *Journal of Helminthology*, **59**, 61–9.

Boag, B. (1987). Reduction in numbers of the wild rabbit (*Oryctolagus cuniculus*) due to changes in agricultural practices and land use. *Crop Protection*, **6**, 347–51.

Boag, B. (1988). Observations on the seasonal incidence of myxomatosis and its interactions with helminth parasites in the European rabbit (*Oryctolagus cuniculus*). *Journal of Wildlife Diseases*, **24**, 450–5.

Boag, B. (1989). Population dynamics of parasites of the wild rabbit. In *Mammals as pests* (ed. R. J. Putnam), pp. 186–95. Chapman and Hall, London.

Boag, B., MacFarlane Smith, W. H., and Griffiths, D. W. (1990). Effects of grazing by wild rabbits (*Oryctolagus cuniculus*) on the growth and yield of oilseed and fodder rape (*Brassica rapus* sub sp. *oleifera*). *Crop Protection*, **9**, 155–9.

Boyd, I. L. (1985). Investment in growth by pregnant wild rabbits: relation to litter size and sex in the offspring. *Journal of Animal Ecology*, **54**, 137–47.

Boyd, I. L. (1986). Effect of daylength on the breeding season in male rabbits. *Mammal Review*, **16**, 125–30.

Boyd, I. L. and Myhill, D. G. (1987). Seasonal changes in condition, reproduction and fecundity in the wild European rabbit (*Oryctolagus cuniculus*). *Journal of Zoology*, **212**, 223–33.

Brambell, F. W. R. (1942). Intra-uterine mortality in the wild rabbit, *Oryctolagus cuniculus*. *Proceedings of the Royal Society* (B), **130**, 462–79.

Brambell, F. W. R. (1944). The reproduction of the wild rabbit, *Oryctolagus cuniculus*. *Proceedings of the Zoological Society, London*, **114**, 1–45.

Brambell, F. W. R. (1948). Prenatal mortality in mammals. *Biological Review*, **23**, 370–405.

Buckley, W. H. (1935). Report on a solution of the rabbit problem: cyanide gassing. *Veterinary Journal*, **91**, 210–15.

Buckley, W. H. (1958). The wild rabbit industry in West Wales. *The UFAW Courier* No. 15, 11–21.

Bull, L. B. and Mules, M. W. (1944). An investigation of *Myxomatosis cuniculi* with special reference to the possible use of the disease to control rabbit populations in Australia. *Journal of the Council of Scientific and Industrial Research, Australia*, **17**, 79–93.

Chitty, D. H. and Southern, H. N. (ed.) (1954) *Control of Rats and Mice*. Clarendon Press, Oxford.

Church, B. M., Jacob, F. H., and Thompson, H. V. (1953). Surveys of rabbit damage to wheat in England and Wales, 1950–52. *Plant Pathology*, **2**, 107–12.

Church, B. M., Westmacott, M. H., and Jacob, F. H. (1956). Surveys of rabbit damage to winter cereals in England and Wales 1953–1954. *Plant Pathology*, **5**, 66–9.

Cobham (1992). *Countryside sports their economic and conservation significance*. Summary report. Standing Conference on Countryside Sports. College of Estate Management, Reading.

Committee on Cruelty to Wild Animals (1951). Report G.B. Parl. Command 8266. HMSO, London.

Corbet, G. B. and Harris, S. (ed.) (1991). *The handbook of British mammals* (3rd edn). Blackwell Scientific Publications, Oxford.

Cowan, D. P. (1984). The use of ferrets (*Mustelo furo*) for the study and management of the European wild rabbit (*Oryctolagus cuniculus*). *Journal of Zoology*, **204**, 570–4.

Cowan, D. P. (1985). Coccidiosis in rabbits. Population dynamics and epidemiology of terrestrial animals: ITE Merlewood research and development paper no. 106 (ed. D. Mollison and P. Bacon), pp. 25–7. Cumbria.

Cowan, D. P. (1987). Aspects of the social organisation of the European wild rabbit (*Oryctolagus cuniculus*). *Ethology*, **75**, 197–210.

Cowan, D. P. and Bell, D. J. (1986). Leporid social behaviour and social organisation. *Mammal Review*, **16**, 169–79.

Cowan, D. P. and Garson, P. J. (1985). Variations in the social structure of rabbit populations: causes and demographic consequences. In *Behavioural ecology and the ecological consequences of adaptive behaviour* (ed. R. M. Sibley and R. H. Smith), pp. 537–55. Blackwell, Oxford.

Cowan, D. P. and Roman, E. A. (1985). On the construction of life tables with special reference to the European wild rabbit *Oryctolagus cuniculus*. *Journal of Zoology*, **207**, 607–9.

Cowan, D. P., Vaughan, J. A., Prout, K. J., and Christer, W. G. (1984). Markers for measuring bait consumption by the European wild rabbit. *Journal of Wildlife Management*, **48**, 1403–9.

Cowan, D. P., Vaughan, J. A., and Christer, W. G. (1987). Bait consumption by the European rabbit in southern England. *Journal of Wildlife Management*, **51**, 386–92.

Cowan, D. P., Hardy, A. R., Vaughan, J. A., and Christer, W. G. (1989). Rabbit ranging behaviour and its implications for the management of rabbit populations. In *Mammals as pests* (ed. R. J. Putman), pp. 178–86. Chapman and Hall, London.

Crawley, M. J. (1989). Rabbits as pests of winter wheat. In *Mammals as pests* (ed. R. J. Putman), pp. 168–77. Chapman and Hall, London.

Crawley, M. J. (1990). Rabbit grazing, plant competition and seedling recruitment in acid grassland. *Journal of Applied Ecology*, **27**, 803–20.

Delibes, M. and Hiraldo, F. (1981). The rabbit as prey in the Iberian mediterranean ecosystem. In *Proceedings of the World Lagomorph Conference* (1979) (ed. K. Myers and C. D. MacInnes), pp. 614–22. University of Guelph. Ontario.

Destructive Imported Animals Act. (1932). 22 Geo. 5, Ch. 12. London.

Dunn, A. M. (1969). The wild ruminant as reservoir host of helminth infection. *Symposium of the Zoological Society, London* (1968) (ed. A. McDiarmid) **24**, 221–48.

Eden, A. (1940*a*). Coprophagy in the rabbit. *Nature*, **145**, 36–7.

Eden, A. (1940*b*). Coprophagy in the rabbit: origin of 'night' faeces. *Nature*, **145**, 628–9.

Elton, C. S. (1958). *The ecology of invasions by animals and plants*. Methuen, London.

Elton, C. S. (1966). *The pattern of animal communities*. Methuen, London.

Errington, P. L. (1946). Predators and vertebrate populations. *Quarterly Review of Biology*, **21**, 144–77.

Errington, P. L. (1963). The phenomenon of predation. *American Scientist*, **51**, 180–92.

European Communities (1991). Council Regulation (EEC) No. 3254/91. *Official Journal*, **34**, L 308/1.

Evans, W. M. R. (1942*a*). Observations on the incidence of some nematode parasites of the common rabbit, *Oryctolagus cuniculus*. *Parasitology*, **32**, 67–77.

Evans, W. M. R. (1942*b*). Observations on the incidence of some common cestode parasites of the wild rabbit, *Oryctolagus cuniculus*. *Parasitology*, **32**, 78–90.

Fairley, J. S. (1966). An indication of the food of the fox in Northern Ireland after myxomatosis. *Irish Naturalists' Journal*, **15**, 149–51.

Farrow, E. P. (1917). General effects of rabbits on vegetation. *Journal of Ecology*, **5**, 1–18.

Farrow, E. P. (1925). *Plant life on East Anglian heaths*. Cambridge University Press.

Fenner, F. and Ratcliffe, F. N. (1965). *Myxomatosis*. Cambridge University Press.

Fenton, E. W. (1940). The influence of rabbits on the vegetation of certain hill-grazing districts of Scotland. *Journal of Ecology*, **28**, 438–49.

Flux, J. E. C. and Fullager, P. J. (1992). World distribution of the rabbit *Oryctolagus cuniculus* on islands. *Mammal Review*, **22**, 151–205.

Forestry Commissioners. (1943). *Post-war forest policy*. Cmd. 6447. HMSO, London.

Fraser Darling, F. (1947). *Natural history in the Highlands and islands*. Collins, London.

Game Act. (1831). 1 and 2 Will., 4, Ch. 32. London.

Gillham, M. E. (1955). Ecology of the Pembrokeshire islands III. The effects of grazing on the vegetation. *Journal of Ecology*, **43**, 172–206.

Gin Traps (Prohibition) Bill (1935).

Gough, H. C. and Dunnett, F. W. (1950). Rabbit damage to winter corn. *Agriculture*, *57*, 374–8.

Ground Game Act. (1880). 43 and 44 Vict., Ch. 47. London.

Harting, J. E. (1898). *The rabbit*. Longmans Green, London.

Hewson, R. and Kolb, H. H. (1973). Changes in the numbers and distribution of foxes (*Vulpes vulpes*) killed in Scotland from 1948–70. *Journal of Zoology*, **171**, 345–65.

Hudson, J. R. and Mansi, W. (1955). Attenuated strains of myxomatosis virus in England. *Veterinary Record*, **67**, 746.

Hume, C. W. (1939). Instructions for dealing with rabbits. UFAW Monograph and Report, 4E. London.

Hume, C. W. (1954). *Instructions for dealing with rabbits*. Universities Federation for Animal Welfare, London.

Hume, C. W. (1958). The gin trap: UFAW's long battle. *The UFAW Courier* No. 15, 1–10.

Hume, C. W. (1962). *Man and beast*. Universities Federation for Animal Welfare, London.

Hurrell, H. G. (1979). The little known rabbit. *Countryside*, **23**, 501–3.

Jefferies, D. J. and Pendlebury, J. B. (1968). Population fluctuations of stoats, weasels and hedgehogs in recent years. *Journal of Zoology*, **156**, 513–17.

Jenkins, D., Watson, A., and Miller, G. R. (1964). Predation and red grouse populations. *Journal of Applied Ecology*, **1**, 183–95.

King, C. M. (1980). Population biology of the weasel *Mustela nivalis* on British game estates. *Holarctic Ecology*, **3**, 160–8.

King, C. M. (1989). *The natural history of weasels and stoats*. Christopher Helm, London.

Kolb, H. H. (1985). The burrow structure of the European rabbit (*Oryctolagus cuniculus*). *Journal of Zoology*, **206**, 253–62.

Kolb, H. H. (1986). Circadian activity in the wild rabbit (*Oryctolagus cuniculus*). *Mammal Review*, **16**, 145–50.

Larceny Act. (1861). Abolished by Theft Act (1968). London.

Leiper, J. W. G. (1937). Natural helminthiasis of the goat involving infection with *Trichostrongylus retortaeformis* of the rabbit. *Veterinary Record*, **49**, 1411–42.

Lever, R. J. A. W. (1959). The diet of the fox since myxomatosis. *Journal of Animal Ecology*, **28**, 359–75.

Lever, R. J. A. W., Armour, C. J., and Thompson, H. V. (1957). Myxomatosis and the fox. *Agriculture*, **64**, 105–11.

Lloyd, H. G. (1962). Humane Traps. *The Review* (June), pp. 15–16. Royal Agricultural Society of England.

Lloyd, H. G. (1963*a*). Spring traps and their development. *Annals of Applied Biology*, **51**, 329–33.

Lloyd, H. G. (1963*b*). Intra-uterine mortality in the wild rabbit (*Oryctolagus cuniculus*) in populations of low density. *Journal of Animal Ecology*, **32**, 549–63.

Lloyd, H. G. (1970*a*). Variation and adaptation in reproductive performance. *Symposium of the Zoological Society, London*, **26**, 165–88.

Lloyd, H. G. (1970*b*). Post-myxomatosis rabbit populations in England and Wales. *EPPO Public Series A*, **58**, 197–215.

Lloyd, H. G. (1981). Biological observations on post-myxomatosis wild rabbit populations in Britain 1955–1979. In *Proceedings of the World Lagomorph Conference* (1979) (ed. K. Myers and C. D. MacInnes), pp. 623–8. University of Guelph, Ontario.

Lockie, J. D. (1956). After myxomatosis: notes on the food of some predatory animals in Scotland. *Scottish Agriculture*, **36**, 65–9.

Lockley, R. M. (1940). Some experiments in rabbit control. *Nature*, **145**, 767.

Lockley, R. M. (1954). The European rabbit flea, *Spilopsyllus cuniculi*, as a vector of myxomatosis in Britain. *Veterinary Record*, **66**, 434–5.

Lockley, R. M. (1955). Failure of myxomatosis in Skokholm Island. *Nature*, **175**, 906–7.

MAFF (1978). Pest Infestation Laboratory Report 1974–76. HMSO, London.

MAFF (1981). Agricultural Science Service Research and Development Reports, *Mammal and bird pests*. HMSO, London.

Marchington, J. (1978). *Pugs and drummers: ferrets and rabbits in Britain*. Faber, London.

Martin, C. J. (1936). Observations on *Myxomatosis cuniculi* (Sanarelli) made with a view to the use of the virus in the control of rabbit plagues. *Bulletin of the Council for Scientific and Industrial Research, Australia*, No. 96.

Matheson, C. (1941). The rabbit and the hare in Wales. *Antiquity*, **15**, 371–81.

McCutchan, J. C. (1980). *Electric fence design and principles*. University of Melbourne.

McDiarmid, A. (1962). Diseases of free-living wild animals. *FAO Agricultural Studies*, No. 57.

McKillop, I. G. (1988). The operation of co-ordinated rabbit control organisations in England and Wales. In *Proceedings of the 13th Vertebrate Pest Conference, California*, pp. 174–9.

McKillop, I. G. and Sibly, R. M. (1988). Animal behaviour at electric fences and the implications for management. *Mammal Review*, **18**, 91–103.

McKillop, I. G. and Wilson, C. J. (1987). Effectiveness of fences to exclude European rabbits from crops. *Wildlife Society Bulletin*, **15**, 394–401.

McKillop, I. G., Pepper, H. W., and Wilson, C. J. (1986). Specifications for wire mesh fences to exclude the European wild rabbit from crops. In *Proceedings of the 12th Vertebrate Pest Conference* (ed. L. R. Davis and R. E. Marsh), pp. 147–52. University of California, Davis.

McKillop, I. G., Pepper, H. W., and Wilson, C. J. (1988). Improved specifications for rabbit fencing for tree protection. *Forestry*, **61**, 359–68.

McKillop, I. G., Sylvester-Bradley, R., Pugh, B. D., Dagnall, J. L., Fox, S. M., and Fox, A. P. (1991). Assessment of damage by rabbits to winter wheat. Working document for Australian Vertebrate Pest Control Conference.

McKillop, I. G., Phillips, K. V., and Ginella, S. G. V. (1992). Effectiveness of two types of electric fence for excluding European wild rabbits. *Crop Protection*, **11**, 279–85.

McNamara, B. P. (1976). Estimates of the toxicity of HCN vapours in man. US Department Army, *Edgewood Arsenal Technical Report* EB-TR-76023.

Mead-Briggs, A. R. (1964*a*). Records of rabbit fleas (*Spilopsyllus cuniculi*), from every county in Great Britain with notes on infestation rates. *Entomological Monthly Magazine*, **100**, 8–17.

Mead-Briggs, A. R. (1964*b*). The reproductive biology of the rabbit flea *Spilopsyllus cuniculi* and the dependence of this species upon the breeding of its host. *Journal of Experimental Biology*, **41**, 371–402.

Mead-Briggs, A. R. (1977). The European rabbit, the European rabbit flea and myxomatosis. *Applied Biology*, **2**, 183–261.

Mead-Briggs, A. R. and Page, R. J. C. (1975). Records of anoplocephaline cestodes from wild rabbits and hares collected throughout Britain. *Journal of Helminthology*, **49**, 49–56.

Mead-Briggs, A. R. and Rudge, A. J. B. (1960). Breeding of the rabbit flea, *Spilopsyllus cuniculi*; requirement of a 'factor' from a pregnant rabbit for ovarian maturation. *Nature*, **187**, 1136–7.

Mead-Briggs, A. R. and Vaughan, J. A. (1969). Some requirements for mating in the rabbit flea, *Spilopsyllus cuniculi*. *Journal of Experimental Biology*, **51**, 495–511.

Mead-Briggs, A. R. and Vaughan, J. A. (1973). The incidence of anoplocephaline cestodes in a population of rabbits in Surrey, England. *Parasitology*, **67**, 351–64.

Meldrum, K. C. (1992). Viral haemorrhagic disease of rabbits. *Veterinary Record*, **130**, 407.

Merton, D. V. (1987). Eradication of rabbits from Round Island, Mauritius: a conservation success story. *Dodo. Journal of Jersey Wildlife Preservation Trust*, **24**, 19–44.

Middleton, A. D. (1932). Syphilis as a disease of wild rabbits and hares. *Journal of Animal Ecology*, **4**, 274–6.

Middleton, A. D. (1942). *The control and extermination of wild rabbits*. Bureau of Animal Population, Oxford.

Monk, K. A. (1989). Effects of diet composition on intake by adult wild European rabbits. *Appetite*, **13**, 201–9.

Moore, N. W. (1956). Rabbits, buzzards and hares: two studies on the indirect effects of myxomatosis. *Terre vie*, **103**, 220–5.

Moore, N. W. (1957). The past and present status of the buzzard in the British Isles. *British Birds*, **50**, 173–97.

Moore, N. W. (1987). *The bird of time*. Cambridge University Press.

Muirhead-Thomson, R. C. (1956*a*). Observations on the European rabbit flea (*Spilopsyllus cuniculi*) in relation to myxomatosis in England. Unpublished Report to Myxomatosis Advisory Committee. MAFF.

Muirhead-Thomson, R. C. (1956*b*). The part played by woodland mosquitoes of the genus *Aedes* in the transmission of myxomatosis in England. *Journal of Hygiene, Cambridge*, **54**, 461–71.

Muirhead-Thomson, R. C. (1956*c*)p. Field studies of the role of *Anopheles atroparvus* in the transmission of myxomatosis in England. *Journal of Hygiene, Cambridge*, **54**, 472–7.

Mykytowycz, R. (1958). Social behaviour of an experimental colony of wild rabbits, *Oryctolagus cuniculus* (L.). **I**. Establishment of the colony. *CSIRO Wildlife Research*, **3**, 7–25.

Mykytowycz, R. (1959). Social behaviour of an experimental colony of wild rabbits, *Oryctolagus cuniculus* (L.). **II**. First breeding season. *CSIRO Wildlife Research*, **4**, 1–13.

Mykytowycz, R. (1960). Social behaviour of an experimental colony of wild rabbits, *Oryctolagus cuniculus* (L.). **III**. Second breeding season. *CSIRO Wildlife Research*, **5**, 1–20.

Mykytowycz, R. (1961). Social behaviour of an experimental colony of wild rabbits, *Oryctolagus cuniculus* (L.). **IV**. Conclusion: outbreak of myxomatosis, third breeding season, and starvation. *CSIRO Wildlife Research*, **6**, 142–55.

Night Poaching Act. (1828). 9 Geo., 9, Ch. 69. London.

Night Poaching Act. (1844). 7 and 8 Vict., Ch. 29. London.

Non-Indigenous Rabbits (Prohibition of Importation and Keeping) Order. (1954). Statutory Instruments No. 927.

Pests Act 1954 2 & 3 Eliz. 2, Ch. 68. London.

Phillips, W. M. (1953). The effect of rabbit grazing on a reseeded pasture. *Journal of the British Grassland Society*, **8**, 169–81.

Phillips, W. M. (1955*a*). The effect of commercial trapping on rabbit populations. *Annals of Applied Biology*, **43**, 247–57.

Phillips, W. M. (1955*b*). An experiment on rabbit control. *Annals of Applied Biology*, **43**, 258–64.

Phillips, W. M., Stephens, M. N., and Worden, A. N. (1952). Observations on the rabbit in West Wales. *Nature*, **169**, 869–70.

Poaching Prevention Act. (1862). 25 and 26 Vict., Ch. 114. London.

Potts, G. R. (1986). *The partridge: pesticides, predation and conservation*. Collins, London.

Prevention of Damage by Rabbits Act (1939) 2 & 3, Geo., 6, Ch. 43. London.

Protection of Animals Act (1911). 11 & 12, Geo., 5, Ch. 14. London.

Protection of Animals (Scotland) Act (1912). 2 & 3, Geo., 5, Ch. 14. London.

Protection of Animals (Amendment) Act (1927). 17 & 18, Geo., 5, Ch. 27. London.

Rees, W. A., Ross, J., Cowan, D. P., Tittensor, A. M., and Trout, R. C. (1985). Humane control of rabbits. In *Humane control of land mammals and birds*, (ed. D. P. Britt), pp. 96–102. Universities Federation for Animal Welfare, Potters Bar, England.

Ritchie, J. (1920). *The influence of man on animal life in Scotland*. Cambridge University Press.

Ritchie, J. N., Hudson, J. R., and Thompson, H. V. (1954). Myxomatosis. *veterinary Record*, **66**, 796–804.

Rolls, E. C. (1969). *They all ran wild*. Angus and Robertson, Sydney.

Ross, J. (1986). Comparison of fumigant gases used for rabbit control in Great Britain. In *Proceedings of the 12th Vertebrate Pest Conference*, pp. 153–7. San Diego, California.

Rothschild, M. (1953). Notes on the European rabbit flea. Report to the Myxomatosis Advisory Committee, Ministry of Agriculture, Fisheries and Food. (Private circulation.)

Rothschild, M. (1958). A further note on the increase of hares (*Lepus europaeus*) in France. *Proceedings of the Zoological Society, London*, **131**, 328–9.

Rothschild, M. (1961). Increase of hares at Ashton Wold. *Proceedings of the Zoological Society, London*, **137**, 634–5.

Rothschild, M. (1963). A rise in the flea-index on the hare (*Lepus europaeus*) with relevant notes on the fox (*Vulpes vulpes*) and wood-pigeon (*Columba palumbus*) at Ashton, Peterborough. *Proceedings of the Zoological Society, London*, **140**, 341–6.

Rothschild, M. (1991). Arrangement of sperm within the spermatheca of fleas, with remarks on sperm displacement. *Biological Journal of the Linnean Society*, **43**, 313–23.

Rothschild, M. and Ford, B. (1964*a*). Breeding of the rabbit flea (*Spilopsyllus cuniculi*) (Dale) controlled by the reproductive hormones of the host. *Nature*, **201**, 103–4.

Rothschild, M. and Ford, B. (1964*b*). Maturation and egg-laying of the rabbit flea (*Spilopsyllus cuniculi*) induced by the external application of hydrocortisone. *Nature*, **203**, 210–11.

Rothschild, M. and Ford, B. (1965). Observations on gravid rabbit fleas (*Spilopsyllus cuniculi*) parasitising the hare (*Lepus europaeus*) together with further speculations concerning the course of myxomatosis at Ashton, Northants. *Proc. Roy. ent. Soc. Lond.* (A), **40**, 109–17.

Rothschild, M. and Ford, B. (1966). Hormones of the vertebrate host controlling ovarian regression and copulation of the rabbit flea. *Nature*, **211**, 261–6.

Rothschild, M. and Ford, B. (1969). Does a pheromone-like factor from the nestling rabbit stimulate impregnation and maturation in the rabbit flea? *Nature*, **221**, 1169–70.

Rothschild, M. and Ford, B. (1973). Factors influencing the breeding of the rabbit flea (*Spilopsyllus cuniculi*): a spring-time accelerator and a kairomone in nestling rabbit urine, with notes on *Cediopsylla simplex*, another 'hormone bound' species. *Journal of Zoology, London*, **170**, 87–137.

Rothschild, M. and Marsh, H. (1956). Increase of hares (*Lepus europaeus*) at Ashton Wold with a note on the reduction in numbers of the brown rat (*Rattus norvegicus*). *Proceedings of the Zoological Society, London*, **127**, 441–5.

Schwenk, S. (1986). The history and spread of the rabbit in Europe. In *The rabbit in hunting, agriculture and forestry*, pp. 6–12. Conseil International de la Chasse, Paris.

Select Committee of the House of Lords on Agriculture (Damage by Rabbits) (1937). HMSO, London.

Shanks, P. L., Sharman, G. A. M., Allan, R., Donald, L. G., Young, S., and Marr, T. G. (1955). Experiments with myxomatosis in the Hebrides. *British Veterinary Journal*, **111**, 25–36.

Sharpe, R. (1918). The taking of wild rabbits. *Journal of the Board of Agriculture*, **25**, 281–98 and 435–44.

Sheail, J. (1971). *Rabbits and their history*. David and Charles, Newton Abbot.

Sheail, J. (1978). Rabbits and agriculture in post-medieval England. *Journal of Historical Geography*, **4**, 343–55.

Sheail, J. (1991). The management of an animal population: changing attitudes towards the wild rabbit in Britain. *Journal of Environmental Management*, **33**, 189–203.

Sibley, R. M., Monk, K. A., Johnson, I. K., and Trout, R. C. (1990). Seasonal variation in gut morphology in wild rabbits (*Oryctolagus cuniculus*). *Journal of Zoology, London*, **221**, 605–19.

Simpson, J. (1908). *The wild rabbit in a new aspect: or rabbit warrens combined with poultry farming and fruit culture*. Parson and Brailsford, Sheffield.

Smith, C. J. (1980). *Ecology of the English chalk*. Academic Press, London.

Sobey, W. R. and Menzies, W. (1969). Myxomatosis: the introduction of the European rabbit flea *Spilopsyllus cuniculi* (Dale) into Australia. *Australian Journal of Science*, **31**, 404–6.

Soriguer, R. C. and Rogers, P. M. (1981). The European wild rabbit in Mediterranean Spain. In *Proceedings of the world Lagomorph Conference* (1979) (ed. K. Myers and C. D. MacInnes), pp. 600–13. University of Guelph. Ontario.

Southern, H. N. (1940*a*). Coprophagy in the wild rabbit. *Nature*, **145**, 262.

Southern, H. N. (1940*b*). The ecology and population dynamics of the wild rabbit. *Annals of Applied Biology*, **27**, 509–26.

Southern, H. N. (1942*a*). Periodicity of refection in the wild rabbit. *Nature*, **149**, 553.

Southern, H. N. (1942*b*). *Electric fencing as a rabbit barrier*. Report 25. Bureau of Animal Population, Oxford University.

Southern, H. N. (1948*a*). Sexual and aggressive behaviour in the wild rabbit. *Behaviour*, **1**, 173–94.

Southern, H. N. (1948*b*). The persistence of hydrogen cyanide in rabbit burrows. *Annals of Applied Biology*, **35**, 331–46.

Southern, H. N. (1956*a*). Ecologists are excited by England's 'Rabbit Disease'. *Animal Kingdom*, **59**, 116–23.

Southern, H. N. (1956*b*). Myxomatosis and the balance of nature. *Agriculture*, **63**, 10–13.

Southern, H. N. (1970). The natural control of a population of Tawny owls. *Strix aluco*. *Journal of Zoology, London*, **162**, 197–285.

Southern, H. N. and Watson, J. S. (1941). Summer food of the red fox (*Vulpes vulpes*) in Great Britain: a preliminary report. *Journal of Animal Ecology*, **10**, 1–11.

Spring Traps Approval Order. (1975). Statutory Instruments No. 1647.

Spring Traps Approval (Variation) Order. (1982). Statutory Instruments No. 53.

Spring Traps Approval (Variation) Order. (1988). Statutory Instruments No. 2111.

Stead, D. G. (1935). *The rabbit in Australia*. Winn, Sydney.

Stephens, M. N. (1952). Seasonal observations on the wild rabbit (*Orygtolagus cuniculus cuniculus L.*) in West Wales. *Proceedings of the Zoological Society, London*, **122**, 417–34.

Stevens, W. F. and Weisbrod, A. R. (1981). The biology of the European rabbit on San Juan Island, Washington, USA in *Proceedings of the World Lagomorph Conference* (1979) (ed. K. Myers and C. D. MacInnes), pp. 870–9. University of Guelph, Ontario.

Sumption, K. J. and Flowerdew, J. R. (1985). The ecological effects of the decline in rabbits (*Oryctolagus cuniculus*) due to myxomatosis. *Mammal Review*, **15**, 151–86.

Swan, C. and Thompson, H. V. (1981). Aquatic rabbits. *Shooting Times*, **5122**, 31.

Tansley, A. G. (1949). *The British Islands and their vegetation*. Cambridge University Press.

Tapper, S. C. (1977). The national game census—a review. *The Game Conservancy Annual Review for 1976*, **8**, 28–35.

Tapper, S. C. (1982). Using estate records to monitor population trends in game and predator species, particularly weasels and stoats. *Transactions of the International Congress of Game Biologists*, **14**, 115–20.

Tapper, S. C. (1992). *Game Heritage*. Game Conservancy, Fordingbridge.

Tapper, S. C., Brockless, M., and Potts, G. R. (1991). The Salisbury Plain predation experiment: the conclusion. *The Game Conservancy Review of 1990*, **22**, 87–91.

Taylor, E. L. (1940*a*). The demonstration of a peculiar kind of coprophagy normally practised by the rabbit. *Veterinary Record*, **52**, 259–62.

Taylor, E. L. (1940*b*). Pseudo-rumination in the rabbit. *Proceedings of the Zoological Society, London*, **110**, 159–63.

Tee, L. A., Rowe, J. J., and Pepper, H. W. (1984). Mammal and bird damage questionnaire report. HMSO, London.

The Times (1992). Issue 18 July 1992.

Thomas, A. S. (1956*a*). Botanical effects of myxomatosis. *Bulletin of the Mammal Society of the British Isles*, **5**, 16–17.

Thomas, A. S. (1956*b*). Biological effects of the spread of myxomatosis among rabbits. *Terre et la Vie*, **103**, 239–42.

Thomas, A. S. (1960). Changes in vegetation since the advent of myxomatosis. *Journal of Ecology*, **48**, 287–306.

Thomas, A. S. (1963). Further changes in vegetation since the advent of myxomatosis. *Journal of Ecology*, **51**, 151–86.

Thomas, J. A. (1980*a*). The extinction of the Large Blue and the conservation of the Black Hairstreak butterflies (a contrast of failure and success). *Annual Report of the Institute of Terrestrial Ecology*, **1979**, 19–23.

Thomas, J. A. (1980*b*). Why did the Large Blue become extinct in Britain? *Oryx*, **15**, 243–7.

Thompson, H. V. (1953*a*). The grazing behaviour of the wild rabbit (*Oryctolagus cuniculus*). *British Journal of Animal Behaviour*, **1**, 16–19.

Thompson, H. V. (1953*b*). The use of repellents for preventing mammal and bird damage to trees and seed. *For. Abstr.* **14**, 129–36.

Thompson, H. V. (1954). The rabbit disease: myxomatosis. *Annals of Applied Biology*, **41**, 358–66.

Thompson, H. V. (1955). The wild European rabbit and possible dangers of its introduction into the USA. *Journal of Wildlife Management*, **19**, 8–13.

Thompson, H. V. (1956). Myxomatosis: a survey. *Agriculture*, **63**, 51–7.

Thompson, H. V. (1961). Getting rid of the rabbit. *Co-operation*, **6**, 34, 36, 38, 40.

Thompson, H. V. (1973). Wildlife conservation and the control of pests. *British Veterinary Journal*, **129**, 202–6.

Thompson, H. V. (1981). Management of lagomorph populations. In *Proceedings of the World Lagomorph Conference* (1979) (ed. K. Myers and C. D. MacInnes), pp. 816–21. University of Guelph, Ontario.

Thompson, H. V. (1985). Control and Conservation. In *humane control of land mammals and birds* (ed. D. P. Britt), pp. 1–6. UFAW London.

Thompson, H. V. (1990). Animal welfare and the control of vertebrates. In *Proceedings of the 14th Vertebrate Pest Conference* (ed. L. R. Davis and R. E. Marsh), pp. 5–7. University of California, Davis.

Thompson, H. V. and Armour, C. J. (1951). Control of the European rabbit *Oryctolagus cuniculus* (L.). An experiment to compare the efficiency of gin trapping, ferreting and cyanide gassing. *Annals of Applied Biology*, **38**, 464–74.

Thompson, H. V. and Armour, C. J. (1952). Rabbit repellents for fruit trees. *Plant Pathology*, **1**, 18–22.

Thompson, H. V. and Armour, C. J. (1953). Power gassing of rabbits. *Agriculture*, **60**, 383–6.

Thompson, H. V. and Armour, C. J. (1954). Methods of marking wild rabbits. *Journal of Wildlife Management*, **18**, 411–13.

Thompson, H. V. and Peace, T. R. (1962). The grey squirrel problem. *Quarterly Journal of Forestry*, **56**, 33–42.

Thompson, H. V. and Thompson, R. H. (1966). Rabbit control by cyanide gassing. *Agriculture*, **73**, 383–8.

Thompson, H. V. and Worden, A. N. (1956). *The rabbit*. Collins New Naturalist, London.

Trout, R. C. and Sunderland, J. C. (1988). A radio transmitter package for the wild rabbit (*Oryctolagus cuniculus*). *Journal of Zoology*, **215**, 377–9.

Trout, R. C., Tapper, S. C., and Harradine, J. (1986). Recent trends in the rabbit population in Britain. *Mammal Review*, **16**, 117–23.

Trout, R. C. and Tittensor, A. M. (1989). Can predators regulate wild rabbit (*Oryctolagus cuniculus*) population density in England and Wales? *Mammal Review*, **19**, 153–74.

Twigg, G. I., Cuerden, C. M., and Hughes, D. M. (1969). Leptospirosis in British wild mammals. *Symposia of the Zoological Society of London* (1968) (ed. A. McDiarmid). No. 24, 75–98.

Vaughan, H. E. N. and Vaughan, J. A. (1968). Some aspects of the epizootiology of myxomatosis. *Symposium of the Zoological Society, London*, **24**, 289–309.

Veale, Elspeth M. (1957). The rabbit in England. *Agricultural History Review*, **5**, 85–90.

Wallis, A. (1904). *The handbook to the natural history of Cambridgeshire*. (ed. J. E. Marr and A. E. Shipley). Cambridge University Press.

Watt, A. S. (1981). Further observations on the effects of excluding rabbits from grassland A in East Anglian Breckland: the pattern of change and factors affecting it (1936–73). *Journal of Ecology*, **69**, 509–36.

Weedon, F. R., Hartzell, A., and Settestrom, C. (1940). Toxicity of ammonia, chlorine, hydrogen cyanide, hydrogen sulphide and sulphur dioxide gases. Animals. *Contribution Boyce Thompson Institute*, **11**,365–85.

Weisbroth, S. H., Flatt, R. E., and Kraus, A. L. (1974). *The biology of the laboratory rabbit*. Academic Press, New York.

Wildlife and Countryside Act. (1981). HMSO, London.

Williams, O. B., Wells, T. C. E., and Wells, D. A. (1974). Grazing management of Woodwalton Fen: seasonal changes in the diet of cattle and rabbits. *Journal of Applied Ecology*, **11**, 499–516.

Wilson, C. J. and McKillop, I. G. (1986). An acoustic scaring device tested against European rabbits. *Wildlife Society Bulletin*, **14**, 409–11.

Worrall, V. (1956). Legal Aspects. In *The rabbit* H. V. Thompson and A. N. Worden). Collins New Naturalist, London.

5

The rabbit in Australia

K. Myers, I. Parer, D. Wood, and B. D. Cooke

5.1 Introduction and spread

Rabbits have been present in Australia since the first European settlement. Five went out with the First Fleet to Sydney in 1788, and numerous rabbits were reported breeding around the houses of the settlement in 1825. In Tasmania in 1827 'common' rabbits were reported as 'running about on large estates in thousands'. From 1850 onwards their importation increased, with government support, under the stimulus of the numerous acclimatization societies set up to 'enrich the country and enhance its attractions'. All of the early importations were domestic breeds unfit for wild existence.

The acclimatization movement in Victoria included Mr Thomas Austin, who introduced rabbits several times, culminating in the release in the Western District of Victoria of a group of wild-type rabbits from England in 1859. There were further releases in other states. In about 1870 rabbits were introduced to Kapunda in South Australia. By 1879 the South Australian and Victorian populations had amalgamated, and farmers began abandoning their properties in the western regions of Victoria. The ensuing spread throughout the continent is now becoming well-documented (Rolls 1969; Long 1972; Strong 1983; Stodart and Parer 1988; Fig. 5.1).

Rabbits spread along the Murray–Darling River system in New South Wales at an estimated rate of 125 km/yr, to reach the Queensland border by 1866 and the Gulf of Carpentaria in northern Australia by 1910. In eastern and southern Australia, denser woodlands slowed their progress towards the coast to about 15 km/yr. For example, they did not appear in the Armidale District on the eastern highlands until 1902. In South Australia they reached Lake Eyre about 1886, and probably moved up the Finke river to invade the MacDonnell ranges and salt lake systems in central Australia, and along drainage channels in the Simpson Desert, at a rate estimated by Strong (1983) of approximately 300 km/yr. The Western Australian border was crossed in 1894, and the west coast was reached near Geraldton in 1906. During the following years the rabbit continued to spread slowly, filling suitable environments along the coast and in the highlands. Today (Fig. 5.2) the rabbit is common on mainland Australia at latitudes higher than 25°S wherever vegetation and soils are favourable, and becomes rare at latitudes less than 20°S; but it is still progressing northwards along the highlands inland from Cairns in north Queensland where woodland is being cleared for agriculture; it has also been introduced on to 48 islands off the Australian coast (Fullagar 1978).

The spread of the rabbit was fastest across the dry southern savannas and along the arid watercourses, and slowest in the mountains, and in coastal woodlands and forests. The greater proportion of the rabbit's present range was covered in 60 years. Although the animal was carried around the country by humans, the main wave of colonization and establishment nevertheless had a strong natural component, influenced by environment; this is illustrated by the fact that in Britain the rate of spread of the rabbit was much slower than in Australia, where, despite widespread introductions, it took the rabbit seven centuries to colonize the country (Lloyd 1981).

Reasons for the successful spread of the rabbit can now only be conjectured about. The pastoral areas of eastern Australia were largely developed between 1830 and 1870. During this activity, European

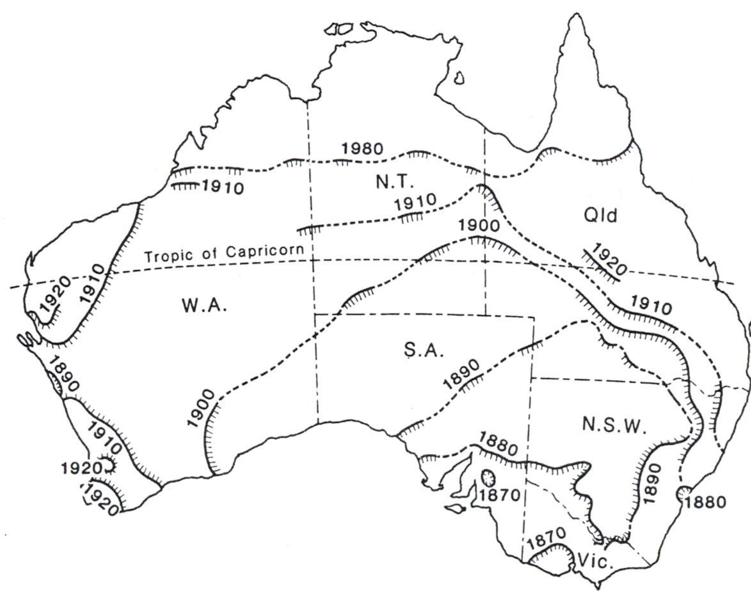

Fig. 5.1 Spread of the rabbit over the mainland of Australia based on reports in newspapers and periodicals (after Stodart and Parer 1988).

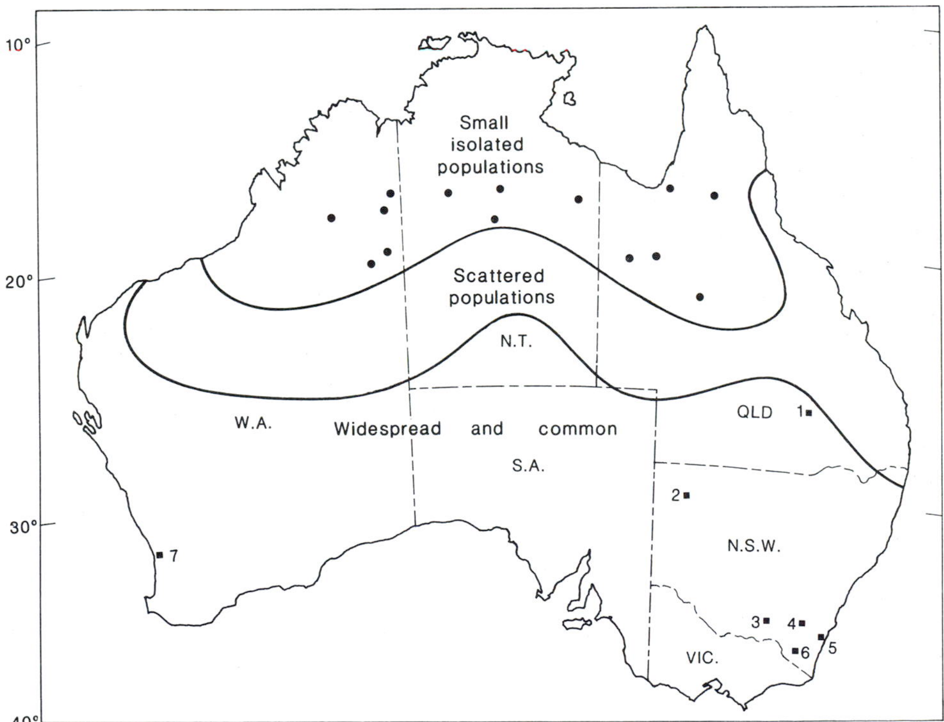

Fig. 5.2 Map of present known distribution of rabbits in mainland Australia, and locations of study sites mentioned in the text: (1) Mitchell, Qld. Subtropical. (2) Tero Creek, NSW. Arid. (3) Urana, NSW. Temperate Mediterranean. (4) Canberra, ACT. Southern Tablelands. (5) Mogo, NSW. Coastal. (6) Snowy Plains, NSW. Subalpine. (7) Chidlow, WA. Mediterranean.

settlers and their animals set in train a general pasture succession throughout southern Australia, in which palatable trees and shrubs and tall native perennial grasses with warm season growth were replaced by unpalatable shrubs and exotic annual herbs and grasses with cool season growth. This succession is excellently depicted in Adamson and Fox (1982), and most workers today accept it as the single most important factor which aided the spread of the rabbit.

In Victoria and New South Wales prior to and during the early years of the rabbit's spread, predators were slaughtered in large numbers, by use of poison, trap, and gun (Marshall 1966; Rolls 1969), especially the dingo, native cats (*Dasyurus* spp.) and wedgetail eagles (*Aquila audax*). This would also have favoured rabbit survival.

Following the removal of trees, the introduction of domestic stock, and the control of predators, native and introduced herbivores on the eastern slopes and highlands of New South Wales staged a series of irruptions, probably reflections of the successional changes in pastures referred to above. Usually kangaroos (*Macropus giganteus*) irrupted first, followed by rat-kangaroos, hares (*Lepus europaeus*) and wallabies (*Macropus* spp.), and finally pademelons (*Thylogale* spp.) and rabbits; but whereas most of the earlier species went through one or two brief irruptive phases, the rabbit persisted at high densities (Jarman and Johnson 1977).

Several species of burrowing marsupials left behind a large number of burrow systems across the southern half of Australia (*Vombatus* spp., *Bettongia leseur*, *Macrotis lagotis*) which in places were very numerous. These burrows undoubtedly aided the spread of the rabbit by affording shelter in semiarid and arid environments.

Over large areas of Australia, important environmental changes which favoured the rabbit before and during its colonizing spread thus coincided with reductions in numbers of its principal native predators and some of its competitors.

5.2 Biology of the rabbit in different environments

A better understanding of the reasons for the success of the rabbit in Australia comes from the findings of modern research.

5.2.1 Distribution

Throughout Australia patterns of numbers and distribution of the rabbit are markedly different in different environments, directly reflecting the impact of the environment on the species.

The most irregular patterns in distribution and numbers are found in arid habitats (Myers and Parker 1975*a*, 1975*b*; Cooke 1982*a*). After a succession of years with good rains, rabbit numbers and distribution increase throughout sand-dune areas (29 warrens km^{-2}). With the onset of drought there is a dramatic fall in numbers (<0.05 warrens km^{-2}) and rabbits become extinct over hundreds of square kilometres except for those populations adjacent to swamps and drainage channels.

At the other climatic extreme, rabbit populations living on subantarctic Macquarie Island also exhibit large fluctuations in numbers, but almost all the rabbits live in, or close to, herb-fields, which contain most food and the best soils for burrowing in. Few rabbits live in tussock grassland, bog, fen, and feldmark formations, except where they abut herb-fields (Copson *et al.* 1981). Rabbit populations living near the snow-line in the Eastern Alps do not irrupt and are almost completely restricted to open grassy valleys with scattered granite outcrops (193 warrens km^{-2}); but they are found in low numbers in adjacent forested habitats (18 warrens km^{-2}) (Myers *et al.* 1975).

In temperate subtropical Queensland rabbits are most common on deposits of sand along drainage channels (114 warrens km^{-2}) (Parker *et al.* 1976). Neighbouring low-lying black soil plains support 2 warrens km^{-2}. The same patterns exist in arid southwestern Queensland, where rabbits inhabit dissected dunes bordering the Bulloo and other large, seasonal rivers. In the humid tropics of Queensland rabbits exist mainly in agricultural clearings at higher altitudes, as on the Atherton Tablelands, inland from Cairns. In the arid tropics of the Northern Territory sparse, scattered populations live in groups of from two to several hundred warrens on suitable limestone or granite outcrops, or where there are sandy soils close to the water table (Low and Strong 1983), for example along fringing sand-dunes and sandy-stream frontage country (Foran *et al.* 1985). Rabbits in pastoral regions fringing the coast north of Carnarvon, Western Australia, however, show less restricted patterns of distribution, living in large areas in adjacent habitats, probably a reflection of the ameliorating influence of a coastal climate (King *et al.* 1983).

Patterns of distribution of the rabbit throughout the southern agricultural regions are less easy to explain. Myers (1962) showed that even in the most favourable climatic environments, regional rabbit populations with densities as high as 3.5 warrens ha^{-1} are discontinuously distributed. Parker (1977) later explained this by showing that in an agricultural region on the eastern tablelands warren densities are lowest where the ground regularly becomes flooded or waterlogged, or where agricultural practices have led to the clearing of harbour, and cultivation of crops or pastures. Densities are greatest in well-drained areas where it is not possible to cultivate. This is usually rough country—creeks and river banks, erosion gullies, rocky outcrops, and the forest–grassland edge.

Further complications in measuring distribution in agricultural and coastal areas are posed by rabbits which live on the surface for some or all of their lives. This is clearly due to seasonal climatic events in some circumstances, for example when heavy snows in mountain areas force warren-dwelling rabbits to vacate their homes and seek refuge in woodland and scrub (Myers *et al.* 1975). In coastal areas of southern New South Wales a large population of rabbits is usually to be found on the surface in clumps of blackberry and bracken fern bordering dairy flats, which are usually too damp to burrow in (Dunsmore 1971). Wheeler *et al.* (1981) present evidence of a similar distribution at Cape Naturaliste, south-western Western Australia, where large numbers of rabbits live above-ground in a sandy woodland–tall shrubland–heath vegetation complex abutting cleared grazing areas.

5.2.2 Physiology and development in different environments

The data described here (and in Section 5.2.5) were collected from a large series of more than 100 shot samples totalling approximately 10 000 rabbits from populations in six different geographical regions in eastern Australia (Fig. 5.2). The collection of each of approximately half of the samples involved up to 38 hours driving and 5 nights of all-night shooting by two workers, and it took a further 2 weeks to process each sample in the laboratory. Hormone analyses from field samples were not feasible when this work was done, and in their place, organ weights and morphologies were measured as indices of physiological function; kidney weights and ear lengths were measured to indicate thermoregulatory responses, adrenal gland, and kidney morphology to indicate water, mineral, stress, and metabolic functions, body weight, liver weight, fat reserves, spleen weight, and blood parameters to indicate general nutritional condition, and weights and morphology of gonads, pituitary and adrenal glands for changes in reproductive physiology. Historically this study was the first serious attempt to understand the nature of ecological and biological variability in wild-rabbit populations in Australia, with the twin objectives of a better understanding of the complexity of the rabbit problem and the framing of pertinent hypotheses for future work. Other studies followed, and are mentioned elsewhere in this chapter.

Rabbits in different geographical regions in eastern Australia vary significantly in body size, physiological indices, and rates of development (Stodart 1965; Casperson 1968; Myers 1971; Myers *et al.* 1981; Tables 5.1–5.3). Many of these differences are large, and are highly significant. Females are on average slightly smaller than males (although the work clearly showed that females in populations with extended breeding seasons were heavier than males) and show similar morphological differences between different regions, but practice different physiological strategies, as evidenced by measurements of the weights and morphology of organs specifically chosen to demonstrate variations in physiological processes (Table 5.2; Figs. 5.3 and 5.4). Most of these environment-specific differences become apparent in development during the first three months of life, as shown by increase of body and organ weights expressed as regression slopes on eyelens weights, used for estimating relative age (Myers and Gilbert 1968) (Table 5.3). Accompanying analyses, not included here, show that although the within-site differences between the sexes in young

Table 5.1 Physical differences between adult male rabbits older than 6 months from four different climatic regions of eastern Australia. Values in this and following tables adjusted to a mean lens weight of 225 mg (Myers and Gilbert 1968)

Rainfall (mm)	Subtropical Q 571	Subalpine NSW 1400	Arid NSW 205	Temperate mediterranean 439	Significance of differences
N	485	429	655	311	3/1876 D.fr.
Paunched weight (g)	1367	1329	1261	1328	$p < .001$
Ear (mm)	81.87	77.35	80.82	78.23	$p < .001$
Foot (mm)	91.95	89.35	90.93	89.38	$p < .001$
Tibia (mm)	96.64	95.05	94.13	93.67	$p < .001$
Pituitary (mg)	34.90	38.60	32.40	38.72	$p < .001$
Adrenals (mg)	345	367	228	263	$p < .001$
Spleen (mg)	322	420	273	410	$p < .001$
Kidney fat (Index 1–4)	1.29	1.59	1.92	1.35	$p < .001$
Testes (g)	2.95	3.00	2.22	3.41	$p < .001$
**N*	104	111	153	99	3/463 D.fr.
Liver (g)	39.16	39.24	32.46	36.14	$p < .001$
Kidneys (g)	10.64	9.04	11.54	9.81	$p < .001$
Packed cell vol. (%)	34.90	33.00	25.50	37.30	$p < .001$
Serum protein (g/100 ml)	5.98	6.39	5.68	6.45	$p < .01$

* After Casperson (1968).

Table 5.2 Physical differences between male and female rabbits older than 6 months from four different climatic regions of eastern Australia

	Male	Female	Significance of difference
N	1880	1810	1/3688 D.fr.
Paunched weight (g)	1315	1307	$p < .001$
Ear (mm)	79.87	78.91	$p < .001$
Foot (mm)	90.58	89.90	$p < .001$
Tibia (mm)	94.91	94.09	$p < .001$
Pituitary (mg)	35.53	51.17	$p < .001$
Adrenals (mg)	296	217	$p < .001$
Spleen (mg)	344	420	$p < .001$
Kidney fat (Index 1–4)	1.58	1.67	$p < .001$
**N*	467	415	1/880 D.fr.
Liver (g)	36.35	41.54	$p < .001$
Kidneys (g)	10.37	11.30	$p < .001$
Packed cell vol. (%)	31.86	28.90	$p < .001$
Serum protein (g/100 ml)	6.02	5.70	$p < .001$

* After Casperson (1968).

Table 5.3 Developmental rates during first three months of life expressed as regression slopes (b) of body measurements and organ weights on lens weight (mg)

	Subtropical Q		Subalpine NSW		Arid NSW		Temperate mediterranean NSW	
	male	female	male	female	male	female	male	female
Paunched weight (g)	7.37	7.40	7.45	7.15	6.54	6.45	6.82	6.98
Ear (mm)	0.54	0.54	0.52	0.51	0.52	0.52	0.52	0.51
Foot (mm)	0.61	0.61	0.60	0.59	0.59	0.58	0.59	0.59
Tibia (mm)	0.65	0.63	0.62	0.62	0.60	0.59	0.60	0.61
Pituitary (mg)	0.22	0.25	0.24	0.27	0.19	0.22	0.25	0.27
Adrenals (mg)	1.25	1.16	1.58	1.35	0.84	0.75	0.98	0.84
Spleen (mg)	2.60	2.88	3.20	3.31	2.12	2.15	3.17	3.59
Testes	8.28	—	7.10	—	6.21	—	11.03	—
Ovaries (mg)	—	0.03	—	0.03	—	0.03	—	0.03
N	162	148	398	189	218	191	193	158

rabbits are not all significant, except for those physiological indices which are directly sex related (pituitary, adrenal), between-site differences are all significant.

Arid-zone rabbits have large kidneys, small livers, low levels of serum proteins, and a low-packed cell volume (Table 5.1). They are slow-growing (Table 5.3) and are small-bodied as adults, but possess long ears and large feet (Table 5.1), and there is a significant widening of the yellow band in the agouti pattern of the pelt (Stodart 1965), indicating adaptation to high temperatures. In rabbits, and other lagomorphs, the length of the ears is of particular significance, since temperature control is by panting and through heat radiation from the ears (Kluger 1975). Other indices point to depression of endocrine and reproductive functions (small adrenals and pituitary, small testes) and to problems relating to shortages of food and water (low body-weight, small liver and spleen, low serum protein, low-packed cell volume, and large kidneys) (Table 5.1), and excretion of sodium (small *zona glomerulosa* in adrenal gland) (Myers 1967; Myers *et al.* 1981). These findings are evidence that the rabbit is reacting to a hot, dry, sodium-rich, and food-poor environment (Blair-West *et al.* 1968; Cooke 1974, 1982*b*; Myers and Bults 1977) to which it is not well-adapted physiologically (Wood and Lee 1985).

The ability of the rabbit kidney to concentrate urine is poor. A mean urine-urea concentration of 1.5 M (max 2.0 M) has been recorded for rabbits confined in a 2-acre enclosure on dry pasture (Hayward 1961). This is much lower than the mean of 3.6 M for the arid-adapted kangaroo rat (Schmidt-Neilsen *et al.* 1948). Sodium concentrating ability is also poor. A maximum mean value of 476 mmol l^{-1} has been recorded for caged rabbits on a diet of lucerne hay containing 10 per cent NaCl (added) and allowed water at the rate of 50 per cent total food and water intake (Wood and Lee 1985). Several arid-adapted small mammals can produce sodium concentrations in their urine of 1200–1900 mmol l^{-1} (Gordon 1968). A maximum mean urine osmolality of 2719 mosm l^{-1} for free-living rabbits was recorded in drought conditions in arid habitat (Wood and Lee 1985) which is below that of jack rabbits (3700 mosm l^{-1}) from the Mojave Desert (Nagy *et al.* 1976) and well below that of the arid-adapted Australian hopping mouse (6550 mosm l^{-1}) (MacMillan and Lee 1969).

In subtropical populations where high temperatures are also experienced, long ears and feet, large kidneys, and a yellower pelage also occur, but unlike

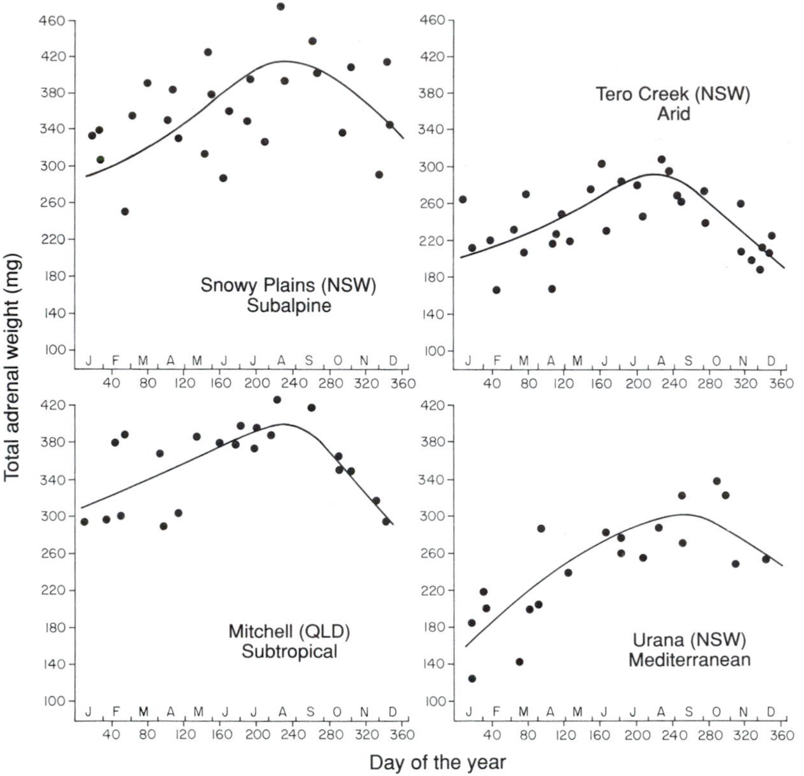

Fig. 5.3 Seasonal changes in weights of adrenal glands (mg) in male rabbits over 9 months of age in four climatically diverse populations in eastern Australia.

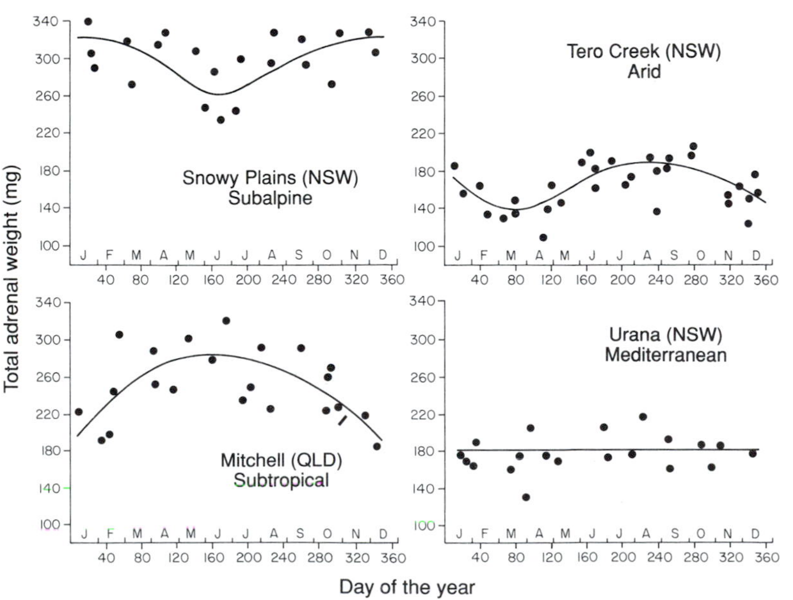

Fig. 5.4 Seasonal changes in weights of adrenal glands (mg) in female rabbits older than 9 months in four climatically diverse populations in eastern Australia.

rabbits in arid environments they grow fast and attain heavy body-weights, indicating good nutrition, and have large adrenal glands with a broad *zona glomerulosa* (Myers *et al.* 1981) demonstrating a need to conserve sodium rather than water, and possibly phosphorus (Parer 1977). Deficiencies of sodium are widespread in tropical pastures and have important effects on lactating cows and their young (Gartner *et al.* 1980). Rabbits living in such areas appear to be similarly affected.

The subalpine rabbit is also large-bodied and grows fast, but in strong contrast to those mentioned above, has a pelage in which the black band is longer than the yellow, shorter ears and feet, and a small kidney. Like the subtropical rabbit it also has a very large adrenal gland. Sufficient work has now been done to show that rabbits living in this environment are extremely sodium deficient (Myers 1958; Blair-West *et al.* 1968), and the lactating female is very susceptible to stress (Myers *et al.* 1981) and reproduces poorly (see later). The addition of sodium to the diet increases growth rates in the young (Myers *et al.* 1981; Richardson and Osborne 1982), but the role of sodium deficiency in limiting reproduction in subalpine areas has yet to be explained.

We attempted to identify some of the reasons for the differences between sexes and populations (Tables 5.4–5.6) using regression analyses, adjusted for age differences and for curvilinearity where present, of the relationships between the biological attributes measured and an array of standard meteorological statistics collected by the Commonwealth Bureau of Meteorology, and variables derived from them. Evaporation summed for the previous 8 weeks was included essentially as an index of temperature on or near the surface of the ground and for calculation of the ratio $P/E^{0.7}$ (Hounam 1961). As an index of nutrition we used the plant growth index of Fitzpatrick and Nix (1970) which combines the effects of light, temperature, and moisture on plant

Table 5.4 Summary of regression ($p < .001$) of climatic and plant growth variables on body weight and reproductive indices, adjusted for age, showing differences between male and female rabbits older than 6 months

Variable	Male	Female
N	1880	1810
Body weight (g)	Plant growth index 12*+	Evaporation 8−
Kidney fat (Index 1–4)	—	Evaporation 8−
Spleen (mg)	Rainfall 12+	Rainfall 12+
Pituitary (mg)	—	Rainfall 12+
Adrenals (mg)	Evaporation 8− $P/E^{0.7}$+ Day length−	—
Ovaries (mg)	—	Rainfall 12+
No. pregnant	—	Rainfall 12+ Plant growth index 4, 8, 12+
Testes (g)	Evaporation 8− $P/E^{0.7}$+ Day length−	—
Seminal vesicle diameter (mm)	Evaporation 8−	—

* Numerals refer to the summation of the variable over the 4, 8, or 12 weeks preceding sample.

Table 5.5 Summary of regressions ($p < .001$) of climatic and plant growth variables on body measurements and reproductive indices, adjusted for age, in female rabbits older than 6 months showing differences between populations in four climatic regions in eastern Australia. Reproductive variables of males also included

Variable	Subtropical Q	Subalpine NSW	Arid NSW	Temperate mediterranean NSW	All sites
N	411	411	723	265	1810
Body weight (g)	—	—	Evaporation 8−*	—	Evaporation 8−
Kidney fat (Index 1–4)	—	Plant growth index 4+ Day length−	—	Evaporation 8− Day length−	Evaporation 8− Day length−
Spleen (mg)	—	—	Rainfall 12+	Rainfall 12+ $P/E^{0.7}$+	Rainfall 12+
Pituitary (mg)	—	Day length−	—	Evaporation 8− $P/E^{0.7}$+	Rainfall 12+
Adrenals (mg)	—	—	—		—
Ovaries (mg)	—	Day length−	—	Rainfall 12+ $P/E^{0.7}$+	Rainfall 12+
No. pregnant	—	Plant growth index 8+	—	Rainfall 4, 8, 12+ Evaporation 8−	Rainfall 12+ Plant growth index 4, 8, 12+
Testes (mg)	Evaporation 8−	Plant growth index 12+	Evaporation 8− Day length−	Evaporation 8− $P/E^{0.7}$+	Evaporation 8− $P/E^{0.7}$+ Day length−
Seminal vesicle diameter (mm)	—	—	Evaporation 8− Day length−	$P/E^{0.7}$+	Evaporation 8−

* Numerals refer to the summation of the variable over the 4, 8, or 12 weeks preceding sample.

Table 5.6 Summary of regressions ($p < .001$) of climatic and plant growth variables on development of body weight, ear length, and glands in female rabbits less than 3 months old in four populations in eastern Australia. Reproductive organs of males also included.

Variable	Subtropical Q	Subalpine NSW	Arid NSW	Temperate mediterranean NSW	All sites
N	148	189	191	158	726
Body weight (g)	Rainfall 12*+ Plant growth index 8+	Plant growth index 8+	Evaporation 8−	—	—
Spleen (g)	—	Plant growth index 8+	Rainfall 4, 8, 12+ $P/E^{0.7}$+	Rainfall 12+ Plant growth index 4, 8, 12+	Rainfall 12+ Plant growth index 4, 8+
Ear (mm)	Evaporation 8−	Plant growth index 12+	**Evaporation 8−	**Evaporation 8−	Evaporation 8− Plant growth index 4, 8+
Pituitary (mg)	—	—	—	$P/E^{0.7}$+	Rainfall departure from normal−
Adrenals (mg)	—	—	Rainfall 12+ $P/E^{0.7}$+	Evaporation− $P/E^{0.7}$+	Evaporation 8− Rainfall departure from normal−
Ovaries (mg)	—	—	Evaporation 8−	Evaporation 8− $P/E^{0.7}$+	Evaporation 8
Testes (mg)	—	—	—	Evaporation 8−	Evaporation 8− Plant growth index 12+
Seminal vesicle diameter (mm)	—	$P/E^{0.7}$+	—	Evaporation 8− Day length−	Evaporation 8− Plant growth index 12+

* Numerals refer to the summation of the variable over the 4, 8 or 12 weeks preceding sample.
** $p < .01$.

performance in different climatic regimes. To allow for lag effects, rainfall, and plant growth indices were summed for the previous 4, 8, and 12 weeks.

The differences between the sexes shown in Table 5.2 and Figs. 5.3 and 5.4 were examined in relation to climatic and nutritional variables (Table 5.4), but only those regressions significant at the $p < .001$ level have been listed. Body condition (weight and fat reserves) are best predicted by evaporation (negative correlation) in females, and by plant growth during the previous 12 weeks (positive correlation) in males. Seasonal endocrine rhythms are very different between the sexes (see, for example, Figs. 5.3 and 5.4), with metabolism and reproduction (pituitary gland, adrenal gland, gonads) responding mainly to day-length and temperature in males (negative correlation) and to rainfall and plant growth indices in females (positive correlation).

Differences between populations in different geographic environments (Table 5.1) are shown by important between-site differences in relationships between biology and climate and nutritional indices (Table 5.5). In a given population, although ear length, adjusted for age, changes with the season (in hot environments average ear lengths are adaptively longer in summer than in winter) morphological characters of adult rabbits are essentially unaffected by short-term changes in climate in all populations, but differences between sites in the responses of physiological indices and reproduction are marked. Subtropical rabbits show no relationships except those between temperature and testis weight. The temperate Mediterranean population, on the other hand, shows a comprehensive array of significant regressions of biological attributes on climate which closely agrees with the mean regressions for pooled data from all sites. This might be expected for rabbits in an environment to which they are well-adapted.

Regression analyses using data on juveniles (Table 5.3) and climatic variables similar to those described for adults yielded similar results (Table 5.6). In particular, the development of the ear significantly correlated with evaporation (temperature) in populations in hotter environments. In the subtropical and subalpine populations few of the physiological and reproductive variables showed relationships with climate and food as measured. The temperate Mediterranean population, however, again showed strong relationships between development, climate, and indices of nutrition.

Exploratory analyses of this kind, in addition to those cited earlier, and in concert with accompanying work on parasites and disease (Dunsmore 1966*a*, 1966*b*; Dunsmore and Dudzinski 1968; Stodart 1965, 1968*a*, 1968*b*; Williams 1972) lead clearly to the general hypothesis that each environment affects the biology of rabbits in critically different ways (Myers 1971; Myers and Bults 1977; Myers *et al.* 1981; Gilbert and Myers 1981; Myers 1986; Gilbert *et al.* 1987). In the extreme climatic regions represented by subalpine, subtropical, and to some extent the arid region, physiological and reproductive patterns are poorly predicted by climate and food. In temperate Mediterranean regions, biological rhythms are strongly in tune with climatic and nutritional ones. These findings imply the existence in wild rabbits in Australia of genetic pre-adaptation to Mediterranean-type environments at the level of winning combination or balanced selection.

Williams and Moore (1989*a*, 1989*b*, 1990) sought verification of the causes of some of the differences described above, in a critical experiment which examined growth and fecundity in rabbits from arid, subalpine, and temperate Mediterranean field populations, and their progeny. The experimental animals from these different stocks were bred and reared in a standard animal-house environment for two generations; then released into small pens exposed to ambient temperatures in the temperate southern highlands climate of Canberra, ACT. The work clearly demonstrated the dominant impact of environment on development and reproduction, including the important relationship between ambient temperature and growth of ears. Evidence was also presented to suggest that there has been genetic selection of other characters in the wild ancestors of the rabbits we observed, although in ways that could not be explained. Thus rabbits taken from Urana, NSW, a temperate Mediterranean environment, are less fecund in Canberra than rabbits from an arid environment; the latter unexpectedly develop a superior fecundity in the more temperate environment of the Southern Tablelands (cf. Table 5.8).

5.2.3 Climate and food

Throughout many of the areas it inhabits in Australia there are deficiencies in the diet of the rabbit which are due to seasonal impacts on pasture growth, varying soil fertility, droughts, or to a high density of rabbits.

The main nutritional problems the rabbit has had to face have been those caused by high temperatures, mineral imbalances, and a low, variable rainfall. Additionally, towards northern Australia rainfall becomes progressively more summer-dominant, leading to a summer pasture of perennial grasses with a low protein and high fibre content (Parer 1987), much less suitable for rabbits than those supported by winter–spring rains in the southern half of the continent.

The protein requirements of wild rabbits for maintenance have been estimated by Cooke (1974) at about 12 per cent. In subtropical Queensland and western New South Wales the protein content of recently ingested food may be as low as 5.4 per cent and 6.3 per cent respectively (Myers and Bults 1977). In inland Australia, where breeding seasons are usually short, the period of availability of high-quality food in some years may not be long enough to allow any of the young of the year to survive (Cooke 1983*a*; Low and Strong 1984*a*), due to milk starvation or to nutrient deficiencies in the pasture (Myers and Poole 1962; Parer 1977; Cooke 1982*a*; Wheeler and King 1985*c*). Short (1985) presented graphic evidence for this when measuring the functional grazing response of rabbits in a bluebush (*Maireana pyrimidata*) shrubland at Kinchega in western New South Wales, where he showed that in that environment food intake in rabbits levels out at 68 g $kg^{-0.75}$ day^{-1}, at a pasture biomass of 250 kg ha^{-1} or greater. During the preceding 3.5 years the average pasture biomass at Kinchega varied between 9 and 1090 kg ha^{-1}, and was below 180 kg ha^{-1} for most of the time.

During severe food and water shortages, rabbits eat bark, fallen tree leaves and seed pods, tree roots, and termites. In droughts average body weights of adult male rabbits fall by 22 per cent (Myers and Bults 1977; Cooke 1982*a*) and individual rabbits may lose 50 per cent of their body weight (Hayward 1961).

During prolonged droughts rabbit populations may fall to below 1 per cent of peak numbers (Myers and Parker 1975*b*). Young rabbits die first (Cooke 1982*a*, 1982*b*) as they have higher water requirements than older rabbits (Richards 1979). Water turnover rates in free-living rabbits range from 214.3 ml kg^{-1} in cool, wet conditions to 46.1 ml kg^{-1} in a moderate drought. Young rabbits 1–2 months of age have water turnover rates 1.7 times that of non-breeding adults (Richards 1979).

Adult rabbits require food with a water content of more than 55 per cent (Cooke 1982*b*; Wood and Lee 1985). In the arid regions of Australia most of the plants which have a high water and protein content in droughts are chenopods (*Mairiana* spp., *Sclerolaena* spp., *Atriplex* spp.). The inability of rabbits to handle a diet containing 5 per cent salt when total water intake is only 50 per cent of total food and water intake (Wood and Lee 1985) suggests that a diet of chenopod shrubs, many of which have a gross salt content above 5 per cent, would be unsuitable in a drought period; however Dawson and Ellis (1979) recorded a diet of 22 per cent herbs, 25 per cent browse, 53 per cent chenopods in a very dry summer. Young stems of some chenopods contain less than 5 per cent salt and above 50 per cent water (Wilson 1966; Cooke 1974). Catling and Newsome (1992), however, have recently shown that rabbits are very adept at selecting items of diet within the chenopod group itself to maximize protein and water and minimize sodium intake (see Section 5.2.4).

The use of warrens enables the rabbit to colonize dry environments. At Albury, NSW, Hayward (1961) showed that in summer the air in burrows remained at a constant 26°C and was almost saturated with water vapour, whilst the ambient temperature rose to 40°C and relative humidity was as low as 5 per cent. In extremely arid environments burrow temperatures can exceed 30°C and relative humidity may fall to less than 40 per cent (Hall and Myers 1978; Parer and Libke 1985; Cooke 1990*a*).

Inability to produce urine with a high-osmotic concentration or high-sodium content compared with rodents endemic to desert areas suggests that rabbits are ill-equipped physiologically for living in arid habitats. However, use of warrens, plasticity in behaviour patterns leading to lower water losses and selection in the diet of food resources of relatively high water content without intake of large amounts of salt seem to be the strategies enabling them to survive in arid habitats often devoid of free water for long periods.

Further north, summer rains cause high humidities, and the ecological problem shifts from one principally concerning water conservation and sodium excretion to that of temperature regulation. The rabbit is not a good regulator, and core temperatures rise quickly when humidity increases at high temperatures, especially if reproduction is attempted (Cooke 1977).

In south-eastern Queenland leached soils add the problem of sodium and phosphorus deficiency in pastures (Gartner *et al.* 1980; Parer 1987) and rabbits there must also find metabolic energy for sodium retention (Myers *et al.* 1981).

At the other extreme rabbits have little trouble in dealing with extremely low temperatures provided nutrition is adequate (Griffiths *et al.* 1960). In the subalpine region of south-eastern Australia, however, pastures are extremely deficient in minerals, especially sodium, and the rabbit is fully extended to maintain physiological homeostasis (Blair-West *et al.* 1968). Furthermore it has been shown that under such conditions the rabbit cannot withstand stress (Myers *et al.* 1981).

In the subalpine region the number of rabbits declines during periods of unusually prolonged snow-cover, as rabbits are inefficient at foraging in the snow and have to rely on browse or areas of pasture cleared by wombats (*Vombatus ursinus*). The hare (*Lepus europaeus*), which is more adapted to browsing, has a higher altitudinal limit than the rabbit.

5.2.4 Behaviour

Much of the detailed information on behaviour of the rabbit in Australia is derived from observations of confined populations (Rowley 1956; Mykytowycz 1958*a*, 1958*b*, 1959, 1960, 1961; Myers and Poole 1959, 1961, 1962, 1963*b*; Fullager 1981). Among this work Fullagar's (1981 and unpublished) observations are of critical importance. Working in a large 33 ha fenced paddock near Canberra ACT he objectively measured the frequency of association in space and the behavioural interactions between individual rabbits in a small free-living population, and used principal co-ordinate and minimum spanning tree analyses to show the reality of social groupings, of social status in individual rabbits, and of territorial behaviour. His work substantiated and extended the observations of the earlier workers referenced above.

Valuable information on specific aspects of behaviour in natural populations has also been obtained by field workers (Parer 1977, 1982*a*; Fullager 1981; Wheeler *et al.* 1981), often in the course of studies of other aspects of rabbit biology (Daly 1980, 1981). Most aspects of behaviour important in population biology can be included under the headings which follow.

5.2.4.1 Behavioural adaptibility in securing resources and avoiding risks

The rabbit is highly competent in satisfying its nutritional requirements. As pasture plants mature and set seed, rabbits detect the flow of protein, water, and minerals and alter their patterns of intake accordingly (Myers and Poole 1963*a*). Even newly weaned young rabbits quickly learn how to fell seed heads on tall stems to gain access to the food they contain. In sodium-deficient subalpine environments one plant (*Stylidium* sp) actively accumulates sodium in seed heads, which are avidly sought and eaten by rabbits as they appear (Blair-West *et al.* 1968). In the arid zone rabbits avoid sodium-rich plants even though these plants may have the highest water and protein content. During flush periods they utilize green herbage (mainly forbs) and perennial chenopods for water and nitrogen and switch to round chenopods in dry periods (Catling and Newsome, 1992). That this behaviour is obligate is suggested by the finding that it remains unchanged whether or not sheep or cattle are feeding on the same pastures.

In dry times rabbits extend their home ranges, dig for reserves of protein and water in roots, or turn to bark and leaves of perennials (Cooke 1982*a*). In their search for the latter the animals climb shrubs and low trees. Whatever the conditions in which rabbits live, the quality of the food in the cardiac stomach represents the best available to small herbivores (Cooke 1974; Myers and Bults 1977).

The rabbit is equally competent in selecting soils of the right kind and situation to burrow in (Myers and Parker 1965; Parer and Libke 1985). In the cooler and more temperate environments in Australia, rabbits seek out and proliferate in sandy soils bordering grasslands, especially where there are scattered shrubs and trees or outcrops of rock for protection from predators. This is also the preferred living place of rabbits in coastal areas in southern Spain (Rogers and Myers 1979). In arid regions where high soil temperatures become critically important, other factors determine burrowing success. Thus Parer and Libke (1985) measured the quality of soils within the range of a very large natural population of rabbits in semi-arid western New South Wales savanna woodlands and grasslands, and showed that rabbit warrens were concentrated in those areas where impact-penetrometer readings indicated friable soil to a depth of at least 75 cm. Isolated warrens were found only in patches of favourable soil, and were absent from areas with shallow soils.

5.2.4.2 Territoriality

Territorial behaviour in the rabbit is intimately related to reproduction, with males protecting females from other males, and females protecting burrows against other females. This is common to all populations both in enclosures and the field (Parer 1977; Daly 1980). It is seasonal and varies in timing from population to population according to differences in reproductive patterns and breeding intensity (Mykytowycz 1960; Myers and Poole 1961).

Aggressive behaviour in defence of physical resources has been less frequently seen. The only populations in which this has been observed in our studies are those of Mykytowycz (1961), who provided artificial warrens for a confined population of rabbits, and in a free-living subalpine population of NSW (Dunsmore 1974) suffering a serious sodium deficiency. In both populations, both sexes aggressively defended the warren even in the absence of reproductive activities. In the free-living population aggression was most intense in autumn–winter, and was directed mainly towards immigrants and juveniles. With high warren densities, low pasture growth and frequent falls of snow, possession of a warren was critical for survival (Hale *et al.* unpublished data).

Overt physiological differences between dominant and subordinate rabbits do not develop until density rises and the size of social groups increases (Myers 1966; Myers *et al.* 1971). As in other mammals, high densities cause increased rates of aggressive encounters, resorptions of embryos and inhibition of oestrus, and a wide variety of physiological changes in both adults and young indicative of stress. Young rabbits born to stressed mothers show severe stunting, and are behaviourally and physiologically different when adult from rabbits born under more favourable circumstances (Myers *et al.* 1971). The impact of environment on development thus includes the effects of social as well as physical and chemical stimuli.

The maintenance of territorial boundaries is supported by the use of the secretions of skin glands, especially those of the anal glands and the submandibular 'chin' glands. The secretions of the anal glands which coat the surfaces of hard pellets as they pass through the rectum, producing the well-known 'rabbity' smell, are used mainly for territorial defence, especially by the strategic placement of dung-hills on special sites within territory on which faeces are regularly deposited (Mykytowycz and Gambale 1969). The secretions of the chin glands are precisely applied by dominant males onto all objects within the territory and onto the bodies of individual members of the social group, and give superior psychological advantages ('confidence') to the owners of the territory in confrontations with strangers (Mykytowycz *et al.* 1976).

Initial analyses of the chemical composition of the secretions of the glands, using chromatographic, spectroscopic and electrophoretic techniques, have shown that the secretions of the anal, submandibular, and inguinal glands are different. There are also differences between the sexes and between individuals. Gland secretions contain proteins and carbohydrates. Lipid extracts obtained from gland homogenates consist of hydrocarbons, nonglyceryl esters, fatty acids, and cholesterol. Triglycerides, diglycerides, and monoglycerides are also produced by the chin glands and by the sebaceous portions of the inguinal glands (Goodrich and Mykytowycz 1972).

Most rabbits are initially subordinate before they become dominant. In natural populations dominance is a consequence of a rabbit living longer than its colleagues, and is more likely to be a result of ecological efficiency than of behavioural aggressiveness.

5.2.4.3 Changes in size of home range

Home-range size varies in relation to season, sex, and age, and differs markedly from one environment to another. When resources are plentiful, home range contracts during the breeding season (Myers and Poole 1961; Parer 1977) and expands when breeding ceases, or when resources are depleted while breeding, for example in the Snowy Plains, NSW, where the spring flush of pasture is accompanied by an extreme deficiency of Na^+ (Blair-West *et al.* 1968) and the lactating female home range more than doubles in size as rabbits search for plants containing that mineral (Hale *et al.* unpublished). There is also a large expansion of range in the winter months in subalpine populations when pasture growth is minimal and protein in the food supply falls to low levels (Myers and Bults 1977). In Mediterranean climates range increases in the summer when pastures dry out and food quality deteriorates (Parer 1982*a*).

The home range of males is larger than that of females, both in confined and in natural populations (Myers and Poole 1959; Fullagar 1981; Parer 1982*a*). In a subalpine population, the mean home range of males at dusk has been measured as 0.049 ± 0.002 ha and of females 0.041 ± 0.002 ha. The same differences are seen in populations in arid environments. Fullagar (1981) measured mean true home ranges of 4.73 ha for males and 3.34 ha for females in rabbits in north-western NSW, and states that the largest home ranges observed were those of rabbits occupying warrens on dry sparsely vegetated stony slopes. These contrast with the mean home range of 0.98 ha for males and 0.46 ha for females measured in a 33 ha paddock in the Southern Tablelands (Canberra), where water and food were not limiting.

5.2.4.4 *Shifts of home range*

Rabbits shift home for several reasons. Lost rabbits move large distances. Douglas (1969) recorded a 25 km movement in a tagged rabbit which had been inoculated with myxoma virus and released in central Victoria. Young rabbits often move from one warren to another due to unknown disturbances, especially under conditions of high density. Parer (1982*b*) measured a movement of 1.5 km by one young rabbit. There are also strong social pressures on sub-adult young, especially males, to leave the warren when they start to compete with breeding adults (Myers and Poole 1961; Mykytowycz and Gambale 1965; Dunsmore 1974; Parer 1982*a*). This is especially in evidence during group formation at the beginning of the breeding season, when numerous young males are ejected from warrens and are often forced to live on the surface away from the warrens, and also late in the breeding season when warrens become crowded (Daly 1981).

The most important stimulus to large-scale dispersal in rabbits in Australia is depletion of food or water resources. The younger classes in the population are affected first, usually due to a lack of nutrients in dry pastures. Before myxomatosis it was a common phenomenon to observe large numbers of foot-loose young rabbits along the roads in southern NSW, searching for green food in the rapidly maturing pastures of early summer (November–December). Parer (1982*a*) records nighly movements of 300 m from the warren of rabbits suffering food shortage in a temperate Mediterranean environment, and Newsome (1989) describes movements of rabbits of 1500 m to obtain water in an arid habitat in western NSW. Movements of this kind may develop into mass emigrations over large distances when resource shortage (food, water, shelter) develops over a large area, as in a drought year in the Australian arid zone, where rabbits have been observed to pile up a metre deep along rabbit barrier fences.

Adult rabbits rarely shift home range when resources are adequate; those that do are usually the subordinate or unattached animals, and the dispersal is to an area which has already been reconnoitred.

Rabbit behaviour varies from one environment to another, in response to different physiological and ecological stimuli, especially different levels of resource availability. Since the breeding rabbit and its young are anchored to the warren in which they live, we believe that the ultimate function of territoriality and group behaviour is protection of the young and the resources they require during the first weeks of life. A second function of territoriality is the ejection of surplus animals, usually young, which then become exposed to a different suite of environmental hazards, especially predators.

5.2.5 Reproduction

There are large differences in reproductive performance between rabbits living in different Australian environments (Myers 1971, 1986; Parer 1977; Wood

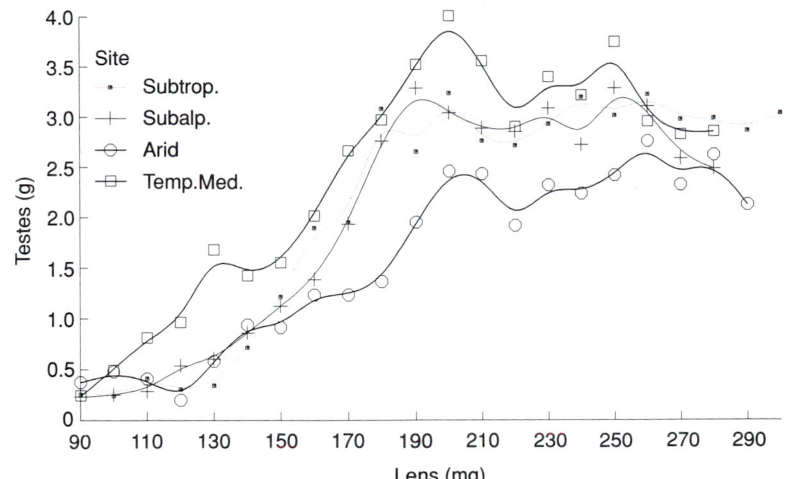

Fig. 5.5 Changes in weights of testes with age (expressed as weights of crystalline eye lens in mg) in four climatically diverse populations in eastern Australia.

1980; Cooke 1983*a*; Wheeler and King 1985*b*; Gilbert *et al.* 1987; Myers *et al.* 1989). The main causes are undoubtedly of environmental origin, affecting the rate of maturation in the young and the frequency and success of reproduction in adults.

It takes the testes of arid zone rabbits about 11 months to reach sexual maturity at a size of 2.5 g (Fig. 5.5). In temperate Mediterranean environments maturity is reached in 7 months and the testes continue to grow to more than 3.5 g. Sperm are present in the tubules of young males as early as 4 months of age at Urana (temperate Mediterranean), but not until 10 months at Tero Creek (arid). Sperm appear at 7 months in the subalpine and subtropical populations. In the arid population the testes atrophy to a degree not measured elsewhere (Hughes 1965), at times in the whole male population, if high summer temperatures and drought coincide for long periods. As shown earlier, the single best predictor of testis weight is evaporation, followed by day length and $P/E^{0.7}$ (Tables 5.4–5.6). In the cooler subalpine (and coastal) populations evaporation is replaced by plant growth indices summed for the preceding 12 weeks as the best predictor.

Changes in the weight and morphology of the male adrenal glands are also significantly related to evaporation, day length, and $P/E^{0.7}$ (Myers *et al.* 1981). Adrenal weight and testis weight are themselves closely and positively correlated, but the male pituitary shows no relationship with reproduction, weather, or plant growth indices, except in the arid zone where changes in phyhsiology in relation to rain are more or less global due to large changes in nutritional status (Myers and Bults 1977; Cooke 1982*a*). The main differences in indices of reproduction measured in adult males in four different populations are shown as mean values, adjusted for age, in Table 5.7. Reproduction in the male rabbit appears to be a mechanical kind of process driven mainly by seasonal climatic rhythms.

Judging from the maturation of ovaries, females usually become fully mature a little later than males, about 9 months in Mediterranean and 12 months in arid environments, although the ability to conceive precedes adulthood by many months (Figure 5.6). The youngest females found pregnant were 3.5 to 4 months old from Urana (temperate Mediterranean), Mogo (coastal) and Canberra (Southern Tablelands), 5 months old (from Tero Creek (arid) and Snowy Plains (subalpine), and 6 months old from Mitchell (subtropical), but high rates of pregnancy in females younger than 6 months of age were recorded only in the Mediterranean environment (Table 5.8).

Pregnancy rates increase to a peak between 9 and 12 months of age, and fluctuate about the level reached, which varies between populations for the rest of life. The main reproductive period in the life of the female wild rabbit lies between 9 months and 2 years of age.

The numbers of eggs shed and fertilized usually

Table 5.7 Mean differences between gonadal weights and related organs in male rabbits in four populations in eastern Australia, adjusted for age

Variable	Mitchell Q	Snowy Plains NSW	Tero Creek NSW	Urana NSW	Significance of difference
Adult					D. fr. 3/1876
Testis weight R (g)	1.27	1.40	0.89	1.33	p < .001
Testis weight L (g)	1.70	1.55	1.30	2.02	p < .001
Total relative testis weight (g/kg)	2.20	2.28	1.73	2.65	p < .001
Seminal vesicle diameter (mm)	8.01	7.87	6.85	8.80	p < .001
Mean sperm stage (Index 1–8)*	7.02	7.12	6.40	7.51	p < .001
Adrenal weight (mg)	34.54	36.66	22.76	26.30	p < .001
Pituitary weight (mg)*	3.50	3.86	3.24	3.87	p < .001
Juvenile					D.fr. 3/967
Testis weight R (g)	0.50	0.45	0.41	0.65	p < .001
Testis weight L (g)	0.66	0.61	0.65	0.91	p < .001
Total relative testis weight (g/kg)	0.80	0.75	0.57	1.23	p < .001
Seminal vesicle diameter (mm)	5.97	5.61	5.30	6.38	p < .001
Adrenal weight (mg)	17.80	22.02	11.99	13.99	p < .001
Pituitary weight (mg)*	3.38	3.43	2.86	3.83	p < .001

* Hughes (unpublished data). Mean sperm stage: 1–5 = no sperm present; 6 = few sperm present; 7 = moderate numbers of sperm present; 8 = large numbers of sperm present.

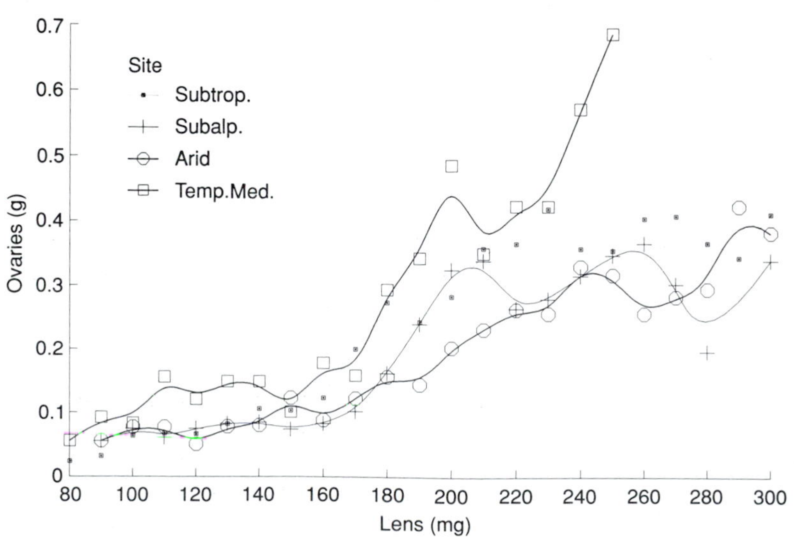

Fig. 5.6 Changes in weights of ovaries with age (expressed as weights of crystalline eye lens in mg) in four climatically diverse populations in eastern Australia.

Table 5.8 Age specific productivity in two sites in eastern Australia

Age (months)	Number of females >3 months old	Number of females (% total)	Females pregnant (%)	Mean litter size (*in utero*)	Young per female per year* (observed)	Relative production (%)
Temperate mediterranean, Urana, NSW						
3–6	73	17.72	23.29	4.38	2.90	10.38
6–12	148	35.92	43.24	5.60	8.12	29.05
12–18	125	30.34	52.80	5.89	12.07	43.18
18–24	44	10.68	45.45	5.86	3.62	12.95
>24	22	5.34	54.54	6.18	1.24	4.44
Totals and means	412	100.00	43.45	5.65	27.95* 29.07**	100.00
Arid, Tero Creek, NSW						
3–6	130	13.86	4.62	3.80	0.55	3.25
6–12	250	26.65	14.40	4.66	3.21	18.96
12.18	203	21.64	31.53	4.56	4.57	26.99
18.24	180	19.19	37.22	4.33	4.42	26.11
>24	175	18.66	34.29	4.53	4.18	24.69
Totals and means	938	100.00	24.84	4.49	16.93* 18.62**	100.00

* Estimate based on observed pregnancies.
** Adjusted to allow for recent ovulations and resorptions.

increase with age, but there are significant differences between populations in this respect. In Mediterranean environments mean embryo counts continue to increase beyond the second year of life. In populations from the arid zone there is no increase beyond the first year (Table 5.8). Although females aged two years or older have the largest litter sizes in Mediterranean environments, they contribute little to the annual crop of young as they constitute only 5 per cent of the population. In the arid zone females aged two years or older constitute a much higher proportion (19 per cent) of the population.

There is a surprisingly constant pattern of seasonal production of young rabbits at all sites (Figs. 5.7, 5.8). More than 75 per cent of the young are born in the second half of the year with peaks of production between August and November (spring), skewed towards the summer in the subalpine population and towards winter in the arid population. The differences between the timing of production of the annual crop of young in each population can be represented by an average birth date (day of year) at each site:

Arid NSW	205 ± 75
Temperate Mediterranean NSW	220 ± 69
Subtropical Q.	232 ± 86
Southern Tablelands ACT	247 ± 64
Coastal NSW	251 ± 78
Subalpine NSW	262 ± 99

In the strongly Mediterranean climate of south-western Western Australia, reproduction peaks earlier still (Wheeler and King 1985*a*, 1985*b*; Gilbert *et al.* 1987). In the Northern Territory, on the other hand, with a progression from south to north of an unpredictable summer rainfall, breeding is usually confined to short bursts, and reproductive failures are higher than measured elsewhere in Australia, mirroring responses to infrequent and inadequate

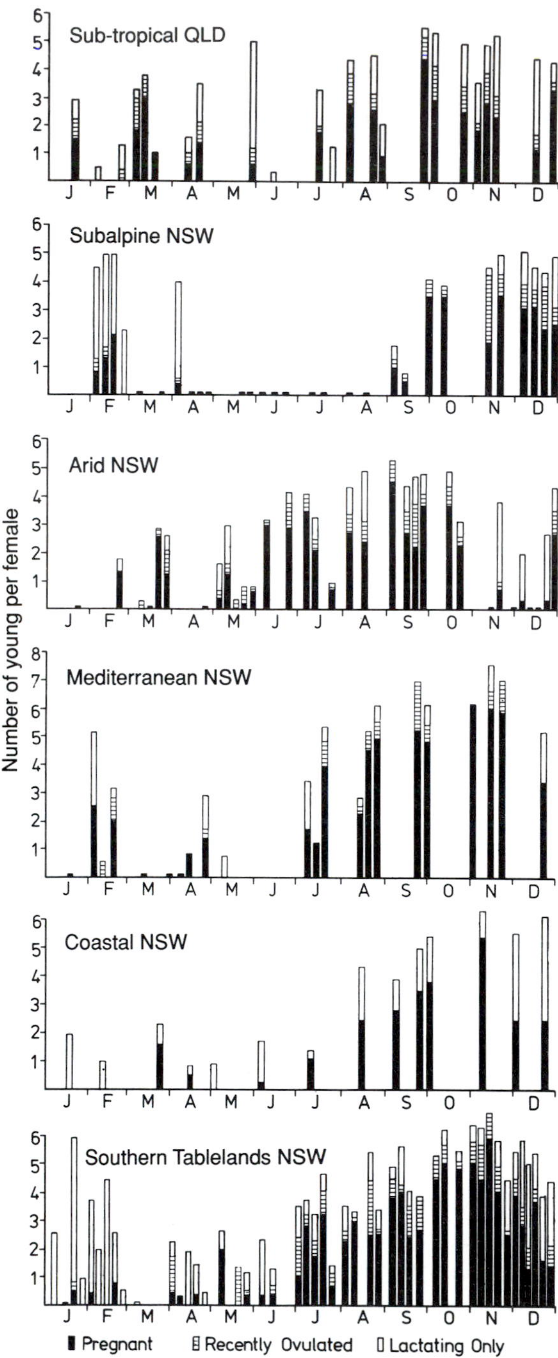

Fig. 5.7 The breeding season in female rabbits in six climatically diverse environments in eastern Australia. (Graphs 1–4 (from top) after Myers 1971; Coastal NSW after Dunsmore 1971; Southern Tablelands after Parer and Libke 1991).

rains in a harsh and variable, arid environment (Low *et al.* 1983). A more comprehensive examination of the seasonal timing of reproduction in Australian populations is presented fully in Gilbert *et al.* (1987).

There is also a clear annual pattern of the combined effects of resorptions and/or failures of *post partum* conceptions ('lactating only', Fig. 5.7). Most rabbit populations show low rates of interruptions to reproduction in the spring and higher rates in summer and autumn. In Queensland and the Northern Territory rates are high in every month of the year. The population from the Mediterranean region (females older than 3 months) shows the lowest rates of 'lactating only' (5.6 per cent) and the highest estimated production (27.95 young per year, based on actual numbers of pregnancies observed, or 29.07 young per year when pregnancy rates are increased by numbers of recent ovulations (on the assumption that all the recently ovulated eggs implant), and then decreased by partial resorptions (estimated as the differences between numbers of functional corpora lutea on the ovaries and healthy embryos).

Estimates of numbers of young born per female based on data from dead samples are of value for comparative purposes only, and then only when the estimated age groups used are stated, and the basis for estimates given. For example, Gilbert *et al.* (1987) quote annual production at Urana of 38 young per year. Their analyses refer to rabbits older than 12 months, in their breeding prime. After corrections of this kind are made, most populations approach their maximum reproductive rates in the spring, while lower rates are recorded in summer and autumn.

A summary of differences in reproduction in female rabbits older than 3 months of age, in six diverse climatic regions in eastern Australia, is presented as weighted means of samples, adjusted for age where necessary, in Table 5.9.

Differential rates of reproduction in Australian rabbits are not difficult to explain. Reproduction in rabbits is significantly related to climatic variables affecting physiology and food (Poole 1960; Myers and Poole 1962; Parer 1977; Wood 1980; Cooke 1981; Wheeler and King 1985*b*; Tables 5.4–5.6). Unlike the larger mammals, rabbits are induced ovulators. The final luteinizing hormone surge which causes the follicles to rupture and free the ova is caused by nervous stimulation of the hypothalamus during the act of mating itself. The remale rabbit in grazing

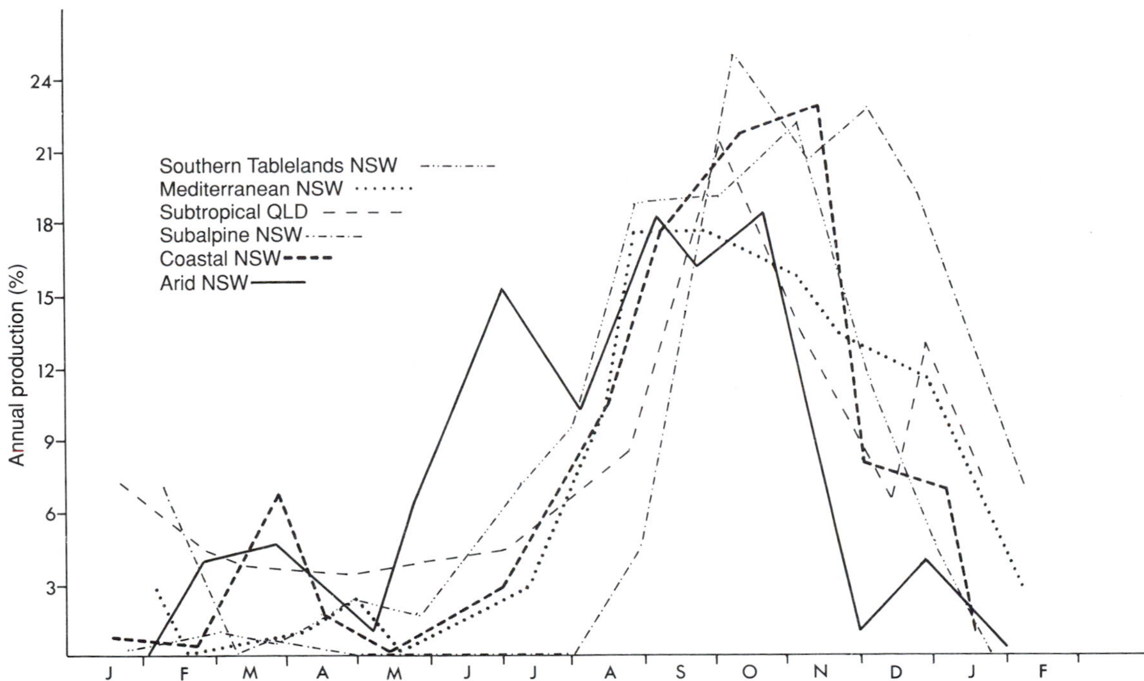

Fig. 5.8 Estimated mean monthly production of young (per cent total mean annual production) in six climatically diverse environments in eastern Australia.

natural pastures does not normally solicit the male to copulate unless her diet contains actively growing grass or forbs which appear to act as an oestrogenic factor (Poole 1960). In natural populations green grass is also necessary for lactating females and for growth of young. After a dry period the first conceptions usually follow within a week of drought-breaking rains (Hughes 1965; Wood 1980). Where body weights have fallen as a result of prolonged feeding on dry pastures, breeding is delayed until body condition is regained (Myers and Poole 1962). A supply of high-quality green food, or a substitute, is necessary for the continued production of litters, and the higher the protein level the larger the litter size and the lower the mortality to weaning (Stodart and Myers 1966; Omole 1982; Wheeler and King 1985*b*). As soil moisture declines so does pasture growth and the conception rate (Cooke 1982*a*; Wheeler and King 1985*b*; Myers *et al.* 1989); the last conceptions usually occur in the month when available soil moisture declines to zero. In a Mediterranean environment almost two-thirds (63 per cent) of the annual crop of kittens that survive to emergence are conceived in the months of July, August, and September (Myers *et al.* 1989).

In the rabbit, pregnancy appears to be terminated by anything which affects the secretion of the luteotrophic complex in the pituitary or stops ovarian oestrogen secretion (Austin and Short 1972). The close relationship between the pituitary and the corpus luteum means that adverse environmental change can act quickly to suppress implantations and pregnancies likely to produce young which will not survive. However, provided the plane of nutrition does not fall too low, the female rabbit remains in a condition similar to dioestrus in other animals, 'lactating only', especially if she has been pregnant before; and in wild populations, she exhibits a 7-day follicular and behavioural cycle (Myers and Poole 1962) which permits her to become pregnant again as soon as conditions improve.

Rainfall and plant growth indices assume a dominant role in predicting reproduction in female rabbits. Except in the arid zone, evaporation is of

Table 5.9 Summary of reproduction in female rabbits older than 3 months from populations in six diverse climatic regions in eastern Australia (weighted means)

	Subtropical Q	Subalpine NSW	Arid NSW	Temperate mediterranean	Coastal* NSW	Southern** Tablelands
Rainfall (mm)	571	1400	205	439	1000	629
Ovaries (g)	0.33	0.29	0.24	0.41	0.45	0.27
Pregnant (%)	32.1	24.1	24.8	43.4	26.6	33.9
Recently ovulated (%)	10.9	7.7	6.7	7.8	—	8.5
Lactating (%)	46.3	33.5	24.9	41.0	36.5	42.4
Lactating only (%)	19.0	12.9	8.1	5.6	—	15.9
No. embryos	4.80	4.52	4.49	5.65	5.23	5.01
Size of egg sets***	4.86	4.91	5.04	6.95	—	—
Partial resorption (%)	15.0	14.7	14.7	12.3	—	11.9
Young per female per year based on:						
i. Observed pregnancies	16.87	14.06	16.93	27.95	15.12	17.67
ii. Observed pregnancies plus recent ovulations less resorptions	19.23	14.76	17.52	29.07	—	20.66
Reproductive inadequacies**** (%)	52.91	45.52	32.68	24.87	40.57	44.54
Number	548	636	937	412	301	889

* After Dunsmore (1971).
** After Parer and Libke (1991).
*** Hughes unpublished data.
**** Difference between estimate of production based on observed pregnancies, and estimate of potential production if all females in sample exhibiting reproductive activity of some kind were pregnant.

relatively minor importance, and there the strengths of the regressions are not marked. The adrenal gland, as in the male, correlates with evaporation and day length, and probably subserves the same metabolic functions, increasing in the cooler and decreasing in the warmer months. The pituitary gland, however, fluctuates significantly in relation to rainfall and reproduction. Two trends are evident. Ovary weight, hence the ability to ovulate, changes markedly in relation to rainfall and short-term changes in plant growth. Reproductive quiescence ('lactating only') is best predicted by day length and shorter-term plant growth indices. Pregnancy, lactation, and production of young, are more readily predicted by longer-term indices of plant growth and evaporation.

Reproduction in the female is apparently a compromise between long-term and short-term stimuli. These responses are best explained in terms of changing body condition on the one hand, which takes time, and short-term behavioural and endocrine responses to changing climatic conditions, which quickly effect nutrition and physiology. In contrast to the male, the female rabbit responds mainly to rain and food, not temperature and day length. The two sexes appear to practice different physiological strategies as far as reproduction is concerned (Myers *et al.* 1981), and the female rabbit is the more labile of the two. Similar findings have been recorded for reproduction of rabbits in southern Spain (Soriguer and Myers 1986) where male reproduction is clearly controlled by radiation and temperature, and female reproduction by nutrition and climate (see Chapter 3).

5.3 The warren

Within the family Leporidae the rabbit has the most complex social system, one of the most complex odour-signalling systems (Mykytowycz *et al.* 1984), the most complex warrens (Parer *et al.* 1987), and its young are born at the least advanced stage of development. Because of this and because it has a short gestation period (30.27 days (Myers and Poole 1962)) the rabbit is capable of having litters in quick succession. Rabbits born in warrens do not need fur or mobility to avoid predators and extremes of climate during their vulnerable infancy.

The rabbit is a burrowing mammal but it has no obvious morphological adaptations for burrowing. In suitable soils it can dig 2 m of burrow in one night (Myers 1958) yet new burrows are not readily constructed in the wild to accommodate increased numbers of rabbits (Parer 1977). The warren represents the collective effort of many rabbits over considerable periods of time, and cannot be replaced easily. For example, an experimental warren in Canberra known to be only six years old had 150 entrances, and it was estimated that 10.35 m^3 of soil had been excavated; the total tunnel length was 517 m (Parer *et al.* 1987).

Female rabbits are responsible for most of the burrowing, although males may assist. Bursts of digging occur after rain and also periodically in relation to pregnancy (Myers and Poole 1962). Mykytowycz *et al.* (1960) have described how the resident female in a small warren dropped five litters in one breeding season, each in a separate breeding chamber at the end of a separate tunnel excavated for that purpose.

One of the consequences of rabbits being born without fur is that nestling rabbits up to 12 days old have a limited ability to thermoregulate (Poczopko 1969). Although the doe lines the nesting chamber with grass and with fur plucked from the abdomen, this nesting material loses its insulating properties if it becomes saturated, and the nestlings may die from hypothermia. Flooding is thus a significant cause of mortality of nestlings (Myers and Poole 1963*b*; Parer *et al.* 1987). For the above reasons most rabbit warrens are constructed in well-drained soils (Myers and Parker 1965). Sandy soils alongside drainage channels provide optimum habitat in Spain, and also in the semi-arid and arid regions of Australia, as the sandy soils are well drained, are readily colonized, and the drainage channels provide food during dry periods (Rogers and Myers 1979; Myers and Parker 1975*a*, 1975*b*). The numbers of entrances in a warren vary from 1 to over 150, but warrens are smaller on sandy soils than on heavy soils and have fewer interconnections between the tunnels (Myers and Parker 1965; Oliver and Blackshaw 1979), undoubtedly because sandy warrens collapse and disappear during droughts and rabbit population crashes (Myers and Parker 1975*b*).

In Australia rabbits never dig warrens on heavy cracking-clay soils, although they may have litters in hollow logs on this soil type. Between sandy soils and cracking-clay soils there is a gradient of favourability of soil types which probably depends on the clay content and soil depth (Parer and Libke 1985; Parer *et al.* 1987). Soils with a high clay content tend to be hard to dig in when dry, and they become waterlogged in winter. Before the introduction of myxomatosis, when rabbits were numerous in southern Australia, they were present in areas of heavy clay soils using logs and shrubs as shelter. However, after myxomatosis most of the populations on the heavy soils died out and the distribution of rabbits is now concentrated on the more favourable soil types (Myers 1962), except in coastal areas with stands of dense low shrubs and tussocks, where rabbits do not always use warrens in the daylight hours (King *et al.* 1984).

In most Australian environments warrens are an obligate resource. Available warren space sets a limit to the number of rabbits that can live in an area, and there is a limit to the number of rabbits which can live in a warren. Large numbers of rabbits in a warren directly affect its gaseous environment and the ability of rabbits to survive (Hayward 1966), and this effect varies between warrens in different soils, and in different seasons. Warrens in sandy soils are generally hotter and dryer than those in heavy soils at the same depth. Warren temperatures as high as 33°C and relative humidities of 40 per cent have been measured in arid Australian environments (Hall and Myers 1978; Parer and Libke 1985; Cooke 1990*a*). Competition between females for nesting sites also affects warren use. When warren space is in short supply, dominant does interfere with litters of subordinate does, digging is suppressed (Myers *et al.* 1971), and very few litters born to subordinate does survive and emerge (Mykytowycz and Fullagar 1973).

Above-ground activities concerned with obtaining food and water and avoiding predators are centred on the warren. Wood *et al.* (1987) measured the amount of biomass eaten per year by a warren population in a semi-arid grassland habitat in central-western New South Wales, and showed that 800 kg ha^{-1} were taken at 12 m from the warren, 220 kg ha^{-1} at 25 m, and 150 kg ha^{-1} at 100 m, reflecting an inverse relationship between intensity of feeding and distance from the warren. We have not observed stable populations in which there were more than 1.3 adult rabbits for each warren entrance.

Aerial photographs have been used to survey warrens, but usually fewer than 60 per cent of warrens are detected (Parker 1979) and the proportion of warrens located is a linear function of warren size (Martin and Zickefoose 1976). The number of rabbits living in an area may be estimated by the number of warren entrances showing evidence of use by rabbits (Myers *et al.* 1975; Parer 1982*b*; Parer and Wood 1986).

5.4 Predators, parasites, and myxomatosis

5.4.1 Predators

The following is a list of the predators of the rabbit in Australia. Mammals and birds of prey are listed in their order of probable importance.

- Fox (*Vulpes vulpes*)
- Cat (*Felis catus*)
- Dingo (*Canis familiaris dingo*)
- Wedgetail eagle (*Aquila audax*)
- Little eagle (*Hieraaetus morphnoides*)
- Australian goshawk (*Accipiter fasciatus*)
- Brown falcon (*Falco berigora*)
- Whistling kite (*Haliastur sphenurus*)
- Spotted harrier (*Circus assimilis*)
- Swamp harrier (*Circus aeruginosus*)
- Black kite (*Milvus migrans*)
- Black-breasted buzzard (*Hamirostra melanosternon*)
- Black falcon (*Falco subniger*)
- Barking owl (*Ninox connivens*)
- Masked owl (*Tyto castanops*)
- Australian raven (*Corvus coronoides*)
- Monitor lizards (*Varanus gouldii, V. varius, V. giganteus, V. rosenbar*
- Brown snakes (*Pseudonaja textilis, P. nuchalis, P. affinis, Pseudechis australis*)
- Tiger snakes (*Notechis ater, N. scutatus*)
- Red-bellied black snake (*Pseudechis porphyriacus*)
- Woma (*Aspitites ramsayi*)
- Carpet python (*Morelia spilotes*)

The two major predators of the rabbit throughout its range in mainland Australia are the introduced fox (*Vulpes vulpes*) and cat (*Felis catus*). Rabbits and, to a lesser extent, small mammals are the main food of cats and foxes in both pastoral and forested country (McIntosh 1963; Coman and Brunner 1972; Coman 1973; Ryan and Croft 1974; Bayly 1978; Croft and Hone 1978; Jones and Coman 1981; Catling 1988). In most pastoral areas rabbits are virtually the only mammal of suitable size available to cats and foxes; native mammals are usually either very rare or absent, and house mice are abundant only at infrequent intervals.

Where rabbits breed in shallow warrens on sandy soils up to 75 per cent of the annual production of young rabbits is dug out by foxes (Wood 1980). In deep, sandy soils and on heavy soils few nests are excavated. After a drought in western NSW rabbit numbers increase much faster in stony soils where the nestlings cannot be dug out by foxes, than on sand-dunes where fox predation on nestlings is common (Myers and Parker 1975*a*). Nestling rabbits are unavailable to cats, but cats are more efficient than foxes in killing young rabbits above-ground (Parer 1977), whilst foxes are better at catching adults. Catling (1988) presents data to show that rabbits younger than 80 days occur as 40 per cent of fox diets and 91 per cent of cat diets. Rabbits older than 80 days form 60 per cent of fox diets and only 9 per cent of cat diets. When rabbit populations crash, fox and cat populations also collapse, after a time lag (Ridpath and Brooker 1986), but they remain active and effective in rabbit refuge areas bordering inland swamps and channels (Myers and Parker 1975*b*; Wood 1980; Ridpath and Brooker 1986).

Rabbits represent 6–13 per cent of the diet of the dingo (*Canis familiaris dingo*) in coastal NSW and 9–29 per cent in the Southern Tablelands (Newsome *et al.* 1983), but in the Northern Territory rabbits are the preferred prey of dingoes and constitute up to 61 per cent of their diet (Corbett and Newsome 1987).

Other mammals which prey on rabbits are of little importance. The native cat (*Dasyurus viverrinus*) is known to eat rabbits but it is now rare or extinct on the Australian mainland. No mustelids are native to Australia. Mongooses were released in NSW to control rabbits but failed to acclimatize. Ferrets (*Mustela furo*) have been used for many years to catch rabbits but despite thousands of accidental releases they have not established feral populations even in those areas where rabbits were very abundant, probably due to lack of alternative prey when rabbits stop breeding.

Almost 50 per cent of the diet of the medium and large diurnal birds of prey in the pastoral districts is rabbit (Cupper and Cupper 1981; Baker-Gabb 1984). Baker-Gabb estimated that hawks and eagles consumed 7–14 per cent of the annual production of young rabbits. The wedgetail eagle mainly kills rabbits over four months old (Brooker and Ridpath 1980) and the other diurnal birds of prey mainly kill kittens (Dunnet 1957). In habitats where alternate prey is scarce the diet of the wedgetail eagle may be 97 per cent rabbit. The eagles do not breed if the density of rabbits is less than 60 km^{-2} (Ridpath and Brooker 1986). It is unlikely that native mammal populations ever reached the densities attained by rabbit populations; the rabbit is probably supporting larger numbers of eagles than existed before it came.

Rabbits are the principal diet of masked and barking owls (Green 1965; Calaby 1951) but as these birds are restricted in distribution they are significant predators only in a few situations. Ravens are not efficient predators. They feed opportunistically on rabbits suffering from myxomatosis and on very young kittens (Mykytowycz *et al.* 1959; Parer 1977). During outbreaks of myxomatosis birds of prey may kill a high proportion of both adult and young rabbits (Williams *et al.* 1973).

Goannas prey on nestlings (Parer 1977). It is unlikely that they kill older rabbits. The snakes listed at the beginning of this section often feed on rabbits, but no dietary studies are available.

Few experimental studies in Australia have focused on predation as an ecological process. Richardson and Wood (1982) protected kittens in a subalpine environment from predation by hawks and eagles using parallel lengths of cord 30 cm apart over the warren area. By comparing survival on protected and unprotected warrens they estimated that 12.5 per cent of the mortality to 80 days of age was due to birds of prey. Using radio-tracking data in an independent study they found a mortality rate of 2.7 per cent per day for the first 14 days after emergence. Cats and foxes were responsible for a mortality rate of 1.5 per cent per day and avian predators for 0.7 per cent per day. More recently, another important study has shown that after a drought had caused the collapse of a rabbit population in semi-arid central NSW, rabbit numbers increased four times faster on an area where cats and foxes were removed than on a control area (Catling 1988; Newsome *et al.* 1989). Rabbit constituted 45.1 per cent in stomachs of foxes and 54.0 per cent in cats. Both predators exhibited a functional response to rabbits during the rabbit-breeding season. Predation was heaviest on an increasing rabbit population on green pastures, and on a decreasing population during drought. When pastures became dry and rabbits stopped breeding, both foxes and cats switched to invertebrates, birds, reptiles, and carrion, but foxes still preyed heavily on adult rabbits.

A comparison of the predators of rabbits in Australia and in Spain where the rabbit originated, shows that in Spain there are 17 species of mammals, 19 species of birds and 4 reptile species which prey on rabbits (Soriguer 1981; Chapter 3), whereas in Australia there are only 3 mammals and 13 birds. Nine of the eleven genera of birds which prey on the rabbit in Australia are also predators of the rabbit in Spain.

If we consider only those predator species for which rabbits comprise 40 per cent or more of the biomass consumed, Spain has six species of mammals and eight species of birds which could be considered as specialist predators of the rabbit (Delibes and Hiraldo 1981). Australia has two mammal species, at least eight bird species and an unknown number of reptiles which could be considered as specialist predators of the rabbit. Five of the six specialized mammalian predators in Spain take nestling rabbits but only one mammalian species in Australia, the fox, preys on nestlings. In Spain and in Australia small mammals are not abundant (Delibes 1975) and in both areas the rabbit is the principal source of energy for most medium to large predators.

Studies in Australia have not yet reached the stage of being able to indict predation as a regulatory factor in rabbit populations, but predators have been shown to increase rates of decline and decrease rates of increase, by exerting strong pressures on numbers of rabbits during the late stages of population crashes and early stages of population recovery. Newsome *et al.* (1989) call this process 'Environmentally Modulated Predation', since it depends on the intervention of a large environmental event, such as drought, for its operation.

5.4.2 Ectoparasites

The principal ectoparasites of the rabbit in Australia are two mites (*Listrophorus gibbus*, *Cheyletiella parasitivorax*), three fleas (*Echidnophaga myrmecobii*, *E. perilis*, *Spilopsyllus cuniculi*) and one louse (*Haemadipsus ventricosus*). *E. myrmecobii* and *E. perilis* are native species which are commonly found on a variety of marsupials (Shepherd and Edmonds 1978) whilst the louse and the mites were introduced with the rabbit. Various other species of ectoparasites such as *Echidnophaga gallicaceae* and *Bdellonyssus bacoti* are occasionally found on rabbits (Mykytowycz 1957*a*; Williams 1972).

Echidnophaga spp. are most abundant in low rainfall areas with light soils. Their numbers are highest in late summer when most rabbits are infested (Shepherd and Edmonds 1978). *H. ventricosus* is most common in semi-arid areas (Williams 1972). *C. parasitivorax* is equally abundant in semi-arid and higher rainfall areas. *L. gibbus* is most abundant in areas where the rainfall exceeds 640 mm.

The European rabbit flea, which was released in Australia in 1968, is now well-established as a vector of myxomatosis in all parts of Australia except the arid zone. Flea populations have not persisted where the annual rainfall is less than 200 mm (Cooke 1984).

E. myrmecobii, *H. ventricosus*, and *C. parasitivorax* have all been shown to be capable of transmitting myxomatosis although not very efficiently (Bull and Mules 1944; Mykytowycz 1958*a*). Williams (1972) and Dunsmore *et al.* (1971) suggested that *L. gibbus* might be the vector responsible for the winter epizootics of myxomatosis, which occur most commonly in those areas where high infestations of *L. gibbus* would be expected. Except as vectors of myxomatosis, ectoparasites have not been shown to adversely affect the rabbit in Australia.

5.4.3 Endoparasites

The major species of endoparasites in Australia (Table 5.10) are the same as those in New Zealand (Bull 1953; Dunsmore 1981). The anoplocephalid cestodes (*Cittotaenia pectinata*, *C. denticulata*, *C. ctenoides*, *Andrya cuniculi*), found in rabbits in England, have not been observed in Australia. Also lacking is the nematode lungworm (*Protostrongylus rufescens cuniculorum*) and the two most pathogenic forms of intestinal coccidiosis, *Eimeria intestinalis* and *E. flavescens* (Coudert 1979).

The intensity of infection with *Eimeria* spp., as measured by oocyst output, is greatest in the higher rainfall areas of eastern Australia and least in the arid zone (Stodart 1968*a*, 1968*b*, 1971). Oocyst counts decline with age of rabbits in all environments. In the arid zone even adult rabbits can be susceptible to severe coccidiosis because of their low degree of exposure as kittens. All species of *Eimeria* except for *E. piriformis* occur over wide areas, even in arid habitats.

Coccidiosis, especially hepatic coccidiosis (*E. stiedae*), can be lethal to young rabbits (Mykytowycz 1962). In a relatively moist coastal environment Dunsmore (1971) demonstrated that it is a very important juvenile mortality factor. In the Southern Tablelands of NSW no young rabbits have liver lesions in a dry year, but over 40 per cent have lesions in a wet year (Parer and Libke 1991).

The stomach nematode (*Graphidium strigosum*) is found in large numbers (> 100 per rabbit) only in areas with high rainfall but it is absent from subtropical Queensland (Dunsmore 1966*a*). *Trichostrongylus retortaeformis*, a small intestinal nematode, is commonly found in high rainfall areas, but is less affected by climate than the stomach worm and is found even in the arid zone (Dunsmore 1966*b*).

Table 5.10 The important species of endoparasites of the rabbit in Australia

Protozoa	Nematoda	Platyhelminthes
Eimeria stiedae	*Graphidium strigosum*	*Taenia pisiformis*
Eimeria media	*Trichostrongylus retortaeformis*	*Taenia serialis*
Eimeria perforans	*Passalurus ambiguus*	*Fasciola hepatica*
Eimeria magna		
Eimeria irresidua		
Eimeria piriformis		
Eimeria exigua		

Table 5.11 The results of an experiment in which 16 female rabbits produced litters when (a) worm-free and (b) paratisized by *Trichostrongylus retortaeformis* (Dunsmore 1981)

	(a)	(b)
Number of young born alive	73	71
Number of young weaned	70	58
Mean weight at birth (g)	45	40
Mean weight at weaning (g)	230	203
Mean milk production (g)	1534	1114
Food intake during lactation	2856	2293

Infestations with *G. strigosum* and *T. retortaeformis* increase with host age, and breeding females have the heaviest parasite burdens. *T. retortaeformis* has been shown to have detrimental effects on the productivity of female rabbits (Table 5.11) and it is likely that *G. strigosum* acts similarly. The distribution of *Passalurus ambiguus* is probably co-extensive with that of the rabbit in Australia. Although very heavy infestations (> 50 000) have been recorded it is not thought that this nematode has adverse effects on rabbits (Dunsmore 1981).

Taenia pisiformis and *T. serialis* are commonly found in rabbits (Dunsmore 1981). The metacestode of *T. serialis* develops in the muscles and is sometimes so large that it can impede mobility, and indirectly lead to the death of its host. *Fasciola hepatica* is common in rabbits in the wetter areas where the parasite is endemic. In the Canberra district the percentage of rabbits with liver damage caused by the fluke varies from 19 per cent in winter to 4 per cent in summer.

5.4.4 Myxomatosis (see also Chapter 7)

Myxomatosis in Australia greatly reduced rabbit numbers in all environments, although less markedly in semi-arid and arid areas away from swamps and drainage channels. In higher rainfall regions the areas of distribution of rabbits were greatly decreased, and open grazing and crop lands were vacated for more favourable and protected sites in sandy areas bordering streams and swamps, and woodlands and rocky terrain bordering grasslands (Myers 1962).

Many of the higher rainfall areas in eastern Australia are well served by large, seasonal populations of several insect species capable of transmitting myxomatosis. Transmission of myxomatosis in the eastern Riverina and south-western slopes of NSW, for example, falls into five distinct seasonal phases. Four of these are dependent on seasonal production of winged insects, mainly mosquitoes, whereas the fifth, designated as winter outbreaks, is basically a product of rabbit behaviour, dependent upon the close social contacts engendered by reproductive activities.

1. *Anopheles annulipes* shows two distinct seasonal peaks in adult activity centred about the late spring (November) and early autumn (April) months, principally reflections of the seasonal distribution of waters available or suitable for breeding (Myers 1955). During these periods *Anopheles* is accompanied by the less important vector *Culex australasiae*.

A. annulipes exhibits an array of behaviour patterns which probably make it the most efficient of all the vectors. It is a wide-ranging species, and feeds avidly on rabbits and transmits myxomatosis within warrens during the daylight hours. During the night, above-ground, the mosquito shows no special preference for rabbit blood (Myers 1956). It is a hardy species, able to live in many semi-arid environments, where the harbour provided by rabbit warrens is probably of critical importance for its survival as well as for disease transmission. *A. annulipes* was undoubtedly the principal agent in disseminating the virus throughout the region in environments away from rivers and swamps during the first three years of spread.

2. *Culex annulirostris* feeds in the high summer months, with a single annual peak between December and February. *C. annuilrostris* is accompanied along major river frontages by the simuliid *Simulium melatum*, which also readily transmits myxomatosis (Mykytowycz 1957*b*).

   Unlike *A. annulipes*, *C. annulirostris* is closely confined to humid habitats along river and creek frontages, and about semi-permanent and permanent swamps. This culicine is an avid feeder on rabbits, but only while they are above ground at night. Its success as a vector is due to its strict confinement to habitat, leading to repeated feeding in the same locality, to a high incidence of interrupted feeding on several hosts before becoming satiated, and to long movements within its habitat. Transmission is also assisted because stock, on which it also feeds heavily, tend to vacate the areas it inhabits during the night, leaving rabbits and birds as its main source of food.

   C. annulirostris was probably the most important agent in the initial long-distance dissemination of the virus along and between the major river systems and swamps during 1950–1. This mosquito was the only species which could have carried a high enough infection rate to account for the widespread movement recorded during the 1950–1 summer. The *Anopheles* populations in between the river systems were virtually free of the virus during the 1950–1 epizootic.

   S. melatum, which accompanies *C. annulirostris* in some habitats, breeds in large rivers following spring floods. It is a day-biter, which reduces its value as a vector, and attacks stock and man, as well as rabbits.

   Between them, the anopheline and culicine phases of transmission of myxomatosis accounted for almost all the great epizootics of the early years, and in most localities occurred in sequence.

3. This phase differs markedly from the first two described above. The aedine fauna of the south-western slopes and plains of NSW is dominated by three species or species groups, *Aedes alboannulatus*, *Aedes sagax*, and *Aedes theobaldi*, with incursions of *Aedes vittiger* from northern areas. Depending on the rainfall regime, the *Aedes* spp. show two annual peaks in adult activity, centred about the autumn (April–May) and early spring (August–September). There is also some opportunistic breeding during the winter and after the less-frequent summer rains.

   Observations have shown that these species transmit myxomatosis sporadically during population peaks, but no epizootic has been known to result from their activities. Their inefficiency as vectors of myxomatosis is explained by their basic behaviour patterns. They are wide-ranging day-biters, and rabbits are exposed to attack by them only during the short period between emergence from their warrens in late afternoon and dusk. The mosquitoes attack rabbits, stock, and other mammals equally. During the evening, however, when rabbits were exposed to their activities, the stock to rabbit feeding ratio on one site was estimated as 10 000:1 (Myers, unpublished data). The chances of an infective mosquito returning to rabbits for blood is low.

4. Numerous species of simuliidae breed in the large rivers of inland Australia, and in the small intermittent streams which drain the catchments in between. In the eastern Riverina, *S. melatum* restricts its breeding to the large rivers after spring flooding, and often accompanies the mosquito *C. annulirostris* during epizootics in high summer. More than seven other species, dominated by *S. nicholsoni*, *S. bancrofti*, *S. ornatipes* and *Austrosimulium furiosum*, utilize smaller streams and rain-filled and spring-fed water courses.

   A. furiosum is the dominant simuliid vector

away from the large rivers. Adult activity in this species commences in early spring (August), rises to a peak in October, and decreases as the small creeks cease running. Transmission of myxomatosis by *A. furiosum* was measured by exposing rabbits to bites in twelve epizootics during October, after the spring Aedine peak, and prior to the commencement of Anopheline activity (Myers unpublished). *A. furiosum*, like *S. melatum*, is a day-biter, and attacks stock, man, and other mammals. It is transported long distances by winds, and congregates in dense woodlands. In such localities stock tend to move away and rabbits remain as the principal source of food to a vector with restricted movements.

5. Winter epizootics, occurring mainly from May to August, have always been the most difficult to explain, since most tend to be inconspicuous and long-lived, and, despite many comprehensive attempts during the early years (Lee *et al.* 1957; Dyce and Lee 1962; Myers, unpublished observations), it has proved generally impossible to incriminate winged, blood-feeding insects caught at the site of an epizootic.

Numerous observations of confined populations have shown clear relationships between social activities and transmission of myxomatosis. In an enclosure experiment in September 1960, in Adelaide, South Australia, for example, a population of wild rabbits consisting of two social groups was accidentally infected by a field strain of myxomatosis when an outside male was added. In territorial defence the dominant males of groups A and B attacked and fought the strange male on numerous occasions between 5 and 12 September, by which date it became obvious that the new animal was infected. Both dominant males exhibited typical symptoms of myxomatosis by 20 September. On this evening one of the three females in group B came into oestrus. The dominant male, in typical manner, chased, nuzzled, licked, and attempted to mount the female repeatedly, and repeatedly attacked and fought with the subordinate male in his group, during which bites and scratches were inflicted. The dominant males were removed on 22 September. The subordinate male and the female involved in the above encounters became infected and exhibited typical symptoms between 30 September and 2 October, when they were removed. No further cases were recorded (Myers unpublished).

Two winter epizootics have been examined in detail in natural populations, at Canberra (Williams *et al.* 1973), and at Snowy Plains (Dunsmore *et al.* 1971). In both populations transmission was very slow and the virus was of grade III virulence, but the case mortality rate was very high. The workers concluded that contact between rabbits was a major factor in transmission, and that ectoparasites were probably involved. Williams (1972) measured the seasonal and regional variations in numbers of ectoparasites of rabbits in NSW and suggested that the mite, *Listrophorus gibbus* (Pagenstecher) was a probable vector in close-contact transmission.

In the eastern Riverina and nearby slopes, during the first two decades after its introduction, transmission of myxomatosis occurred somewhere in every month of the year. Spread of the virus was most active during the warmer months, and it remained active during winter in local populations, undoubtedly caused by social behaviour of rabbits, ectoparasites, and sporadic feeding of aedine mosquitoes.

Parer and Korn (1989) recently analysed patterns of epizootics of myxomatosis throughout NSW from the coast in the east to the arid environments of the west, and showed that although myxomatosis is reported in that state more frequently in summer than winter, the probability of summer epizootics is greater in the western and central plains than on the slopes, tablelands, and coast; and summer epizootics on the slopes and plains are positively related to rainfall. This comprehensive survey adds spatial understanding to the kind of observations described above.

Clear patterns of transmission have not been detected in other Australian environments, because of paucity or absence of vectors (Calaby *et al.* 1960, in Western Australia; Lee *et al.* 1957, and Dyce and Lee 1962, on NSW coast and tablelands; Low and Strong 1984*a*, in Northern Territory) or lack of predictability of vector activity.

Because of the importance of the rabbit flea, *Spilopsyllus cuniculi*, as a vector in Britain, the insect was imported into Australia in 1966 to improve disease transmission in arid and tableland environments. It was freed from quarantine in 1968 (Sobey and Menzies 1969) and introduced into rabbits at a number of sites throughout Australia. Transmission rates were clearly increased in most areas in which it was released, including arid environments in South Australia (Cooke 1983*b*), and were usually accompanied by alterations in the seasonality of the epizootics they caused (see King *et al.* 1985). The flea does not survive for long periods in the severe climates of South Australia and the Northern Territory (Cooke 1984) and Western Australia (King *et al.* 1985) below the 200 mm isohyet, and investigations are under way into the possibility of obtaining more efficient ectoparasitic vectors (Cooke 1990*a*).

The vast area covered by Myxomatosis in Australia is a striking reflection of the numbers of endemic insects which used the introduced rabbit as a source of food, and to the long distances they traversed in carrying the virus throughout the continent and to islands more than 100 km off-shore.

Although predators were numerous before myxomatosis arrived (Calaby 1951) they were unable to breed at the same rate as rabbits, were easily satiated, and had little effect on rabbit population numbers; but since the reduction in numbers and changes in distribution of rabbits after myxomatosis, predators have become a critically important cause of mortality (Parer 1977; Brooker and Ridpath 1980; Richardson and Wood 1982; Baker-Gabb 1984; Catling 1988; Newsome *et al.* 1989; Newsome 1990), sharing with myxomatosis a dynamic control of rabbit numbers over large areas.

One of the major problems in semi-arid areas before the introduction of the European rabbit flea was that inoculations had to be done at times when mosquitoes were abundant; however, when mosquitoes were abundant, field strains of myxomatosis were also present, and the field strains outcompeted the virulent strains. Although field strains of the myxoma virus are now of intermediate virulence, and rabbits have some degree of genetic resistance to the disease, myxomatosis is still preventing a large upsurge in rabbit populations (Parer *et al.* 1985). If myxomatosis were to suddenly become ineffective, the cost of controlling rabbits in the higher rainfall country would be greatly increased, but the application of modern control techniques there should be able to prevent any recurrence of rabbit plagues. In the semi-arid rangelands, however, rabbit control would become necessary where little is practised today. The cost would be high and farmers on some of the smaller properties could be forced off their land. In the arid zone in areas where control measures may not be economic, recurrent plagues would again become a reality, resulting in further degradation of the environment.

It is of some concern, therefore, to note that resistance to myxomatosis appears to be continuing to increase. In their recent studies, Williams *et al.* (1990) observed that mortality rates in rabbits from arid, subalpine, and temperate Mediterranean populations had decreased by 10 per cent in the last 10–20 years when challenged by the virulent Lausanne strain and 50 per cent over the past 40 years with the avirulent Uriarra strain. Parer (1991) discusses modern problems relating to viral strains in field populations, including the problem of differentiating between genetic and acquired resistance.

Many different techniques of introducing myxomatosis into field populations have been devised. Early workers used gin traps with only one jaw, which held needles coated with virus (Anon. 1942), injected freeze-dried virus into rabbits caught by use of ferrets or netted by spotlight, or rubbed freeze-dried virus with an abrasive powder on the eyelids of trapped rabbits (Sobey *et al.* 1967). Other methods included aerosol generators (Conolly and Sobey 1985), air-gun pellets with virus impregnated cotton-wool in a hole in the pellets, fleas coated with virus (Parer *et al.* 1981), and, where access was difficult, arrows tipped with glass vials of infective fleas fired on to warrens.

A fuller account of myxomatosis in Australia is presented in Chapter 7.

5.5 Demography

The interplay between the variable rates of reproduction and mortality described in earlier sections cause marked differences in sex and age structures and in different capacities for increase between populations in different environments.

5.5.1 Sex ratio

In experimental populations, competition for mates and resources have important effects on sex ratio, which is 50:50 at emergence. In enclosures males and females suffer similar mortalities until reproduction commences (Myers and Poole 1963*b*). During their first breeding season, there is an increase in mortality in young subordinate males ejected from breeding groups, caused mainly by avian predation, which results in a 2.5:1 female to male ratio in adults at the end of the breeding season. The young males lie up on the surface and have few avenues for escape, whereas most females and the dominant males possess burrows. Thus under conditions of strong competition a harem behavioural structure is translated into population structure, and the main causal factor is selective predation.

These differences are not so apparent in natural populations, where competition for space may be less severe, and where other environmental factors may exert a strong influence. In shot samples there is a strong trend to female preponderance from month to month during the breeding season in subalpine populations (χ^2 = 49.77, 11 DF), and noticeable trends in temperate tableland (χ^2 = 23.18), and coastal populations (χ^2 = 11.8; Table 12). Similar trends are also evident in data from Victoria (Shepherd *et al.* 1981) and New Zealand (Gibb *et al.* 1985). They are probably due to changes in rabbit behaviour.

Dunsmore (1974) observed a preponderance of females in his careful study of a marked warren-dwelling population on a subalpine site which had a mean male:female ratio of 0.65:1 over four years. Shot samples in the surrounding area, however, disclosed a return to male preponderance in the non-breeding months and an overall male:female ratio of 1.15:1 (Myers 1971; Table 5.12), suggesting that warren populations are not representative of the total population. The simplest explanation is that in the above populations many of the ejected animals survived and they were mostly males and were accessible to shooting.

At the other extreme, in the arid zone the population shows a strong preponderance of females almost all the year. Ejected males in that population apparently died. In the subtropical population where reproduction continues throughout the year at reduced levels, male numbers appear to dominate most of the year, but at Urana, where the breeding season is more clearly defined, the ratio changes during the first months of the breeding season (May–August) in favour of females, and this is significant (χ^2 = 6.86, 11 DF).

An examination of changes in sex ratio in relation to age shows that in populations in cooler environments there is a general increase in the proportion of females in samples during and after their first breeding season (9–18 months) with a return to male preponderance in older age classes. The same phenomenon is obvious in the New Zealand data described in Gibb *et al.* (1985). These patterns do not appear in samples from populations in hotter environments, especially in arid samples, and are probably related to the same phenomena mentioned above.

5.5.2 Age structure

Field populations show a variable age structure (Fig. 5.9; Myers 1971; Parer 1977; Wood 1980; Cooke 1983*b*; Wheeler and King 1985*c*; Gilbert *et al.* 1987; Parer and Libke 1991). Shot examples in eastern Australia suggest that populations in arid and subtropical regions not only have large sex-ratio imbalances but are also top heavy with older individuals. The coastal population on the other hand is composed of predominantly young animals. Differences of this kind clearly point to differential mortality. Generally the more unfavourable the environment, the older the age structure.

5.5.3 Attempts at modelling Australian data

Due to the size and complexity of the Australian rabbit problem, Australian workers approached their initial studies in an analytical manner; this involved

Table 5.12 Seasonal changes in male:female ratio (expressed as males/females/month) in rabbits older than 6 months in shot samples of rabbits in six populations in eastern Australia

Site	J	F	M	A	M	J	J	A	S	O	N	D	Mean	N
Subtropical Q	1.59	1.04	—	1.04	1.00	1.47	1.33	1.25	1.07	1.12	1.09	1.29	1.16	1049
Subalpine NSW	0.73	0.75	0.86	1.58	2.04	1.58	0.98	1.70	1.38	0.68	0.24	0.69	1.15	1083
Arid NSW	0.92	1.00	0.86	0.90	0.84	0.87	0.77	1.32	0.84	0.75	1.11	0.91	0.90	1563
Temperate mediterranean NSW	1.61	1.92	2.44	1.49	0.89	1.07	1.00	0.95	1.38	1.21	1.33	1.33	1.20	743
Coastal NSW*	1.40	1.00	1.15	1.15	2.08	1.84	1.14	0.88	1.00	0.92	0.86	0.92	1.19	423
Southern** Tablelands NSW	0.74	1.63	1.24	1.26	0.93	1.47	1.37	0.86	0.90	0.99	0.70	0.65	1.04	1327

* After Dunsmore (1971).
** After Parer and Libke (1991).
Each 'month' has been calculated as 365/12.

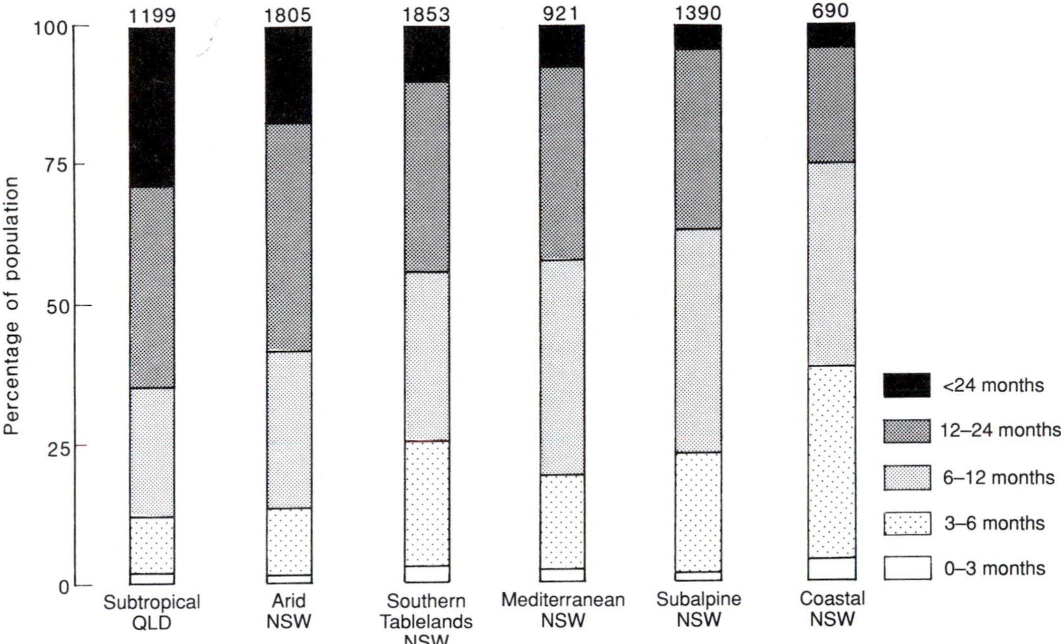

Fig. 5.9 Age distributions estimated by weights of crystalline eye lens in shot samples taken from six populations in eastern Australia.

studies of the rabbit in confined populations to gain fundamental knowledge on reproduction and behaviour, followed by the shooting of a large series of samples of rabbits from climatically different regions to gain an understanding of biological and ecological variability among populations, and a series of live studies in the areas from which samples were collected in an attempt to measure dynamics.

In early years when much of the work described in this chapter was done, most Australian workers accepted the premise that the rifle collected fairly representative samples of rabbits older than 3 months. This was in general accord with published information of the time, but despite this, the relative scarcity of rabbits younger than 6 months in samples, and the strange shifts in sex ratios by age and season, suggested that there was an unknown component of hunting mortality in the collected samples. In the absence of measured survival rates, and aware of the problems of calculating demographic parameters from age frequency data, as a heuristic and comparative exercise, Myers (1971) calculated what he termed 'provisional life-tables', using the shot samples of rabbits older than 3 months as a dx series. He then calculated an array of 'provisional capacities for increase' (Laughlin 1965) by varying the possible values of dx of the 0–3 months age group from 0 to 1. Because of the close relationship between the $1x$ and dx, the absolute demographic differences between the populations as sampled remained unchanged. Although the exercise underestimated mortality in the young and overestimated adult mortalities, yielding calculations of capacities for increase which were too low, it demonstrated the superior capacity of the temperate Mediterranean population to increase in numbers, and the need for very high mortalities among its young (more than 80 per cent of the annual cohort) for the population to remain stable, later confirmed in Parer (1977), Gilbert and Myers (1981), and Gilbert *et al.* (1987).

Gilbert (unpublished data) later looked more critically at the possibility of estimating the survival rates required to give the observed age distributions in successive shot samples, using age-specific reproduction rates as measured, and age estimates based on the weight of the crystalline eye-lens (Myers and Gilbert 1968). Those analyses revealed heavy losses

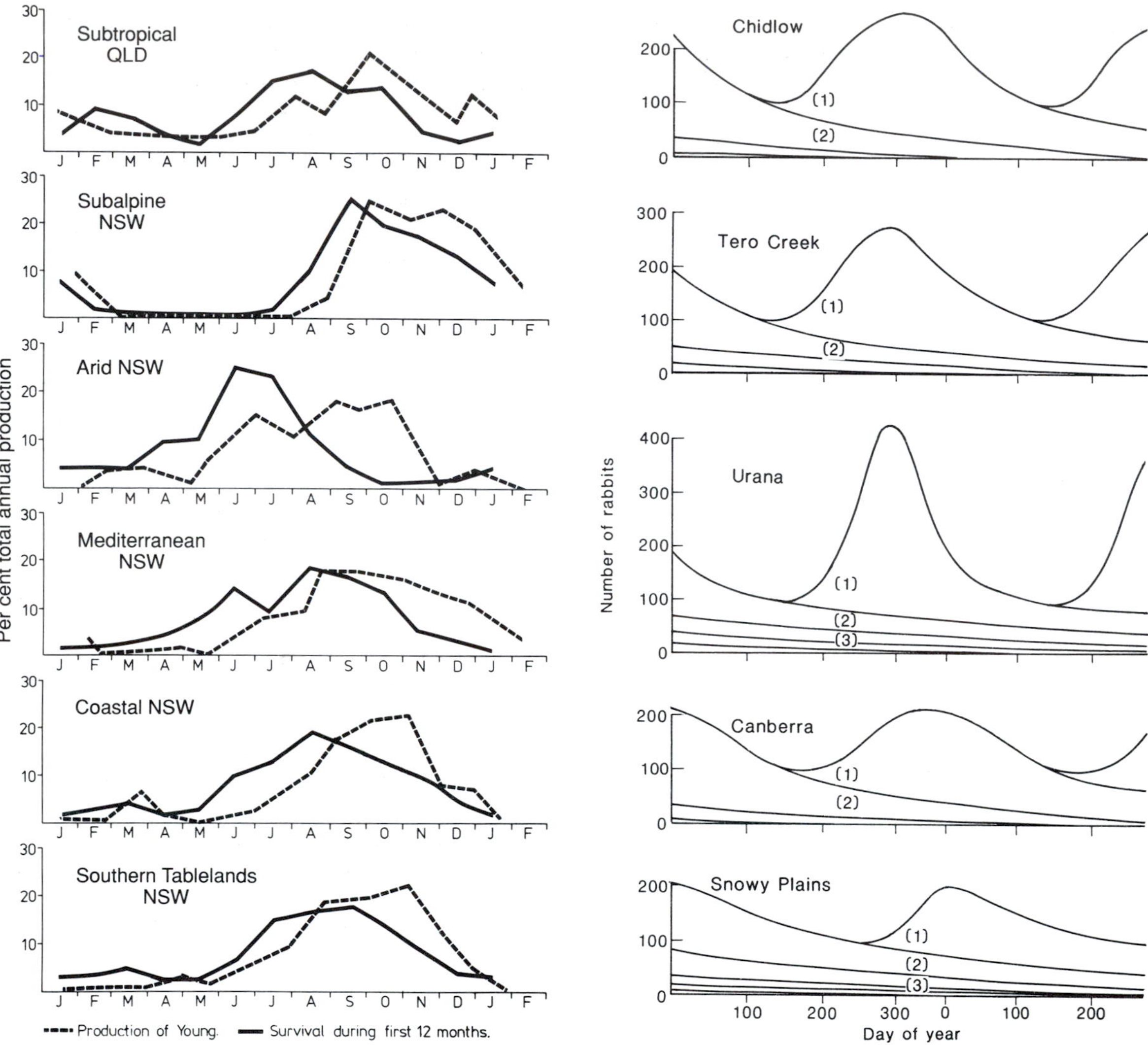

Fig. 5.10 Estimated mean monthly production of young rabbits (per cent total annual production) (solid line) and the percentage of rabbits less than 12 months old in shot samples that were estimated by weights of crystalline eye lens to have been born in each month (dashed line) in six climatically diverse populations in eastern Australia. References as in Fig. 5.7.

Fig. 5.11 Computed age distribution and relative numbers of rabbits (more than 4 weeks of age) throughout the year in five populations of rabbits in Australia. Curves at each site are scaled to an arbitrary minimum population size of 100. The curves repeat from one year to the next. 1 = rabbits born in current season, 2 = those born in previous season, etc. (After Gilbert *et al*. 1987.)

(around 15 per cent per week) in young rabbits during the first four months of life, and rates between the 10th and 12th month equal to adults (defined here as older than 12 months). But the survival rates between 5 and 9 months on an average exceeded one, indicating that rabbits in that age group, and especially from 4–6 months of age, were undersampled by the shooting. In order to obtain accurate survival rates from shot samples, Gilbert concluded that each sample would need to be increased to about 200 animals, an impact which few populations could absorb if samples were taken frequently. It would also be necessary to calibrate the shooting biases by shooting samples from populations of rabbits with ages to 9 months precisely known, which was not feasible, on a logistic basis alone.

In experimental populations longevity is markedly longer at low density, due to an amelioration of the effects of malnutrition, unfavourable environmental conditions, parasites, disease, predation, and social stress, especially during the first 3 months of life for both sexes, and for females during reproduction (Mykytowycz 1961; Myers and Poole 1963*b*). At low densities in confinement the main mortality in young rabbits affects the last born of the season, which suffer milk starvation when the does cease lactating. At higher densities longevity markedly decreases, especially among young, but young born early in the season still survive better than those born later. Irrespective of date of birth, high mortality is common among rabbits, due to malnutrition, if their food changes from green to dry when they are still on the steep part of the growth curve. The growth of older young is retarded and they remain stunted for life.

Similar demographic patterns were measured in natural populations, both in shot samples (Fig. 5.10) and in live studies (Parer 1977; Wood 1980; Wheeler and King 1985*c*; Gilbert *et al.* 1987). In every population sampled, after allowing for the small numbers born during the summer months in all populations, rabbits born during the early months of breeding survive better than those born during and after the spring peak, when predation is heavy and food quality is usually decreasing rapidly.

With the completion of several live studies it became possible to combine the measurements of reproduction in shot samples with rates of survival measured in cohorts of marked individuals, averaged from site to site. Thus Gilbert and Myers (1981) and Gilbert *et al.* (1987) constructed annual sine curves, fitted by regression, to the annual cycles in pregnancy and litter size *in utero* (see Figs 5.7 and 5.8), corrected for age of mother, and linked them with the survival curves to make 'variable life tables', representing dynamic changes in numbers through an average year.

It is not easy to collect comprehensive data on rabbits in natural populations subjected to frequent epizootics of myxomatosis, to human interference, and to complex patterns of predation. The model represents the best that is presently possible with the data available. The work is based on studies at 10 sites spanning the major habitats of the rabbit in Australia, and one in New Zealand, and represents the result of an enormous amount of labour and application of field workers in different Australasian states, each travelling large distances in the collection of comparable data.

The computed age composition of each population as it varies throughout the year is shown in Fig. 5.11. The effects of the various factors described in earlier sections are apparent in the different dynamics shown in the graph. In particular it emphasizes the overall favourability of temperate Mediterranean environments like Urana, NSW, and the wide differences between populations in different environments.

5.6 Evolutionary change

Historically the fossil record shows a conservatism in evolutionary change within the order Lagomorpha and in particular the family Leporidae (Dawson 1981; Chapter 1). *Oryctolagus* appears to have a high capability of maintaining heterozygosity (Richardson 1981).

In Australia the rabbit colonized an array of environments differing widely in climate and vegetation. It also exhibited the boom/bust population dynamics typical of a fecund species invading a generally favourable environment. In some environments it still exhibits this demographic pattern. Together these facts suggest that a relatively rapid rate of evolutionary change may have occurred in Australian rabbits.

We have mentioned that data from rabbits shot at sites in a wide range of habitats occupied by the species have shown regional differences in fecundity and seasonal breeding patterns and in body and appendage sizes, as well as differences in internal organs important to physiological processes. Recent experiments by Williams and Moore (1989*a*, 1989*b*, 1990) have shown that much of the variability between sites is phenotypic adaptation, but there is a small, detectable genetic component of differences in ear size relative to body size. This genetic difference corresponds to the thermal environments of the respective free-living forebears; relatively large ears in hot environments, relatively small ears in cold environments. Ear lengths of Australian rabbits (77–82 mm) (Table 5.1) are longer than English (60–70 mm) or French (71 mm) rabbits (Lloyd 1977; Soriguer 1980).

Variation in coat colour between field populations also has been examined for evidence of evolutionary changes. In northern Tasmania there is a cline in the percentage of black rabbits in populations between the coast and the crest of an escarpment about 1000 m altitude (Barber 1954). Between 20 per cent and 30 per cent of the rabbits are black above 500 m in cleared rain-forest areas where rainfall is over 1250 mm. Below 500 m, where rainfall is less than 1250 mm and the original vegetation was eucalypt forest, 10 per cent are black. Less than 2 per cent are black where the rainfall is about 625 mm and the vegetation is eucalypt savannah or the rich farming land of the coastal plain.

Examination of skins from sites in arid, subalpine, subtropical and Mediterranean climatic zones on the Australian mainland has shown that there has been selection in arid sites towards yellowness and in wet, subalpine sites towards blackness compared to the other sites with intermediate rainfall and temperature regimes (Stodart 1965). This evolutionary divergence seems to have been carried out by selection of quantitative genes rather than qualitative genes as appears to have been the case in Tasmania, and has taken place in under 80 years.

There are significant morphological differences in the skulls of rabbits from five localities in Australia (McCluskey *et al.* 1974) but whether these differences are due to selection or to founder effects is unknown. Richardson (1981) found significant differences in the gene frequencies of electrophoretically detectable enzyme polymorphisms between populations in Australia, but there was no clear geographical pattern and there was no correlation between genetic distance and geographical distance. The genetic distance within Australian populations is similar to the distance between Australian and British populations, suggesting that there has been little change since the rabbit arrived in Australia.

5.7 Impact on flora and fauna

Until recently, information on the impact of the rabbit on the Australian environment has come mainly from circumstantial observations and reports. Some of them (Anon. 1901; Ratcliffe 1936, 1959; Jessup 1951) describe the destruction by rabbits of vegetation in arid habitats. Ratcliffe's measured assessment of the problem identified the stripping and killing of seedling trees and perennial shrubs as the most serious form of damage.

Similar findings have been presented by Johnson and Baird (1970) who described changes in vegetation between 1930 and 1955 in parts of the Nullarbor Plain grazed only by kangaroos and rabbits. Large areas of bluebush were destroyed and regeneration of *Acacia* spp. was prevented wherever rabbits lived.

Long-term studies in the arid zone in South Australia have also emphasized the competence with which rabbits suppress regeneration of trees and shrubs, especially *Acacia* spp. (Lange and Graham 1983). Serious absence of regeneration of woody plants in central Australia is associated with rabbit densities as low as 3 ha^{-1} (Friedel 1985). The impact of rabbit grazing on herbaceous plants in arid regions is less noticeable, as the pasture has already been degraded, both by the grazing of domestic stock and previous generations of rabbits. Large numbers of rabbits do not now degrade the pasture, but rather restrict its development in response to improving seasonal conditions (Foran 1986). In areas where intensive grazing has destroyed seed beds, no change in species composition results when rabbits are denied access to sample plots (Leigh *et al.* 1989).

Graphic descriptions of the depredations of the rabbit and the soil erosion caused in agricultural areas are presented in a large pre-myxomatosis literature in State Agricultural Journals, but the actual effect of rabbits on pastures received little attention until Gooding (1955) measured the large quantities of pastures eaten by rabbits in Western Australia.

The finding in England that rabbits grazing rye–clover pastures reduced the density of rye grass and clover and increased the content of less palatable grasses and weeds (Phillips 1953) was confirmed by Myers and Poole (1963*a*) in eastern Australia. In the subalpine high plains Leigh *et al.* (1987) showed that the feeding activities of rabbits led to substantial reductions in cover, biomass, and species diversity of forbs. Over a 7-year period there was a net loss of nine palatable forb species where rabbits grazed.

Rabbits have been introduced to 48 islands off the Australian coast, and where no predators and competitors are present have often completely destroyed or transformed the vegetative cover (Fullagar 1978; Armstrong 1982; Norman 1988; Davey 1990).

Little information on the direct impact of the rabbit on native fauna exists. No native mammal has gone extinct north of the range of the rabbit, where there are foxes but no domestic stock (Brooker 1977). Most extinctions have been of small- and medium-sized mammals, and throughout much of the area invaded by rabbits, man and stock had already caused major changes in the environment before the rabbit came. It is generally believed, however, that rabbits played a role in eliminating two small burrowing marsupials, the boodie rat (*Bettongia lesueur*) and bilby (*Macrotis lagotis*), and the sticknest rat (*Leporillus* spp.) perhaps by direct competition for food but, in the case of the former two, more likely for the burrows they both inhabited (Stodart 1966). Old residents in Broken Hill, western NSW, still recount that trappers killed innumerable bilbies when trapping large warrens for rabbits along the South Australian–NSW border. The last bilbie was caught in that region in about 1900.

In South Australia common wombats (*Vombatus ursinus*) disappeared from the north-western part of their range within 20 years of the arrival of rabbits, except on one pastoral lease which had been kept free of rabbits (Mallet and Cooke, personal communication). Rabbits undoubtedly hastened successional changes in the pasture and the loss of the wombat was probably related to the disappearance of perennial native plants.

The vegetation changes and heavy erosion that have followed the introduction of rabbits onto coastal islands have posed serious threats to the existence of seabirds which use the islands as resting and breeding sites. This process has been exacerbated where predators have also been carried onto islands, where seabirds form their only source of food. Seabirds no longer use Macquarie Island as a winter resting place because of heavy predation by skuas and cats.

The effects on native fauna of the massive trapping and poisoning campaigns against the rabbit in past years, described by Rolls (1969), must have been particularly damaging to the small herbivorous marsupials. Today, the widespread use of 1080 poison throughout Australia is accompanied by serious research efforts to minimize its effect on native fauna (McIlroy 1986).

5.8 Economic aspects

Prior to myxomatosis there was always a steady domestic trade in rabbits for the table, especially during years of depression. An active export trade in rabbit skins and carcases also existed. The main market for carcases was the United Kingdom, and the United States was the main buyer of skins. Much of the fur was used to make hats, especially during the First World War. By the late 1940s up to 6.3 million kg in skins worth $15 million per year were exported. However, this market collapsed due to changes in fashion, falling to less than 1 million kg per year by 1970.

During the three years 1983–5, exports of skins registered 37 564 kg (1983), 39 132 kg (1984) and 63 855 kg (1985), and in 1990 5079 kg, all to Belgium–Luxembourg. In 1990, 142 309 kg of rabbit meat were exported, mainly to the USA (106 429 kg) and Norway (27 420 kg). The skin and carcass trade earned $300–$400 000 per year in 1983–5, and in 1990 this earned $142 309 for skins, and $25 620 for meat.

Counterbalancing this profit are the costs of control and losses in production whcih farmers have had to bear. Most of these have never been addressed, especially the many indirect flow-on effects which would not normally be considered. There have been numerous estimates, however, of the large and direct economic effects of rabbits on agriculture. One of the most pertinent was presented by the Commonwealth Bureau of Agricultural Economics giving figures of the five-year average production before and after the introduction of myxomatosis. In NSW the greasy wool shorn from adult sheep increased by 26 per cent in 1950–5, mainly because of an increase in the number of sheep, despite the fact that over the same period the number of sheep slaughtered went up 25 per cent. Numbers of beef cattle increased by 10 per cent after myxomatosis, at a time when the cattle slaughtered increased by 26 per cent (Waithman 1979). Figures collected on a property basis by Fennessy (1966) in South Australia and NSW demonstrated increases of 26–50 per cent in stock-carrying capacity following removal of rabbits.

Wood *et al.* (1987, personal communication) have estimated from available data that the mean herbage loss due to rabbit grazing in southern Australia is now 87 kg ha^{-1} year $^{-1}$. This is equivalent to a lowering of carrying capacity by 1 sheep per 5 ha, assuming a sheep consumes 450 kg yr^{-1} dry matter. In areas affected by rabbit plagues herbage losses may exceed 800 kg ha^{-1}. Short (1985) recently showed in small enclosure grazing trials in an arid environment in western NSW that the maximum food intake of rabbits of 58 g $kg^{-0.75}$ day^{-1} in that environment is marginally higher than that of sheep and kangaroos, and that 16 rabbits eat approximately the same amount as one sheep. This differs greatly from the subjective estimate of 10–12 rabbits per sheep often used for rabbits grazing sown pastures in agricultural areas.

A comprehensive and modern overview of the economic and scientific aspects of rabbit control in Australia and New Zealand is presented in Williams *et al.* (in press).

5.9 Management

Government and landholders became aware of the threat posed by rabbits to agriculture as early as 1880, and hastily constructed thousands of miles of rabbit-proof fences to try and halt their spread. By 1890, the rabbit population in the southern half of south-eastern Australia was out of control. Journals and papers of the time are full of accounts of the desperate measures landholders took to try to safeguard crops and stock. The countryside was subjected to mass campaigns of poisoning with strychnine and phosphorus in pollard baits. Watering points were poisoned to kill rabbits when they came to drink. Large concerted rabbit drives were held, the rabbits being directed by wing-nets into yards where thousands could be slaughtered with sticks. Most control techniques were inefficient, and their effects were often negated by the immigration of rabbits from areas that had not been treated.

Some farmers effectively controlled rabbits by netting their boundaries, removing all harbour, digging out warrens with pick and shovel, regular inspection of fences, and hunting down any rabbits as soon as they appeared. This strategy was time-consuming and expensive, but was made possible by the high prices for primary products and the availability of cheap labour at the time.

After the advent of myxomatosis, effective rabbit control was possible on most farms in the higher rainfall areas because of the reduced number of rabbits and their lower potential for increase, the decay and disappearance of large numbers of warrens in difficult habitats, the availability of tractors for ripping warrens in farmlands, and the use of new poisons. Recommendations on the use of control techniques based on scientific research also made control more effective.

Compound 1080 (sodium monofluoroacetate) is the main poison used to control rabbits in Australia today, but since 1080 is very toxic to the rabbit's main predators, the cat and fox, rabbit populations usually recover quickly after poisoning and controls have to be constantly repeated. Evidence is gathering that the rabbit is becoming behaviourally resistant to 1080 (Oliver *et al.* 1982). To overcome this problem the anticoagulant poisons are being investigated in Western Australia (Oliver *et al.* 1982). Fumigation of warrens and ripping of warrens by tractor continue to be used as the principal supplements to poisoning, usually in an integrated campaign (Cooke 1981; Burley 1986).

Very little rabbit control is practised in low rainfall areas because of the high costs involved. In an arid area south-west of Alice Springs, NT, Foran *et al.* (1985) showed that rabbit grazing qualitatively changed the species composition of grasses and trees, but suggested that the changes did not justify the expense of rabbit control since seasonal effects were much more important. On smaller holdings in southern Australia, however, Wood (1985) and Cooke and Hunt (1987) present data to support economic advantages of rabbit control.

Modern attempts to control rabbits in arid environments are focusing on the strategic destruction of 'refuge' warrens (warrens which support rabbit populations during drought years) (Myers and Parker 1975*b*; Parker *et al.* 1976; Wood 1985), the search for arid adapted ectoparasites which might act as vectors for myxomatosis (Cooke 1990*b*), and the better understanding of the important role of predation in control strategies (Newsome 1990) so that it can be included in integrated campaigns.

One of the more important advances in rabbit management in Australia in the past decade has been a determined effort by government authorities and research workers to initiate socio-economic studies in rabbit control. Control of rabbits in agricultural areas lies essentially in the hands of farmers and their local governing bodies. Australian farmers are strong individualists, with attitudes shaped by hardship and individual effort in an unpredictable environment, and decisions on rabbit control are often given less weight, for good reasons, than decisions on ploughing, seeding, harvesting, and fencing. The rabbit problem has its basis in sociological and economic as well as ecological processes, which the new studies are addressing (Nolan 1981; Low and Strong 1984*c*; Parer and Pech 1988; Williams *et al.* in press).

An entirely novel concept in mammalian pest control has been launched by CSIRO scientists, aimed at provoking an auto-immune reaction in the breeding rabbit to specific proteins essential for successful gametogenesis, fertilization, or implantation, using a recombinant myxoma virus as the vector. Molecular and immunological studies to identify sperm acrosomal membrane-specific antigens, oocyte-specific antigens, proteins associated with implantation, and genes expressed during the onset of puberty and seasonal gametogenesis are being made with the objective of suppressing their expression. The use of a recombinant myxoma virus as a gene vector is a brilliant idea, because of the species-specific nature of the virus, the availability of insect viral vectors in natural populations, and because recombinant myxoma viruses have already been successfully prepared. It seems likely that the first objectives are achievable. Beyond that, however, it is not possible to predict, since final success depends upon ecological processes not under research control.

Investigations are also under way in the CSIRO Australian Animal Health Laboratory, Geelong, Victoria, into the possible use of rabbit haemorrhagic disease as a biological control agent in Australian wild rabbits. The disease, first described in China in 1984, has steadily spread throughout the rabbit farming industry and in wild rabbit populations in continental Europe, causing large mortalities (see Chapter 3).

5.10 Discussion

Rabbit populations in Australia live in a broad range of habitats, each with its own distinct environment and its own characteristic dynamics.

Climate is the principal factor affecting rabbit population biology in Australia. It has a strong influence on the physiology of rabbits, especially as it relates to metabolism and reproduction, on the nature of the plant food upon which rabbit reproduction, growth and survival depend, and on the predators, parasites, and disease organisms which utilize rabbits as food. It sets the primary limits of a rabbit population's capacity for increase. Rabbit populations within the same climatic region living in similar environments have similar dynamics (Parer and Libke 1991).

In the light of Australian work it makes little sense to promote general theories of population regulation based on single factors. No such theory can be applied to the rabbit throughout its range in Australia, where the same factors exert different effects in different environments. Despite this it is obvious that, as a small grazing mammal, the rabbit practises similar ecological strategies wherever it lives, with nutrition having a prime place in reproduction and development, and predation (including myxomatosis) in mortality.

The central importance of food as a factor influencing rabbit populations is shown most clearly in arid areas where irregular rainfall dramatically changes the timing of plant growth and the abundance of pasture from one year to the next (Myers and Bults 1977; Wood 1980; Cooke 1982*a*; Low *et al.* 1983). In years of high rainfall plants may continue to grow for several months, permitting the production of several litters of young. In dry years rabbits may not reproduce at all.

The demonstration that some populations of rabbits in Australia are relatively stable (Wood 1980; Gilbert *et al.* 1987) suggests the possible operation of density-dependent regulation. In confined populations in Australia and New Zealand there are large effects of density on physiology and behaviour, on food resources, and on rates of predation (Myers and Poole 1963*b*; Myers *et al.* 1971; Gibb *et al.* 1978). Field populations, however, do not usually live in closed systems, and it has been difficult to obtain evidence of the presence of true density-dependence in free-living rabbit populations despite the frequently observed relationships between rabbit numbers and predators and rabbit numbers and their food supply.

Social behaviour, operating within the confined system of a relatively fixed number of warrens is one density-related factor common to all natural populations, but whether behaviour can be said to be regulating numbers or not depends upon the fates of the young ejected. If all or most of the rabbits ejected are killed by predators, or some other mortality factor, then behaviour can be indicted as a major regulatory factor. If sufficient young rabbits live, however, to permit total population numbers to increase, behaviour cannot be said to be regulatory even though it continues to govern numbers in warren-dwelling populations at a density favourable for the production of young (the deme level, Richardson 1981). The history of the rabbit in Australia is mute evidence of the failure of social restraints on increase in numbers, left to themselves. In the lower density and spatially restricted populations of today, however, the activities of predators, including myxomatosis, are such as to suggest that density dependent regulation of numbers is occurring in refuge areas such as those described by Dunsmore (1974), Parer (1977) and Wood (1980).

Work in Australia has progressed to the stage where sensible models have been constructed, but much remains to be done in their application to predicting and understanding long-term population changes, especially in relation to ecological processes, for example patterns of rainfall and numbers (Cooke 1981, 1982*a*); and the impact of predation (Newsome *et al.* 1989).

As has been said elsewhere (Myers 1971; Gilbert *et al.* 1987) there is no such thing as 'the' population dynamics of the rabbit in Australia. Reproduction rates vary widely from habitat to habitat, being ultimately determined by the affects of climate on local environments and populations. There is also a varied and patchy distribution across the continent of predators, parasites, and diseases, often operating in unknown ways, to exert large influences on rabbit populations. Ecological processes of this kind need to be further studied before a full understanding of the rabbit in Australia is obtained and for models of the kind described to become useful tools for management.

The rabbit has succeeded as a colonizer of Australia for a number of reasons:

1. It appears to have been physiologically pre-adapted to the climate of much of the southern half of the continent.
2. The rabbit carried a depauperate parasite fauna when it arrived, when compared with rabbits in Spain, where it originated.
3. It is essentially a colonizer of disturbed regions. The alteration to the Australian landscape by humans and domestic stock changed a predominantly perennial vegetation into one containing a large proportion of annual grasses and forbs, presenting a small grazing herbivore such as the rabbit with a source of high-quality food. The clearing of large areas of woodland resulted in abundant surface harbour.
4. As a result of human activities the numbers of native predators were depleted and the ecology of

small endemic herbivores was disrupted, leaving the rabbit with fewer predators and competitors, except stock, during its primary irruption. The regions where the rabbit has thrived are those areas (Australia, New Zealand, small islands) which had a depauperate mammalian fauna when the rabbit was introduced, or (Chile) where the native predators rejected the introduced rabbit as a source of food (Jaksic and Soriguer 1981). In Australia, predators added later to the system (the fox and the cat), possessed far lower rates of reproduction than the rabbit, and came too late to be able to influence events. It was not until 1950, with the addition of myxomatosis to the system and the large reduction in rabbit numbers it caused, that predation could exert a controlling effect.

5. Several of the native grazing mammals left behind them extensive burrow systems which were taken over by rabbits, thereby permitting rapid colonization, especially in arid environments.
6. The rabbit is an excellent ecological generalist, and has not had to call on special genetic qualities for its survival.

These conclusions are not meant to intimate that the rabbit never penetrated new environments without human assistance. It did, and on a large scale, but wherever this has been done it has usually been from a position of ecological strength assisted by human activities.

We find it most useful to think of the rabbit in Australia in terms of Caughley's (1977) strategic model of plant–herbivore–predator relationships. European settlers set up the first trophic level by altering the environment and excluding predators, then added new herbivores, and finally new predators (including myxomatosis) to form second and third trophic levels. The evolutionary outcome of these events remains to be measured.

5.11 Summary

Rabbit populations in Australia live in a broad range of habitats from the arid to the subalpine, each with its own distinct environment and its own characteristic dynamics.

Evidence is presented to show that climate is the principal factor affecting rabbit population biology in Australia. It has a strong influence on the physiology of rabbits, especially as it relates to metabolism and reproduction, on the nature of the plant food upon which rabbit reproduction, growth and survival depend, and on the predators, parasites, and disease organisms which utilize rabbits as food. It sets the primary limits of a rabbit population's capacity for increase. Rabbit populations within the same climatic region living in similar environments have similar dynamics. In the light of Australian work it makes little sense to promote general theories of population regulation in rabbits based on single factors. As a small grazing mammal, the rabbit nevertheless practises similar ecological strategies wherever it lives, with nutrition having a prime place in reproduction and development, and predation (including myxomatosis) in mortality.

The demonstration that some populations of rabbits in Australia are relatively stable suggests the possible operation of density-dependent regulation, especially in areas where there are limited numbers of warrens, and where all or most of the young ejected are killed by predators, or some other mortality factors.

Work in Australia has progressed to the stage where sensible demographic models are being constructed, but much remains to be done in their application to predicting and understanding long-term rabbit population changes, especially in relation to ecological processes, for example patterns of rainfall and numbers, and the impact of predation.

The rabbit has succeeded as a colonizer of Australia for several reasons. It appears to have been physiologically pre-adapted to the climate of much of the southern half of the continent. It carried a depauperate parasite fauna when it arrived. It is essentially a colonizer of disturbed regions. As a result of human activities the numbers of native predators were depleted and the ecology of small endemic herbivores was disrupted, leaving the rabbit with fewer predators and competitors, except stock, during its primary irruption. Predators added later to the system (the fox and the cat), possessed far lower rates of reproduction than the rabbit, and came too late to be able to influence events. It was not until 1950, with the addition of myxomatosis and the large reduction in rabbit numbers it caused, that predation could exert a controlling effect. Several of the native grazing mammals had extensive burrow systems which were taken over by rabbits, thereby permitting rapid colonization, especially in arid environments. The rabbit is an excellent ecological generalist, and has not had to call on special genetic qualities for its survival.

Acknowledgements

We thank our many colleagues in Australian State and Federal agencies who took part in this work, for their cooperation, and the use of their data. We also thank the late H. G. Andrewartha and L. C. Birch for showing us a useful way to approach the study of large ecological problems, and Neil Gilbert for wise council and continued support.

References

Adamson, D. A. and Fox, M. D. (1982). Change in Australasian vegetation since European settlement. In *A history of Australasian vegetation* (ed. J. M. B. Smith), pp. 109–60. McGraw-Hill, Sydney.

Anon. (1901). *Royal Commission to inquire into the condition of Crown Tenants, Western Division*. NSW Government Printer, Sydney.

Anon. (1942). A mechanical device for the spread of disease agents amongst rabbits. *Journal of the Commonwealth Scientific and Industrial Research Organisation, Australia*. **15**, 82–3.

Armstrong, P. (1982). Rabbits (*Oryctolagus cuniculus*) on islands: a case study of successful colonisation. *Journal of Biogeography*, **9**, 353–62.

Austin, C. R. and Short, R. V. (ed.) (1972). *Reproduction in mammals. 3. Hormones and reproduction.* Cambridge University Press, London.

Baker-Gabb, D. J. (1984). The breeding ecology of twelve species of diurnal raptor in north-western Victoria. *Australian Wildlife Research*, **11**, 145–60.

Barber, H. N. (1954). Genetic polymorphism in the rabbit in Tasmania. *Nature*, **173**, 1227–9.

Bayly, P. C. (1978). A comparison of the diets of the red fox and feral cat in an arid environment. *South Australian Naturalist*, **53**, 20–8.

Blair-West, J. R., Coghlan, J. P., Denton, D. A., Nelson, J. F., Orchard, E., Scoggins, B. A., Wright, R. D., Myers, K., and Junqueira, C. L. (1968). Physiological, morphological and behavioural adaptation to a sodium deficient environment by wild native Australian and introduced species of animals. *Nature*, **217**, 922–8.

Brooker, M. G. (1977). Some notes on the mammalian fauna of the western Nullabor Plain, Western Australia. *West Australian Naturalist*, **14**, 2–15.

Brooker, M. G. and Ridpath, M. G. (1980). The diet of the wedge-tail eagle *Aquila audax* in Western Australia. *Australian Wildlife Research*, 7, 433–52.

Bull, P. C. (1953). Parasites of the wild rabbit, *Oryctolagus cuniculus* (L.) in New Zealand. *New Zealand Journal of Science and Technology B*, **34**, 341–72.

Bull, L. B. and Mules, M. W. (1944). An investigation of *Myxomatosis cuniculi*, with special reference to the possible use of the disease to control rabbit populations in Australia. *Journal of the Scientific and Industrial Research Organisation, Australia*, **17**, 79–93.

Burley, J. R. W. (1986). Advances in the integrated control of the European rabbit in South Australia. In *Proceedings of the 12th Vertebrate Pest Control Conference*, pp. 140–6, University of California, Davis, California.

Calaby, J. H. (1951). Notes on the little eagle; with particular reference to rabbit predation. *Emu*, **51**, 33–57.

Calaby, J. H., Gooding, C. D., and Tomlinson, A. R. (1960). Myxomatosis in Western Australia. *CSIRO Wildlife Research*, **5**, 89–101.

Casperson, K. (1968). Influence of environment upon

some physiological parameters of the rabbit *Oryctolagus cuniculus* (L.) in natural populations. *Proceedings of the Ecological Society of Australia*, **3**, 113–19.

Catling, P. C. (1988). Similarities and contrasts in the diets of foxes, *Vulpes vulpes*, and cats, *Felis catus*, relative to fluctuating prey population and drought. *Australian Wildlife Research*, **15**, 307–17.

Catling, P. C. and Newsome, A. E. (1992). A new technique to determine the seasonal changes in the quality of food selected by free ranging rabbits. In *Proceedings of the Sixth International Conference on Mediterranean-type Ecosystems, Meleme, Crete, Sept 23–27, 1991* (ed. Costas A. Thanos), pp. 177–82. University of Athens, Greece.

Caughley, G. C. (1977). *Analysis of vertebrate populations*. Wiley, London.

Coman, B. J. (1973). The diet of red foxes, *Vulpes vulpes* (L.) in Victoria. *Australian Journal of Zoology*, **21**, 391–401.

Coman, B. J. and Brunner, H. (1972). Food habits of the feral house cat in Victoria. *Journal of Wildlife Management*, **36**, 848–53.

Conolly, D. and Sobey, W. R. (1985). Myxomatosis: use of an aerosol generator to introduce a virulent strain of myxoma virus for the control of populations of wild rabbits, *Oryctolagus cuniculus*. *Australian Wildlife Research*, **12**, 249–55.

Cooke, B. D. (1974). *Food and other resources of the wild rabbit,* Oryctolagus cuniculus (L.). Unpublished Ph.D. thesis. University of Adelaide.

Cooke, B. D. (1977). Factors limiting the distribution of the wild rabbit in Australia. *Proceedings of the Ecological Society of Australia*, **10**, 113–20.

Cooke, B. D. (1981). Food and dynamics of rabbit populations in inland Australia. In *Proceedings of the World Lagomorph Conference* (1979) (ed. K. Myers and C. D. MacInnes), pp. 633–47. University of Guelph, Ontario.

Cooke, B. D. (1982*a*). A shortage of water in natural pastures as a factor limiting a population of rabbits. *Oryctolagus cuniculus* (L.), in arid north-eastern South Australia. *Australian Wildlife Research*, **9**, 465–76.

Cooke, B. D. (1982*b*). Reduction of food intake and other physiological responses to a restriction of drinking water in captive wild rabbits *Oryctolagus cuniculus* (L.). *Australian Wildlife Research*, **9**, 247–52.

Cooke, B. D. (1983*a*). Population dynamics in the arid lands. In *Rabbit biology and control in semi-arid lands*. (ed. W. A. Low), pp. 5–7. Conservation Commission of the Northern Territory, Alice Springs.

Cooke, B. D., (1983*b*). Changes in age-structure and size of populations of wild rabbits in South Australia, following the introduction of European rabbit fleas, *Spilopsyllus cuniculi* (Dale), as vectors of myxomatosis. *Australian Wildlife Research*, **10**, 105–20.

Cooke, B. D. (1984). Factors limiting the distribution of the European rabbit flea, *Spilopsyllus cuniculi* (Dale) (Siphonaptera), in inland South Australia. *Australian Journal of Zoology*, **32**, 493–506.

Cooke, B. D. (1990*a*). Rabbit burrows as environments for European rabbit fleas, *Spilopsyllus cuniculi* (Dale), in arid South Australia. *Australian Journal of Zoology*, **38**, 317–25.

Cooke, B. D. (1990*b*). Notes on the comparative reproductive biology and the laboratory breeding of the rabbit flea, *Spilopsyllus cunicularis* Smit (Siphonaptera: Pulicidae). *Australian Journal of Zoology*, **38**, 527–34.

Cooke, B. D. and Hunt, L. P. (1987). Practical and economic aspects of rabbit control in hilly semi-arid South Australia. *Australian Wildlife Research*, **14**, 219–23.

Copson, G. R., Brothers, N. P., and Skira, I. J. (1981). Distribution and abundance of the rabbit *Oryctolagus cuniculus* (L.) at subantarctic Macquarie Island. *Australian Wildlife Research*, **8**, 597–612.

Corbett, L. K. and Newsome, A. E. (1987). The feeding ecology of the dingo. III. Dietary relationships with widely fluctuating prey populations in arid Australia: an hypothesis of alternation of predation. *Oecologia*, **74**, 215–27.

Coudert, P. (1979). Comparison of the pathology of several rabbit coccidia species and their control with robenidine. *International Symposium on Coccidia November 1979*, pp. 159–63, Czechoslovak Science and Technology, Prague.

Croft, J. D. and Hone, L. J. (1978). The stomach contents of foxes, *Vulpes vulpes*, collected in New South Wales. *Australian Wildlife Research*, **5**, 5–92.

Cupper, J. and Cupper, L. (1981). *Hawks in focus*. Jaclin Enterprises, Mildura.

Daly, J. C. (1980). Age, sex, and season: factors which determine trap response of the European wild rabbit, *Oryctolagus cuniculus. Australian Wildlife Research*, **7**, 421–32.

Daly, J. C., (1981). Social organisation and genetic structure in a rabbit population. In *Proceedings of the World Lagomorph Conference* (1979) (ed. K. Myers and C. D. MacInnes), pp. 90–7. University of Guelph, Ontario.

Davey, C. (1990). *A report on the numbers and distribution of Gould's Petrel Pterodroma leucoptera breeding on the John Gould Nature Reserve, N.S.W.* Unpublished report, CSIRO Division of Wildlife and Ecology, Canberra.

Dawson, Mary D. (1981). Evolution of modern Lagomorphs. In *Proceedings of the World Lagomorph Conference* (1979) (ed. K. Myers and C. D. MacInnes), pp. 1–8. University of Guelph, Ontario.

Dawson, J. T. and Ellis, B. (1979). Comparison of the diet of yellow-footed rock wallabies and sympatric herbivores in western New South Wales. *Australian Wildlife Research*, **6**, 245–54.

Delibes, M. J. (1975). Some characteristic features of predation in the Iberian Mediterranean ecosystem. In *Proceedings 12th International Congress of Game Biologists*, pp. 31–6, Lisbon, Portugal.

Delibes, M. and Hiraldo, F. (1981). The rabbit as prey in Iberian Mediterranean ecosystems. In *Proceedings of the World Lagomorph Conference* (1979) (ed. K. Myers and C. D. MacInnes), pp. 614–22. University of Guelph, Ontario.

Douglas, G. W. (1969). Movements and longevity in the rabbit. *Vermin and Noxious Weeds Destruction Board*, Bulletin No. 12, Melbourne.

Dunnet, G. M. (1957). Notes on avian predation on young rabbits, *Oryctolagus cuniculus* (L.). *CSIRO Wildlife Research*, **2**, 66–8.

Dunsmore, J. D. (1966*a*). Nematode parasites of free-living rabbits, *Oryctolagus cuniculus* (L.), in eastern Australia. I. Variations in the number of *Trichostrongylus retortaeformis* (Zeder). *Australian Journal of Zoology*, **14**, 185–99.

Dunsmore, J. D. (1966*b*). Nematode parasites of free-living rabbits, *Oryctolagus cuniculus* (L.), in eastern Australia. II. Variations in the numbers of *Graphidium strigosum* (Dujardin) Railliet and Henry. *Australian Journal of Zoology*, **4**, 625–34.

Dunsmore, J. D. (1971). A study of the biology of the wild rabbit in climatically different regions in Eastern Australia. IV. The rabbit in the south coastal region of New South Wales, an area in which parasites appear to exert a population-regulating effect. *Australian Journal of Zoology*, **19**, 355–70.

Dunsmore, J. D. (1974). The rabbit in subalpine southeastern Australia. I. Population structure and productivity. *Australian Wildlife Research*, **1**, 1–16.

Dunsmore, J. D. (1981). The role of parasites in population regulation of the European rabbit (*Oryctolagus cuniculus*) in Australia. In *Worldwide Furbearer Conference proceedings* (ed. J. A. Chapman and D. Pursley), pp. 654–9. University of Maryland.

Dunsmore, J. D. and Dudzinski, M. L. (1968). Relationship of numbers of nematode parasites in wild rabbits, *Oryctolagus cuniculus* (L.), to host age, sex, and season. *Journal of Parasitology*, **54**, 462–74.

Dunsmore, J. D., Williams, R. T., and Price, W. J. (1971). A winter epizootic of myxomatosis in subalpine south-eastern Australia. *Australian Journal of Zoology*, **19**, 275–86.

Dyce, A. L. and Lee, D. J. (1962). Blood-sucking flies (Diptera) and myxomatosis transmission in a mountain environment in New South Wales. II. Comparison of the use of man and rabbit as bait animals in evaluating vectors of myxomatosis. *Australian Journal of Zoology*, **10**, 84–94.

Fenner, F. and Ratcliffe, F. N. (1965). *Myxomatosis*. Cambridge University Press.

Fennessy, B. V. (1966). The impact of wildlife species on sheep production in Australia. *Proceedings of the Australian Society of Animal Production*, **6**, 148–56.

Fitzpatrick, E. A. and Nix, H. A. (1970). The climatic factor in Australian grassland ecology. In *Australian grasslands* (ed. R. M. Moore), pp. 3–26. Australian National University Press, Canberra.

Foran, B. D. (1986). The impact of rabbits and cattle on an arid calcareous shrubby grassland in central Australia. *Vegetatio*, **66**, 49–59.

Foran, B. D., Low, W. A., and Strong, B. W. (1985). The response of rabbit populations and vegetation to rabbit control on a calcareous shrubby grassland in central Australia. *Australian Wildlife Research*, **12**, 237–47.

Friedel, M. H. (1985). The population structure and density of central Australian trees and shrubs, and relationship to range condition, rabbit abundance and soil. *Australian Rangelands Journal*, **7**, 130–9.

Fullagar, P. J. (1978). *Report on the rabbits on Philip*

Island, Norfolk Island. Unpublished report, CSIRO Division of Wildlife Research, Canberra.

Fullagar, P. J. (1981). Methods for studying the behaviour of rabbits in a 33-ha enclosure at Canberra and under national conditions at Calindary, N.S.W. In *Proceedings of the World Lagomorph Conference* (1979) (ed. K. Myers and C. D. MacInnes), pp. 240–55. University of Guelph, Ontario.

Gartner, R. J. W., McLean, R. W., Little, D. A., and Winks, L. (1980). Mineral deficiencies limiting production of ruminants grazing tropical pastures in Australia. *Tropical Grasslands*, **14**, 266–72.

Gibb, J. A., Ward, C. P., and Ward, G. D. (1978). Natural control of a population of rabbits, *Oryctolagus cuniculus* (L.), for ten years in the Kouraru enclosure. *Bulletin 223*, Department of Scientific and Industrial Research, New Zealand.

Gibb, J. A., White, A. J., and Ward, C. P. (1985). Population ecology of rabbits in the Wairarapa, New Zealand. *New Zealand Journal of Ecology*, **8**, 55–82.

Gilbert, N. and Myers, K. (1981). Comparative dynamics of the Australian rabbit. In *Proceedings of the World Lagomorph Conference* (1979) (ed. K. Myers and C. D. MacInnes), pp. 648–53. University of Guelph, Ontario.

Gilbert, N., Myers, K., Cooke, B. D., Dunsmore, J. D., Fullagar, P. J., Gibb, J. A., King, D. R., Parer, I., Wheeler, S. H., and Wood, D. H. (1987). Comparative dynamics of Australasian rabbit populations. *Australian Journal of Wildlife Research*, **14**, 491–503.

Gooding, C. D. (1955). Rabbit damage to pastures. *Journal of Agriculture, Western Australia*, **4**, 753–5.

Goodrich, B. S. and Mykytowycz, R. (1972). Individual and sex differences in the chemical composition of pheromone-like substances from the skin glands of the rabbit. *Journal of Mammalogy*, **53**, 540–8.

Gordon, M. S. (1968). Water and solute metabolism. In *Animal function: principles and adaptations* (ed. M. S. Gordon), pp. 230–89. Macmillan, New York.

Green, R. H. (1965). Impact of rabbit on predators and prey. *Tasmanian Naturalist*, **3**, 1–2.

Griffiths, M. E., Calaby, J. H., and McIntosh, D. L. (1960). The stress syndrome in the rabbit. *CSIRO Wildlife Research*, **5**, 134–48.

Hall, L. S. and Myers, K. (1978). Variations in the microclimate in rabbit warrens in semi-arid New South Wales. *Australian Journal of Ecology*, **3**, 187–94.

Hayward, J. S. (1961). The ability of the wild rabbit to survive conditions of water restriction. *CSIRO Wildlife Research*, **6**, 160–75.

Hayward, J. S. (1966). Abnormal concentrations of respiratory gases in rabbit burrows. *Journal of Mammalogy*, **47**, 723–4.

Hounam, C. E. (1961). Evaporation in Australia. A critical survey of the network and methods of observation together with a tabulation of the results of observations. *Commonwealth Bureau of Meteorology Bulletin*, **44**, 18–22.

Hughes, R. L. (1965). Reproductive potential of the wild rabbit in the Australian arid zone. In *Australian Arid Zone Conference, Alice Springs, Northern Territory*, pp. 1312–14. CSIRO, Melbourne.

Jaksic, F. and Soriguer, R. (1981). Predation upon the European rabbit (*Oryctolagus cuniculus*) in the Mediterranean habitats of Chile and Spain: a comparative analysis. *Journal of Animal Ecology*, **50**, 269–81.

Jarman, P. J. and Johnson, K. A. (1977). Exotic mammals, indigenous mammals and land-use. *Proceedings of the Ecological Society of Australia*, **10**, 6–66.

Jessup, R. W. (1951). The soils, geology and vegetation of north-western South Australia. *Transactions of the Royal Society of South Australia*, **74**, 189–273.

Johnson, E. R. L. and Baird, A. M. (1970). Notes on the flora and vegetation of the Nullarbor Plain at Forrest, Western Australia. *Journal of the Royal Society of Western Australia*, **53**, 46–61.

Jones, E. and Coman, B. J. (1981). Ecology of the feral cat, *Felis catus* (L.), in south-eastern Australia. 1. Diet. *Australian Wildlife Research*, **8**, 537–47.

King, D. R., Wheeler, S. A., and Schmidt, G. L. (1983). Population fluctuations and reproduction of rabbits in a pastoral area on the coast north of Carnarvon, W.A. *Australian Wildlife Research*, **10**, 97–104.

King, D. R., Wheeler, S. H., and Robinson, M. H. (1984). Daytime locations of European rabbits at three localities in south-western Australia. *Australian Wildlife Research*, **11**, 89–92.

King, D. R., Oliver, A. J. and Wheeler, S. H. (1985). The European rabbit flea, *Spilopsyllus cuniculi*, in south-western Australia. *Australian Wildlife Research*, **12**, 227–36.

Kluger,M. J. (1975). Energy balance in the resting and exercising rabbit. In *Ecological studies analysis and synthesis. Vol. 12. Perspectives of biophysiological ecology* (ed. D. M. Cates and R. S. Schmerl), pp. 497–507. Springer-Verlag, New York.

Lange, R. T. and Graham, C. R. (1983). Rabbits and the failure of regeneration in Australian arid zone *Acacia. Australian Journal of Ecology*, **8**, 377–81.

Laughlin, R. (1965). Capacity for increase: a useful population statistic. *Journal of Animal Ecology*, **34**, 77–91.

Lee, D. J., Dyce, A. L., and O'Gower, A. K. (1957). Blood-sucking flies and myxomatosis in a mountain environment in New South Wales. *Australian Journal of Zoology*, **5**, 355–401.

Leigh, J. H., Wimbush, D. J., Wood, D. H., Holgate, M. D., Slee, A. V., Stanger, M. G., and Forrester, R. I. (1987). Rabbit grazing and fire in a subalpine environment. I. Herbaceous and shrubby vegetation. *Australian Journal of Botany*, **35**, 433–64.

Leigh, J. H., Wood, D. H., Holgate, M. D., Slee, A. V., and Stanger, M. G. (1989). Effects of rabbit and kangaroo grazing on two semi-arid grassland communities in central western New South Wales. *Australian Journal of Botany*, **37**, 375–96.

Lloyd, H. G. (1977). Rabbit *Oryctolagus cuniculus*. In *The handbook of British mammals* (ed. G. B. Corbet and H. N. Southern), pp. 130–9. Blackwell Scientific Publications, Oxford.

Lloyd, H. G. (1981). Biological observations on post-myxomatosis wild rabbit populations in Britain 1955–79. In *Proceedings of the World Lagomorph Conference* (1979) (ed. K. Myers and C. D. MacInnes), pp. 623–8. University of Guelph, Ontario.

Long, J. L. (1972). Introduced birds and mammals in Western Australia. *Agricultural Protection Board of Western Australia Technical Series*, No. 1.

Low, W. A. and Strong, B. W. (1983). Distribution and density of the European rabbit (*Oryctolagus cuniculus*) in the Northern Territory. *Unpublished Report of the Conservation Commission of the Northern Territory*, Alice Springs.

Low, W. A. and Strong, B. W. (1984*a*). The European rabbit flea and other vectors of myxomatosis in the Northern Territory. *Unpublished Report of the Conservation Commission of the Northern Territory*, Alice Springs.

Low, W. A. and Strong, B. W. (1984*b*). Methods and cost of rabbit control in the Northern Territory. *Unpublished Report of the Conservation Commission of the Northern Territory*, Alice Springs.

Low, W. A., Cooke, B. D., and Strong, B. W. (1983). Reproduction and population dynamics of the rabbit in the Northern Territory at the northern edge of its distribution. *Unpublished Report of the Conservation Commission of the Northern Territory*, Alice Springs.

MacMillan, R. E. and Lee, A. K. (1969). Water metabolism of Australian hopping mice. *Comparative Biochemical Physiology*, **28**, 493–514.

Marshall, A. J. (1966). *The Great Extermination*. Heinemann, London.

Martin, J. T. and Zickefoose, J. (1976). The effectiveness of aerial survey for determining the distribution of rabbit warrens in a semiarid environment. *Australian Wildlife Research*, **3**, 79–84.

McCluskey, J., Olivier, T. J., Freedman, L., and Hunt, E. (1974). Evolutionary divergences between populations of Australian wild rabbits. *Nature*, **249**, 278–9.

McIlroy, J. C. (1986). The sensitivity of Australian mammals to 1080 poison. XI. Comparisons between the major groups of animals and the potential danger non-target species face from 1080-poisoning campaigns. *Australian Wildlife Research*, **13**, 39–48.

McIntosh, D. L. (1963). Food of the fox in the Canberra district. *CSIRO Wildlife Research*, **8**, 1–20.

Myers, K. (1955). The ecology of the mosquito vectors of myxomatosis (*Culex annulirostris* Skuse and *Anopheles annulipes* Walk.) in the Eastern Riverina. *Journal of the Australian Institute of Agricultural Science*, **21**, 250–3.

Myers, K. (1956). Methods of sampling winged insects feeding on the rabbit, *Oryctolagus cuniculus. CSIRO Wildlife Research*, **1**, 45–8.

Myers, K. (1958). Further observations on the use of field enclosures for the study of the wild rabbit, *Oryctolagus cuniculus* (L.). *CSIRO Wildlife Research*, **3**, 40–9.

Myers, K. (1962). A survey of myxomatosis and rabbit infestation trends in the eastern Riverina, New South Wales, 1951–60. *CSIRO Wildlife Research*, **7**, 1–12.

Myers, K. (1967). Morphological changes in the adrenal glands of wild rabbits. *Nature*, **213**, 147–50.

Myers, K. (1971). The rabbit in Australia. In *Proceed-*

ings of the Advanced Study Institute on 'Dynamics of Numbers in Populations', Oosterbeek, 1970. (ed. P. J. den Boer and G. R. Gradwell), pp. 478–506. Centre for Agricultural Publishing and Documentation, Wageningen.

Myers, K. (1986). Introduced vertebrates in Australia, with emphasis on the mammals. In *Ecology of biological invasions: an Australian perspective* (ed. R. H. Groves and J. J. Burdon), pp. 120–36. Australian Academy of Science, Canberra.

Myers, K. and Bults, H. G. (1977). Observations on changes in the quality of food eaten by the wild rabbit. *Australian Journal of Ecology*, **2**, 215–30.

Myers, K. and Gilbert, N. (1968). Determination of age in wild rabbits in Australia. *Journal of Wildlife Management*, **32**, 841–9.

Myers, K. and Parker, B. S. (1965). A study of the biology of the wild rabbit in climatically different regions in eastern Australia. I. Patterns of distribution. *CSIRO Wildlife Research*, **10**, 1–32.

Myers, K. and Parker, B. S. (1975*a*). A study of the biology of the wild rabbit in climatically different regions in eastern Australia. VI. Changes in numbers and distribution related to climate and land systems in semiarid north-western New South Wales. *Australian Wildlife Research*, **2**,11–32.

Myers, K. and Parker, B. S. (1975*b*). Effect of severe drought on rabbit numbers and distribution in a refuge area in semiarid north-western New South Wales. *Australian Wildlife Research*, **2**,103–20.

Myers, K. and Poole, W. E. (1959). A study of the biology of the wild rabbit, *Oryctolagus cuniculus* (L.), in confined populations. I. The effects of density on home range and the formation of breeding groups. *CSIRO Wildlife Research*, **4**, 14–26.

Myers, K. and Poole, W. E. (1961). A study of the biology of the wild rabbit, *Oryctolagus cuniculus* (L.), in confined populations. II. The effects of season and population increase on behaviour. *CSIRO Wildlife Research*, **6**, 1–41.

Myers, K. and Poole, W. E. (1962). A study of the biology of the wild rabbit, *Oryctolagus cuniculus* (L.), in confined populations. III. Reproduction. *Australian Journal of Zoology*, **10**, 225–67.

Myers, K. and Poole, W. E. (1963*a*). A study of the biology of the wild rabbit, *Oryctolagus cuniculus* (L.), in confined populations. IV. The effects of rabbit grazing on sown pastures. *Journal of Ecology*, **51**, 435–51.

Myers, K. and Poole, W. E. (1963*b*). A study of the biology of the wild rabbit, *Oryctolagus cuniculus* (L.), in confined populations. V. Population dynamics. *CSIRO Wildlife Research*, **8**, 166–203.

Myers, K., Hale, C. S., Mykytowycz, R., and Hughes, R. L. (1971). The effects of varying density and space on sociality and health in animals. In *Behaviour and the environment* (ed. H. Essor), pp. 148–87. Plenum Press, New York.

Myers, K., Parker, B. S., and Dunsmore, J. D. (1975). A study of the biology of the wild rabbit in climatically different regions in eastern Australia. X. Changes in the numbers of rabbits and their burrows in a subalpine environment in south-eastern New South Wales. *Australian Wildlife Research*, **2**, 121–34.

Myers, K., Bults, H. G., and Gilbert, N. (1981). Stress in the rabbit. In *Proceedings of the World Lagomorph Conference* (ed. K. Myers and C. D. MacInnes), pp. 103–36. University of Guelph, Ontario.

Myers, K., Parer, I., and Richardson, B. J. (1989). Leporidae. In *Fauna of Australia. Mammalia*. (ed. D. W. Walton and B. J. Richardson), pp. 917–31. Australian Government Publishing Service, Canberra.

Mykytowycz, R. (1957*a*). Ectoparasites of the wild rabbit, *Oryctolagus cuniculus* (L.), in Australia. *CSIRO Wildlife Research*, **2**, 63–5.

Mykytowycz, R. (1957*b*). The transmission of myxomatosis by *Simulium melatum* Wharton (Diptera: Simuliidae). *CSIRO Wildlife Research*, **2**, 1–4.

Mykytowycz, R. (1958*a*). Contact transmission of infectious myxomatosis of the rabbit, *Oryctolagus cuniculus* (L.). *CSIRO Wildlife Research*, **3**, 1–6.

Mykytowycz, R. (1958*b*). Social behaviour of an experimental colony of wild rabbits, *Oryctolagus cuniculus* (L.). I. Establishment of the colony. *CSIRO Wildlife Research*, **3**, 7–25.

Mykytowycz, R. (1959). Social behaviour of an experimental colony of wild rabbits, *Oryctolagus cuniculus* (L.). II. First breeding season. *CSIRO Wildlife Research*, **4**, 1–13.

Mykytowycz, R. (1960). Social behaviour of an experimental colony of wild rabbits, *Oryctolagus cuniculus* (L.). III. Second breeding season. *CSIRO Wildlife Research*, **5**, 1–20.

Mykytowycz, R. (1961). Social behaviour of an

experimental colony of wild rabbits, *Oryctolagus cuniculus* (L.). IV. Conclusion: outbreak of myxomatosis, third breeding season, and starvation. *CSIRO Wildlife Research*, **6**, 142–55.

Mykytowycz, R. (1962). Epidemiology of coccidiosis (*Eimeria* spp.) in an experimental population of the Australian wild rabbit, *Oryctolagus cuniculus* (L.). *Parasitology*, **52**, 375–95.

Mykytowycz, R. and Fullagar, P. J. (1973). Effect of social environment on reproduction in the rabbit, *Oryctolagus cuniculus* (L.). *Journal of Reproduction and Fertility Supplement*, **19**, 503–22.

Mykytowycz, R. and Gambale, S. (1965). A study of the inter-warren activities and dispersal of wild rabbits, *Oryctolagus cuniculus* (L.), in a 45-ac. paddock. *CSIRO Wildlife Research*, **10**, 111–23.

Mykytowycz, R. and Gambale, S. (1969). The distribution of dunghills and the behaviour of free-living rabbits, *Oryctolagus cuniculus* (L.), on them. *Forma et Functio*, **1**, 333–49.

Mykytowycz, R., Hesterman, E. R., and Purchase, D. (1959). Predation on the wild rabbit by the Australian raven. *Emu*, **59**, 41–3.

Mykytowycz, R., Hesterman, E. R., and Purchase, D. (1960). Technique employed in catching rabbits *Oryctolagus cuniculus* (L.), in an experimental enclosure. *CSIRO Wildlife Research*, **5**, 85–6.

Mykytowycz, R., Hesterman, E. R., Gambale, S., and Dudzinski, M. L. (1976). A comparison of the odors of rabbits, *Oryctolagus cuniculus*, in enhancing territorial confidence. *Journal of Chemical Ecology*, **2**, 13–24.

Mykytowycz, R., Goodrich, B. S., and Hesterman, E. R. (1984). Methodology employed in the studies of odour signals in wild rabbits, *Oryctolagus cuniculus* (L.). *Acta Zoologica Fennica*, **171**, 71–5.

Nagy, K. A., Shoemaker, V. H., and Costa, W. R. (1976). Water, electrolyte, and nitrogen budgets of jackrabbits (*Lepus californicus*) in the Mojave desert. *Physical Zoology*, **49**, 351–75.

Newsome, A. E. (1989). Large vertebrate pests. In *Mediterranean landscapes in Australia* (ed. J. C. Noble and R. A. Bradstock), pp. 406–15, CSIRO, Melbourne.

Newsome, A. E. (1990). The control of vertebrate pests by vertebrate predators. *Tree*, **5**, 187–91.

Newsome, A. E., Catling, P. C., and Corbett, L. K. (1983). The feeding ecology of the dingo. II. Dietary and numerical relationships with fluctuating prey populations in south-eastern Australia. *Australian Journal of Ecology*, **8**, 345–66.

Newsome, A. E., Parer, I., and Catling, P. C. (1989). Prolonged prey suppression by carnivores—predator removal experiments. *Oecologia*, **78**, 458–67.

Nolan, Ivan. (1981). A survey of the rabbit problem in Victoria based on information obtained from Inspectors of Lands. *Unpublished Report of the Department of Crown Lands and Survey, Division of Inspection and Vermin and Noxious Weeds Destruction*. pp. 1–198. Melbourne.

Norman, F. I. (1988). Long term effects of rabbit reduction on Rabbit Island, Wilson's Promontory, Victoria. *Victorian Naturalist*, **105**, 136–41.

Oliver, A. J. and Blackshaw, D. D. (1979). The dispersal of fumigant gases in warrens of the European rabbit, *Oryctolagus cuniculus* (L.). *Australian Wildlife Research*, **6**, 39–55.

Oliver, A. J., Wheeler, S. H., and Gooding, C. D. (1982). Field evaluation of 1080 and pindone oat bait, and the possible decline in effectiveness of poison baiting for the control of the rabbit, *Oryctolagus cuniculus*. *Australian Wildlife Research*, **9**, 125–34.

Oliver, A. J., Wheeler, S. H., Gooding, C. D., and Bell, J. (1982). Changes in bait acceptance by rabbits in Australia and New Zealand. *Proceedings of the Tenth Vertebrate Pest Control Conference*, pp. 101–3, University of California.

Omole, T. A. (1982). The effect of level of dietary protein on growth and reproductive performance in rabbits. *Journal of Applied Rabbit Research*, **5**, 83–8.

Parer, I. (1977). The population ecology of the wild rabbit, *Oryctolagus cuniculus* (L.), in a Mediterranean-type climate in New South Wales. *Australian Wildlife Research*, **4**, 171–205.

Parer, I. (1982*a*). Dispersal of the wild rabbit, *Oryctolagus cuniculus*, at Urana in New South Wales. *Australian Wildlife Research*, **9**, 427–41.

Parer, I. (1982*b*). European rabbit (Australia). In *CRC handbook of census methods for terrestrial vertebrates* (ed. D. E. Davis), pp. 136–8. CRC Press, Florida.

Parer, I. (1987). Factors influencing the distribution and abundance of rabbits (*Oryctolagus cuniculus*) in Queensland. *Proceedings of the Royal Society, Queensland*, **98**, 73–82.

Parer, I. (in press). Epidemiology of myxomatosis. In *Rabbit* (ed. B. J. Coman and J. H. Arundel). Australian Wool Corporation, Melbourne.

Parer, I. and Korn, T. J. (1989). Seasonal incidence of myxomatosis in New South Wales. *Australian Wildlife Research*, **16**, 563–8.

Parer, I. and Libke, J. A. (1985). Distribution of rabbit (*Oryctolagus cuniculus*) warrens in relation to soil types. *Australian Wildlife Research*, **12**, 387–405.

Parer, I. and Libke, J. A. (1991). Biology of the wild rabbit, *Oryctolagus cuniculus* (L.), in the southern tablelands of New South Wales. *Australian Wildlife Research*, **18**, 327–41.

Parer, I. and Pech, R. P. (1988). Rabbit management. In *Vertebrate Pest Management in Australia Project Report No. 5* (ed. G. A. Norton and R. P. Pech), pp. 38–52. CSIRO Division of Wildlife and Ecology, Canberra.

Parer, I. and Wood, D. (1986). Further observations on the use of warren entrances as an index of the number of rabbits, *Oryctolagus cuniculus*. *Australian Wildlife Research*, **13**, 331–2.

Parer, I., Conolly, D., and Sobey, W. R. (1981). Myxomatosis: the introduction of a highly virulent strain of myxomatosis into a wild rabbit population at Urana in New South Wales. *Australian Wildlife Research*, **8**, 613–26.

Parer, I., Conolly, D., and Sobey, W. R. (1985). Myxomatosis: the effects of annual introductions of an immunizing strain and a highly virulent strain of myxoma virus into rabbit populations at Urana, N.S.W. *Australian Wildlife Research*, **12**, 407–23.

Parer, I., Fullagar, P. J., and Malafant, K. W. (1987). The history and structure of a large rabbit (*Oryctolagus cuniculus*) warren at Canberra, A. C. T. *Australian Wildlife Research*, **14**, 505–15.

Parker, B. S. (1977). The distribution and density of rabbit warrens on the Southern Tablelands of New South Wales. *Australian Journal of Ecology*, **2**, 329–40.

Parker, B. S. (1979). Measuring the distribution of rabbit warrens from aerial photographs. In *Aerial surveys of fauna populations*, pp. 75–83. Australian National Parks and Wildlife Service Special Publication, No. 1.

Parker, B. S., Myers, K., and Caskey, R. L. (1976). An attempt at rabbit control by warren ripping in semiarid western New South Wales. *Journal of Applied Ecology*, **13**, 353–67.

Parker, B. S., Hall, L. S., Myers, K., and Fullagar, P. J. (1976). The distribution of rabbit warrens at Mitchell, Queensland, in relation to soil and vegetation characteristics. *Australian Wildlife Research*, **3**, 129–48.

Phillips, W. M. (1953). The effect of rabbit grazing on a re-seeded pasture. *Journal of the British Grassland Society*, **8**, 169–82.

Poczopko, P. (1969). The development of resistance to cooling in baby rabbits. *Acta Theriologica*, **14**, 449–62.

Poole, W. E. (1960). Breeding of the wild rabbit, *Oryctolagus cuniculus*, in relation to the environment. *CSIRO Wildlife Research*, **5**, 21–43.

Ratcliffe, F. N. (1936). Soil drift in the arid pastoral areas of South Australia. *Council for Scientific and Industrial Research Pamphlet*, No. 64.

Ratcliffe, F. N. (1959). The rabbit in Australia. In *Biogeography and ecology in Australia: Monographica Biologica 8* (ed. A. Keast, R. L. Crocker and C. S. Christian), pp. 545–54. Junk, The Hague.

Richards, G. C. (1979). Variation in water turnover by wild rabbits, *Oryctolagus cuniculus*, in an arid environment, due to season, age group and reproductive condition. *Australian Wildlife Research*, **6**, 289–96.

Richardson, B. J. (1981). The genetic structure of rabbit populations. In *Proceedings of the World Lagomorph Conference* (1979) (ed. K. Myers and C. D. MacInnes), pp. 37–52. University of Guelph, Ontario.

Richardson, B. J. and Osborne, P. G. (1982). Experimental ecological studies on a subalpine rabbit population. II. The effect of sodium and nutritional supplementation on breeding and physiological condition. *Australian Wildlife Research*, **9**, 451–63.

Richardson, B. J. and Wood, D. H. (1982). Experimental ecological studies on a subalpine rabbit population. I. Mortality factors acting on emergent kittens. *Australian Wildlife Research*, **9**, 443–50.

Ridpath, M. G. and Brooker, M. G. (1986). The breeding of the Wedgetail-eagle *Aquila audax* in relation to its food supply in arid Western Australia. *Ibis*, **128**, 177–94.

Rogers, P. M. and Myers, K. (1979). Ecology of the European wild rabbit, *Oryctolagus cuniculus* (L.), in Mediterranean habitats. I. Distribution in the landscape of the Coto Doñana, S. Spain. *Journal of Applied Ecology*, **16**, 691–703.

Rolls, E. C. (1969). *They All Ran Wild*. Angus and Robertson, Sydney.

Rowley, Ian. (1956). Field enclosures for the study of the wild rabbit, *Oryctolagus cuniculus* (L.). *CSIRO Wildlife Research*, **1**, 101–5.

Ryan, G. E. and Croft, D. J. (1974). Observations on the food of the fox *Vulpes vulpes* (L.), in Kinchega National Park, Menindie, N.S.W. *Australian Wildlife Research*, **1**, 89–94.

Schmidt-Nielsen, K., Schmidt-Nielsen, B., and Brokaw, A. (1948). Urea excretion in desert rodents exposed to high protein diets. *Journal of Cellular and Comparative Physiology*, **32**, 361–80.

Shepherd, R. C. H. and Edmonds, J. W. (1978). The occurrence of stickfast fleas *Echidnophaga* spp. on wild rabbits *Oryctolagus cuniculus* (L.) in Victoria. *Australian Journal of Ecology*, **3**, 287–95.

Shepherd, R. C. H., Edmonds, J. W., and Nolan, I. F. (1981). Observations on variations in the sex ratios of wild rabbits, *Oryctolagus cuniculus* (L.) in Victoria. *Australian Wildlife Research*, **8**, 361–7.

Short, J. (1985). The functional responses of kangaroos, sheep and rabbits in an arid grazing system. *Journal of Applied Ecology*, **22**, 435–47.

Sobey, W. R. and Menzies, W. (1969). Myxomatosis: the introduction of the European rabbit flea *Spilopsyllus cuniculus* (Dale) into wild rabbit populations in Australia. *Journal of Hygiene*, **69**, 331–46.

Sobey, W. R., Conolly, D., and Adams, K. M. (1967). Myxomatosis: the preparation of myxoma virus for innoculation via the eye. *Australian Journal of Science*, **30**, 233.

Soriguer, R. C. (1980). El conejo, *Oryctolagus cuniculus* (L.), en Andalucia occidental: Parametros corporales y curva de crecimieanto. *Doñana Acta Vertebrata*, **7**, 83–90.

Soriguer, R. C. (1981). Biologia y dinamica de una poblacion de conejos (*Oryctolagus cuniculus* (L.) en Andalucia Occidental. *Doñana Acta Vertebrata*, **8**, 1–379.

Soriguer, R. C. and Myers, K. (1986). Morphological, physiological and reproductive features of wild rabbit populations in Mediterranean Spain under different habitat management. *Mammal Review*, **16**, 197–9.

Stodart, E. (1965). A study of the biology of the wild rabbit in climatically different regions in eastern Australia. III. Some data on the evolution of coat colours. *CSIRO Wildlife Research*, **10**, 73–82.

Stodart, E. (1966). Observations on the behaviour of the marsupial *Bettongia lesueur* (Quoy and Gaimard) in an enclosure. *CSIRO Wildlife Research*, **11**, 91–9.

Stodart, E. (1968*a*). Coccidiosis in wild rabbits, *Oryctolagus cuniculus* (L.), at four sites in different climatic regions in eastern Australia. I. Relationship with the age of the rabbit. *Australian Journal of Zoology*, **16**, 69–85.

Stodart, E. (1968*b*). Coccidiosis in wild rabbits, *Oryctolagus cuniculus* (L.), at four sites in different climatic regions in eastern Australia. II. The relationship of oocyst output to climate and to some aspect of the rabbit's physiology. *Australian Journal of Zoology*, **16**, 619–28.

Stodart, E. (1971). Coccidiosis in wild rabbits, *Oryctolagus cuniculus* (L.), at a site on the coastal plain in eastern Australia. *Australian Journal of Zoology*, **19**, 287–92.

Stodart, E. and Myers, K. (1966). The effects of different foods on confined populations of wild rabbits, *Oryctolagus cuniculus* (L.). *CSIRO Wildlife Research*, **11**, 111–24.

Stodart, E. and Parer, I. (1988). Colonisation of Australia by the rabbit *Oryctolagus cuniculus* (L.). *Project Report No. 6*, CSIRO Division of Wildlife and Ecology, Canberra.

Strong, B. W. (1983). The invasion of the Northern Territory by the wild European rabbit *Oryctolagus cuniculus*. *Technical Report No. 3, Conservation Commission of the Northern Territory*, Alice Springs, Northern Territory.

Waithman, J. (1979). Rabbit control in New South Wales—past, present and future. *Wool Technology and Sheep Breeding*, **27**, 25–30.

Wheeler, S. H. and King, D. R. (1985*a*). The European rabbit in south-western Australia. I. Study sites and population dynamics. *Australian Wildlife Research*, **12**, 183–96.

Wheeler, S. H. and King, D. R. (1985*b*). The European rabbit in south-western Australia. II. Reproduction. *Australian Wildlife Research*, **12**, 197–212.

Wheeler, S. H. and King, D. R. (1985*c*). The European rabbit in south-western Australia. III. Survival. *Australian Wildlife Research*, **12**, 213–26.

Wheeler, S. H., King, D. R., and Robinson, M. H. (1981). Habitat and warren utilisation by the European rabbit, *Oryctolagus cuniculus* (L.), as determined by radio-tracking. *Australian Wildlife Research*, **8**, 581–8.

Williams, C. K. and Moore, R. J. (1989*a*). Genetic divergence in fecundity of Australian wild rabbits, *Oryctolagus cuniculus*. *Journal of Animal Ecology*, **58**, 249–59.

Williams, C. K. and Moore, R. J. (1989*b*). Phenotypic adaptation and natural selection in the wild rabbit, *Oryctolagus cuniculus* (L.), in Australia. *Journal of Animal Ecology*, **58**, 495–507.

Williams, C. K. and Moore, R. J. (1990). Environmental and genetic influences on growth of the wild rabbit, *Oryctolagus cuniculus* (L.), in Australia. *Australian Journal of Zoology*, **367**, 591–8.

Williams, C. K., Moore, R. J., and Robbins, S. T. (1990). Genetic resistance to myxomatosis in Australian wild rabbits, *Oryctolagus cuniculus* (L.). *Australian Journal of Zoology*, **38**, 697–703.

Williams, C. K., Parer, I., and Coman, B. J. (in press). *Management of vertebrate pests: rabbits*. Australian Bureau of Rural Resources, Canberra.

Williams, R. T. (1972). The distribution and abundance of the ectoparasites of the wild rabbit, *Oryctolagus cuniculus* (L.), in New South Wales, Australia. *Parasitology*, **64**, 321–30.

Williams, R. T., Fullagar, P. J., Kogon, C., and Davey, C. (1973). Observations on a naturally occurring winter epizootic of myxomatosis at Canberra, Australia, in the presence of rabbit fleas, *Spilopsyllus cuniculi* (Dale), and virulent myxoma virus. *Journal of Applied Ecology*, **10**, 417–27.

Wilson, A. D. (1966). The intake and excretion of sodium by sheep fed on species of Atriplex (saltbush) and Kochia (bluebush). *Australian Journal of Agricultural Research*, **17**, 155–63.

Wood, D. H. (1980). The demography of a rabbit population in an arid region of New South Wales, Australia. *Journal of Animal Ecology*, **49**, 55–79.

Wood, D. H. (1985). Effectiveness and economics of destruction of rabbit warrens in sandy soils by ripping. *Australian Rangelands Journal*, **7**, 122–9.

Wood, D. H. and Lee, A. K. (1985). An examination of sodium, potassium and osmotic concentrations in blood and urine of arid-zone rabbits in seasonal field conditions and in the laboratory. *Australian Wildlife Research*, **12**, 173–82.

Wood, D. H., Leigh, J. H., and Foran, B. D. (1987). The ecological and production costs of rabbit grazing. *Working papers of the 8th Australian Vertebrate Pest Conference, Coolangatta, May 1987*. Personal communication.

6

The rabbit in New Zealand

John A. Gibb and J. Morgan Williams

6.1 Introduction

New Zealand has the same two lagomorphs introduced from Britain in the nineteenth century as has Australia: the rabbit *Oryctolagus cuniculus* and the hare *Lepus europaeus*. Both are widespread and common. Though the hare has been blamed for damaging high country catchments and newly-planted trees (Flux 1967), the rabbit has been the major pest competing with sheep, depleting native vegetation and aggravating erosion.

New Zealand (Fig. 6.1a) has a wide latitude range from 34° to 48°S and mountain ranges rising to 3760 m (Mt Cook). Cook Strait, separating the North and South islands, is at the same latitude as Tasmania, or Madrid. The climate is generally temperate, moist, and equable, mollified by the proximity of the ocean (but with no Gulf Stream), whereas parts of Australia have a Mediterranean climate broadly resembling that of the Iberian peninsula where the rabbit evolved. In New Zealand (Fig. 6.1b) only Central Otago, in the southern half of the South Island, has a continental climate (see Table 6.3 below); thus rabbits do not have to live under such a varied or harsh climate as in Australia. Consequently the scale of research on rabbits has never matched that in Australia (cf. Chapter 5). Rabbits in New Zealand resemble those in Britain; and there are no obvious differences in their appearance between the north of the North Island and the south of the South Island, though separated by more than 12° of latitude.

In New Zealand spring is taken to include September, October, and November, summer is December, January, and February, autumn is March, April, and May, and winter is June, July, and August. And throughout this chapter, male rabbits are referred to as bucks and females as does. We have refrained from calling young rabbits 'kittens', which can be confusing when predation by cats is being discussed.

6.2 Colonization

Domestic rabbits of French origin were carried regularly on ships sailing the southern oceans in the late eighteenth and nineteenth centuries. Rabbits, pigs, and goats were released on remote islands to supplement provisions cached for castaways. The first rabbits known to have been released in New Zealand, 'unknown to the natives', were two couples set down on Motuara I. in Queen Charlotte Sound by Captain James Cook in 1777 (Beaglehole 1967); they died out.

Rabbits were a normal article of trade at shore-trading stations, as on Mana I. off the Wellington coast, from the 1820s onwards, and some doubtless escaped or were released from semi-confinement on the mainland and elsewhere. Thomson (1922) and Wodzicki (1950) describe some early liberations and their subsequent spread. Flux and Fullagar (1983) have listed over 50 islands with rabbits, around New Zealand (Table 6.1). Acclimatization societies took great trouble to import breeding stock and distribute their progeny; rabbits were mentioned favourably in the diaries of many early settlers before 1850 (B. T. Robertson, personal communication). Most releases of domestic breeds failed to

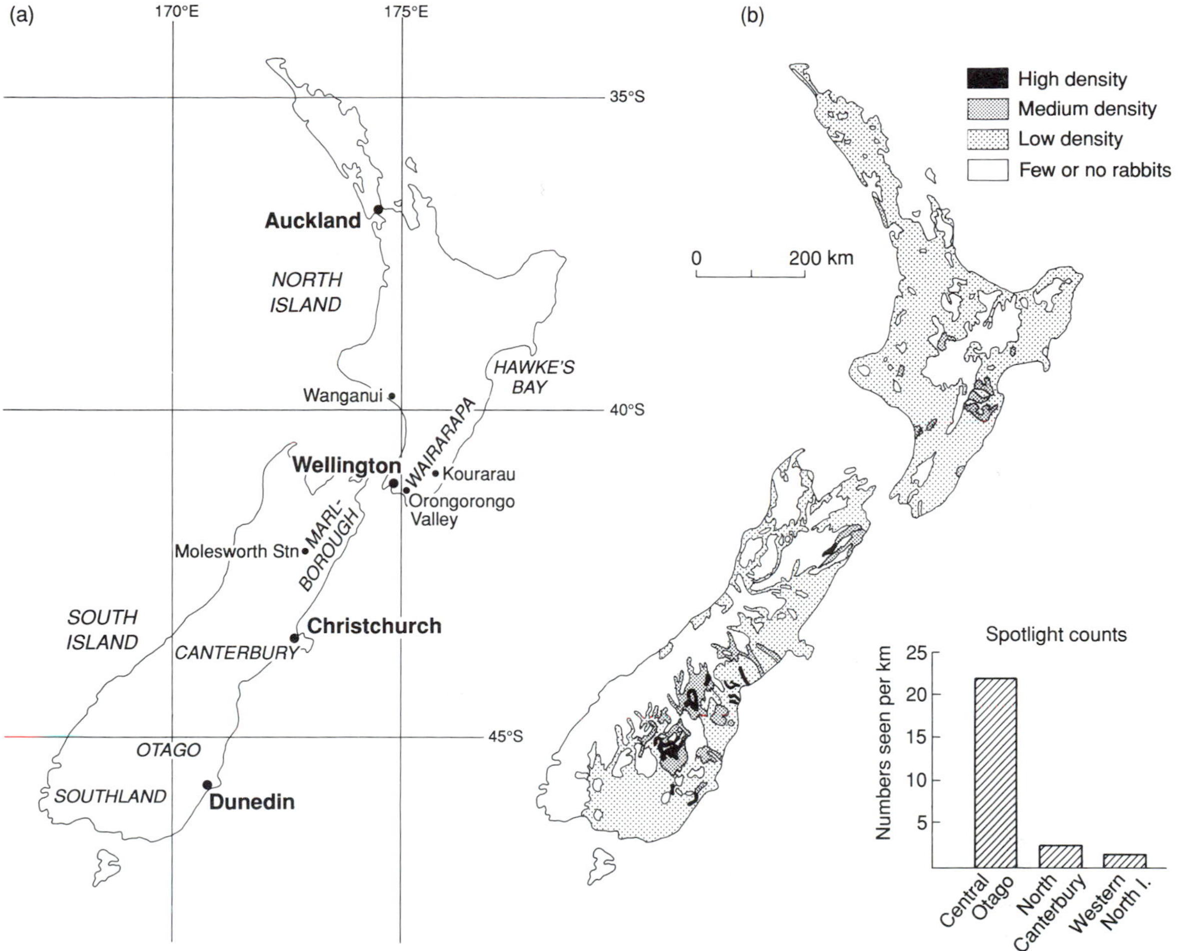

Fig. 6.1 Maps of New Zealand showing (a) places mentioned in the text, and (b) the distribution of rabbits (after Gibb and Williams 1990).

become established in the wild, presumably because of the unsuitable habitat; but some survived and multiplied. It is not known exactly where or when domestic rabbits first became feral, nor when they were eventually supplanted by ordinary wild rabbits; but those released on Rangitata Island (just inland of the mouth of the Rangitata River, 120 km southwest of Christchurch) in 1858 were probably wild-type rabbits and may have been the first to escape and cause serious damage (B. T. Robertson, personal communication).

No time was wasted in the conduct of affairs of State back in 1876. The Journals of the House of Representatives record that on 20 July that year 'a Select Committee [was] appointed to inquire into the Rabbit Nuisance', to report within one month, which it did as follows:

The Select Committee . . . have the honor to report, that they have carefully studied reports . . . in the Australian colonies, in New Zealand, and elsewhere. They have also examined a sufficient number of witnesses [10] to confirm beyond doubt the facts of the case. The mischief already done is most serious, is increasing, and, unless some effectual remedy be adopted, is likely to increase' (Hodgkinson 1876).

Evidence concerning the identity of early rabbits in the Wairarapa, North Island, is contradictory. Mr

Table 6.1 Annotated list of off-shore and outlying islands of New Zealand that have or have had rabbits (updated from Gibb and Williams 1990)

Name	Area (ha)	Latitude	Comment	Present status (✓, ?, ×)
Allports	12	41°14′S	Formerly common	×
Auckland Is	45 975	50°35′S	Introduced by 1847	✓
Bells (Fellows)	150	41°18′S	Silver-greys introduced 1867. Wild-type now	✓
Bests	140	41°18′S	Silver-greys introduced 1867	×
Browns (Motukorea)	60	36°50′S	Introduced 1972–73, increased to 5–7000 by 1978. Still common despite control	✓
Campbell	11 331	52°30′S	Introduced *c*.1883?	×
Chatham	90 650	44°00′S	Introduced but did not survive	×
Clarence		42°10′S	Island in river. Blacks introduced before 1865, then spread	?
D'Urville	16 782	40°50′S	A few reported (1976) 1977–78	?
Enderby	710	50°30′S	French breed(s) probably introduced 1840, 1865. Over-run 1880	✓
Friday	0.5	50°30′S	Rabbits transferred to Rose I. *c*.1850	×
Great Barrier	28 510	36°11′S	Common	✓
Happy Jack (Motukahaua)	*c*.24	36°39′S	Common	✓
Haulashore	0.5	41°16′S		✓
Inner Chetwode	242	40°54′S	White rabbits introduced 1914–18, present to *c*.1935	×
Junction Group (largest of 3)	*c*.7	36°14′S	Present *c*.1936	×
Kawhitihu (Stanley)	120	36°38′S	Introduced 19th century. Probably exterminated 1991	× ?
King Billy	0.4	43°33′S		✓
Korapuki (Rabbit)	11	36°40′S	Introduced 19th century, not common	✓
Leper	0.5	41°15′S	Introduced *c*. 1935, exterminated 1949–50	×
Mahurangi (Goat)	32	36°49′S		?
Mana	217	41°05′S	From 1834, but probably only in captivity	×
Mangere	113	44°17′S	Introduced *c*.1890, exterminated by cats *c*.1895	×
Maud	309	41°02′S	Common *c*.1912	×
Motuara	57	41°05′S	James Cook liberated 2 pr 1777	×
Motuihe	179	36°49′S	Common despite control	✓
Motukawau Group (see Happy Jack, Moturua and Motuwi)				
Motukiore	6.0	35°48′S	Connected to mainland at low tide	✓
Motumaire	5	35°16′S	Present 1974, 1975	✓
Motunau	3.5	43°04′S	Exterminated 1962	×

Name	Area (ha)	Latitude	Comment	Present status (✓, ?, ×)
Moturoa	146	35°12′S	Present 1975	✓
Moturua (Rabbit)	26	36°42′S	Common 1969	✓
Motutapu	1509	36°46′S	Common before 1883, present 1980	✓
Motuwi (Double)	26	36°46′S	Present 1979	✓
Native (Rabbit)	60	46°55′S	Recorded 1872, probably exterminated by 1950	×
Ngawhiti	5	40°49′S	Introduced *c*.1910, exterminated *c*.1912	×
Ohinau	45	36°44′S	Introduced 19th century	✓
Okokewa (Green)	6.0	36°09′S		✓
Otata	22	38°42′S	Present in 1930s	× ?
Penguin	8	37°04′S	Present 1974	✓
Puangiangi	69	40°46′S	Present before 1957	?
Puketutu	200	36°58′S	Present 1940s, 1950s	?
Quail	10	43°38′S	Silver-greys introduced *c*.1851. Few wild-type left after poisoning 1989	✓ ?
Rabbit	1060	41°16′S	Silver-greys introduced 1843, wild-type present now	✓
Rabbit	2.0	44°14′S	Probably never present despite name	×
Rakino	150	36°43′S		?
Rangitata	4000	44°03′S	Island in river; rabbits present before 1864	✓ ?
Rose	75	50°31′S	Introduced *c*.1850 from Friday I.	✓
Rough	140	41°17′	Silver-greys introduced 1867, formerly common	×
Shoe	40	36°59′S		✓
Slipper	210	37°03′S	Doubtful record	×
Stewart	174 600	47°S	First recorded 1860s. Wild-type when exterminated 1946	×
Sugarloaf	<1	39°03′S	Formerly common	×
Tahoramaurea (Browns)	0.8	40°53′S	Present c.1920, exterminated by *c*.1925	×
Taieri	6.9	46°03′S	Connected to mainland at low tide. Common	✓
Takangaroa (Goat)	9.3	36°25′S	Present 1930. Exterminated by 1950	×
Waiheke	9459	36°37′S	Common	✓
Waitaki	10	44°50′S	Island in river. Common	✓
Whale (Motuhora)	140	37°52′S	Introduced *c*.1967. Still common despite control	✓

William Beetham, of Brancepeth Station, reported that there were a few domesticated, lop-eared rabbits about in the years after 1856, but that they never increased 'to a detrimental extent'. Wild grey rabbits turned out by Mr Carter near Carterton had greatly increased since about 1870 and reached Brancepeth (15 km away) in 1875–6. Mr Beetham had seen no French silver-grey rabbits in the Wairarapa. On the other hand, the Hon. G. M. Waterhouse, who farmed in the same general district, when asked by the Chairman, 'Are these wild or tame rabbits you speak of?', replied 'It is a tame rabbit . . . turned out by Mr Carter a few years back'; they had been on his property since about 1871. By about 1890 wild type rabbits lived above the bushline at 1350 m a.s.l. in the Kaimanawa and Tararua Ranges of the North Island.

Evidence from the South Island is not so much contradictory as skimpy. Mr William Smith of Kaikoura, in the Marlborough province, was quite explicit: he had lived there since about 1865 and reported that rabbits were first released on Swyncombe Estate by George and Charles Keene in 1861–2, but had only recently become a nuisance on his own property. Asked what variety they were, he replied: 'The silver grey, I believe they are called Lord Galvin's breed, but they are a French kind, and better known as the silver grey'; and he went on to describe them correctly. Clearly, silver-grey rabbits became sufficiently numerous near Kaikoura to be classed as a pest, having spread from Swyncombe in the late 1860s and early 1870s before wild rabbits arrived. Rabbits forced the Keenes to abandon Swyncombe in 1882 (McCaskill 1970). Pockets of 'coloured' rabbits can still be found near there today (for example in the Clarence Reserve).

In the Clyde district of Central Otago, Captain Fraser reported killing 5000 rabbits in three months in 1876. He was 'under the impression that the rabbits . . . are tame rabbits', and said that they had not yet reached Hawea, 60 km to the north. Mr Cuthbert Cowan had farmed 11 750 ha in the Hokonui district of Southland since 1858, and first saw a rabbit there in 1872. Four years later he had a mixture of English wild rabbits with some silver-greys, the former having come 'up the river from the sand-hills between Invercargill and Riverton' (30 km away) and the latter from east of the Hokonui Range nearer at hand. In 1876 it was 'within the last eighteen months that the nuisance has assumed immense proportions . . .'. Mr Cowan killed 26 000 rabbits on his property between March and July 1876, 'Generally by men with spades, dogs, and one gun'; and he reported that 'My lambs have decreased 20 per cent'—from a flock of 8000 ewes.

Once established, wild rabbits quickly replaced silver-greys over most of the country. They were commonly sold in Southland in 1867 and a year later farmers were complaining of damage to crops. In the eight years 1877–84, 77 sheep runs totalling 627 935 ha were abandoned in Otago, mainly because of rabbits. Newly negotiated rentals were reduced by about 55 per cent as a result (Maitland 1885).

Rabbits reached peak numbers around 1890. In 1891 about 120 000 were taken off the Molesworth and Tarndale runs in Marlborough, by about 40 men dogging and digging (McCaskill 1970). Their subsequent decline, for instance in Otago, was probably due mainly to overgrazing by sheep and rabbits, possibly hastened by trapping and the gassing of warrens, and by the release of many mustelids.

The early spread of rabbits followed the opening up of grazing land for sheep, especially over the dry tussock grasslands of Otago. With its higher rainfall, North Island forest took longer to clear and rabbits spread more slowly. Most suitable country for rabbits in the South Island had been colonized by 1900, but the far north of the North Island was not colonized until the 1950s (Wodzicki 1950). Once established, rabbits seem to have spread at a rate of up to about 16 km a year in New Zealand, much slower than in Australia (Chapters 2 and 5). Wodzicki (1950) described some of the industries based on rabbits before they were 'decommercialized'.

6.3 Brief history of control

New Zealand's unique achievement was to control rabbits without recourse to myxomatosis. This was not done easily, cheaply, or without mistakes; nor even intentionally without myxomatosis since there was an official, but unsuccessful, attempt to introduce it in the early 1950s. Yet the rabbit is now stabilized at low densities almost everywhere (Fig. 6.1b), primarily by natural processes; and pasture production is threatened only in the 'semi-arid' tussock grasslands of the South Island.

Moreover, effective control was achieved by the early 1960s, before the results of early research on the biology of rabbits (for example by Bull 1953, Tyndale-Biscoe and Williams 1955, Watson 1957) could be widely disseminated. This is why, incidentally, we choose to give this brief history of control first, before moving on to describe the biology of rabbits in New Zealand as revealed by subsequent research.

With a land area only 15 per cent greater than that of the Australian state of Victoria, the value of rabbits exported from New Zealand in 1946 was one-third the total value exported from the whole of Australia (Fennessy 1958). By 1946 the rabbit problem was at its worst and runholders were again being forced off the land. Legislation designed to control rabbits had failed, as had primitive attempts at biological control by introducing so-called natural enemies (see Section 6.5.1).

The tables turned against the rabbit with the passing of the Rabbit Nuisance Amendment Act 1947. This innocuous-sounding legislation established a Rabbit Destruction Council (RDC) of eight members appointed by the Government, which adopted what became known as the 'killer policy' for district Rabbit Boards. There were already 108 Boards in existence in 1948, covering nearly 7.5 million hectares of the worst-infested land.

The Council was to promote all aspects of the control of rabbits including the formation of new Boards, and to advise the Minister of Agriculture on the progress of rabbit destruction. By good fortune the first Chairman of the Council, Mr G. B. Baker, was a remarkable man; he held office from 1948 until 1965. A South Island farmer, Bart Baker combined organizational skill with a rare understanding of his fellow farmers. He was absolutely dedicated to the cause of rabbit destruction, particularly to 'getting the last rabbit'. He promoted this belief with religious zeal and expected others to do the same. He had little sympathy for science or scientists unless work was directed strictly towards better methods of killing rabbits. With a less committed man, rabbit control could have foundered again in the 1950s.

The 1947 Act provided a system whereby: (i) landowners paid an annual rabbit rate to locally elected Rabbit Boards, the rate being based on the size, or carrying capacity, of the properties; (ii) rates levied by the boards were to be subsidized £ for £ by the Government; (iii) the whole business of rabbit destruction on public and private land was to be conducted by trained men employed by the boards; (iv) boards were empowered to require farmers to co-operate; and (v) rabbits and rabbit products were to be progressively 'decommercialized'. Though the Minister of Agriculture could create new boards without the support of a majority of local farmers, he rarely did so. This frustrated the RDC's resolve to achieve total coverage of the country by Rabbit Boards.

6.3.1 'Decommercialization'

The Rabbit Nuisance Amendment Act 1947 envisaged the complete devaluation of rabbit skins as articles of commerce. A 10 per cent levy on the sale of skins was imposed for the year ending 31 March 1949, rising to 66.7 per cent in 1952–3. The number of skins exported duly fell from 13.5 million in 1948 to 0.8 million in 1954, and the number of carcases from 4.9 million to 0.4 million.

In its Annual Report for 1953–4 the RDC was pleased that the export of carcases was to be prohibited from 1 June 1954; it hoped that the Government would accept that the export of skins had 'no useful place in our economy and will abolish it completely'. In 1956–7 the Council complained of a trade in carcases from non-board areas. However, the new Rabbit Amendment Act 1956 completely devalued skins and carcases; this was regarded as a major advance. Not content, in 1958 the Council deplored the legal sale of rabbits as pets; but was again pleased in 1960 that live rabbits could now be kept only by hospitals, zoos and research centres. It is said that the Council also wished to prohibit the sale of stuffed rabbits as cuddly toys, but this was never mentioned in its Annual Reports.

6.3.2 Coverage

Between 1948 and 1960 the number of Rabbit Boards more than doubled from about 100 to over 200, while the area covered by boards expanded from 7.5 million hectares in 1948 to 17.2 million hectares in 1962 (Fig. 6.2).

Successive Reports of the Council from 1953 through to 1965 repeat the perceived need for total coverage of the country by boards. They criticize the slow progress made, especially between 1962 and 1970 when the area covered increased by only one million hectares. The Council was annoyed that the Minister still required the support of local farmers before approving new boards.

By March 1963 the Council found it 'most discouraging to see the goal of complete coverage . . . so difficult of attainment'. It was also very embarrassing, for in 1958 the word 'control' had been formally replaced by 'eradication' as the Council's publicized national objective, though this intention had been apparent from the start. In its first Report for the years 1948–53 the Council expressed confidence that the following year would 'bring nearer the goal of complete destruction of the pest'. The Council was even more explicit in 1955, stating that 'the complete eradication of rabbits . . . is not beyond the bounds of possibility'. This was never remotely credible without complete coverage by Rabbit Boards.

In 1964, a wide-ranging Agricultural Development Conference recommended that all areas not by then under Rabbit Board control should be covered by 'County Councils acting as rabbit boards'. The RDC hailed this compromise as a great success, and in its Report for 1964–5 boasted that 'at long last all open areas throughout the country are now under the control of a rabbit board'. But it was really a case of too little, too late; too little in that County Councils, lacking special expertise and having other diverse responsibilities, were a poor substitute for proper rabbit boards; and too late because it was already becoming obvious that total eradication was not possible.

The Agricultural Development Conference further recommended: (i) that the RDC be replaced by a broader Agricultural Pests Destruction Council (APDC), so as to accommodate other introduced vertebrate pests such as possums (*Trichosurus vulpecula*) and rooks (*Corvus frugilegus*); (ii) that a more scientific approach to pest destruction was necessary; and (iii) that an experiment be set up on a substantial scale, lasting at most four years, to test whether or not eradication was really practicable. The old RDC was replaced by the new APDC and a Technical Advisory Committee (TAC) was set up to promote scientific aspects of pest control; but the experiment on eradication lapsed, probably because of administrative difficulties.

6.3.3 Control methods

Early volumes of the *New Zealand Journal of Agriculture* describe some of the methods used to control rabbits at a time when the sale of skins and carcases gave some cash return for the work done. J. Bell (unpublished) listed 36 references to rabbit control between 1917 and 1926, 12 of which included 'strychnine' in the title. This was a popular poison, of course, because it killed rabbits quickly and suddenly, close to the poison line where they could be counted and collected. Phosphorized pollard baits were used on a smaller scale, and warrens were gassed with carbon bisulphide, calcium cyanide, and later chloropicrin; this last released a gas heavier than air which penetrated the burrows without being pressurized.

After rabbits were devalued, strychnine gave way to arsenic as the most commonly used poison; while phosphorized raspberry jam (only the best), dispensed from a 'jam gun' in small amounts on upturned spits of soil, was used for scattered rabbits accessible only on foot. The destruction of large warrens by ripping with a tractor was practical on easy country, but never became as popular as in Australia because of the steep and rocky terrain. The labour-intensive methods of trapping, shooting, and dogging, remained popular but were never cost-effective as the sole method of control, or when applied to sparse populations that in reality needed no control at all.

Reporting for 1953–4, the Council noted the effective use of sodium monofluoroacetate (compound

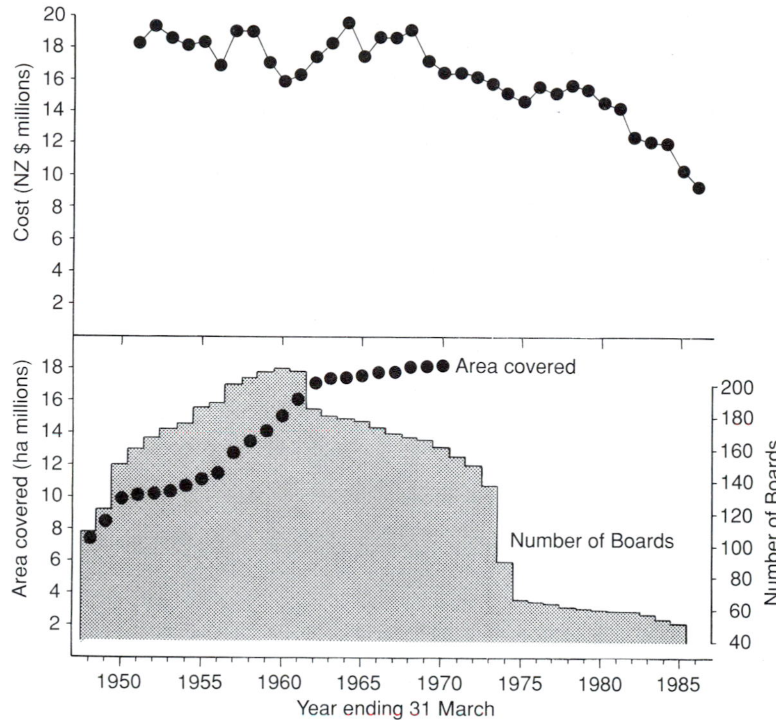

Fig. 6.2 Upper graph. Cost of rabbit destruction since the establishment of the Rabbit Destruction Council in 1948, in 1983. NZ$ adjusted for inflation by the Consumer Price Index (*New Zealand Yearbook*).

Lower graphs. Area covered by Rabbit Boards from 1958 until full coverage effected in 1970 by County Councils acting as Boards; with the number of active Boards, declining since 1970 as a result of amalgamations.

1080) as a new poison for rabbits used in Tasmania. Early trials in New Zealand in 1955–6 found it preferable to other currently-used poisons (Wodzicki and Taylor 1957; McIntosh 1958). Thereafter 1080 became the standard poison used, usually with chopped carrot as bait and often distributed from the air; it remained so despite 'poison-shyness' reported in Central Otago, where 1080 with carrot had been used at least annually for over 10 years (APDC Report for 1968–9).

Fixed-wing aircraft were used increasingly from about 1949 to drop poisoned baits; this revolutionized poisoning on remote country, especially in the South Island. About 2800 tonnes were dropped during nearly 5000 hours' flying in 1953–4; but lesser quantities were needed once rabbits were reduced.

New Zealand has a freer hand than most countries in broadcasting poisoned baits for rabbits, as the risk of killing non-target wildlife on agricultural land is slight. Stock must be removed from paddocks to be poisoned, and dogs are notoriously susceptible to 1080. However, most of New Zealand was forested before being cleared for agriculture during the nineteenth century and is devoid of native mammals. Birds are the most likely fauna to take pollard, grain, or carrot bait, but most of those on agricultural land are of introduced species, usually considered expendable.

There have been problems with native birds, even supposedly insectivorous species, taking finely chopped carrot with 1080 laid for possums in the native forest; and some introduced species, including game-birds and duck, have been killed by eating poisoned carrot on pature, but not on a scale causing loud public protest. In many ways 1080 has proved environmentally safer than some other toxins used for pest or disease control. In its later years the APDC appointed an Environmental Subcommittee to counter public concern over the misuse of poisons generally.

As sparse populations of rabbits, already reduced by poisoning, were difficult to bring any lower by further poisoning, night-shooting came to be accepted as the most likely way to bag the elusive 'last rabbit' in the late 1950s. Enormous effort went into improving the technique, using .22 rifles and 12-gauge shotguns with a spotlight mounted on the cab of 4-wheel-drive vehicles, and latterly using trail-bikes which were more cost-effective. Night-shooting was popular with Board staff and farmers alike, and was used as the standard method of 'controlling' low-density rabbits over vast areas of land (McLean 1966); but scant attention was paid to its real effectiveness in reducing the numbers of rabbits.

There is, in fact, little evidence that night-shooting ever really reduced sparse populations of rabbits at the intensity at which it was normally used, i.e. one or two visits per property each year representing an average of only 1–2 minutes per hectare (Gibb *et al.* 1969, 1985; Williams and Robson 1985). Night-shooting yielded impressive annual tallies, but these often amounted to only a few rabbits shot per square kilometre of land covered (Gibb *et al.* 1985). However, at least the rabbits' predators were spared and the tallies of rabbits shot were used years later to assess the case for or against government funding in the mid-1980s.

6.3.4 Myxomatosis (see also Chapter 7)

Myxomatosis is not present in New Zealand. There was an official, unsuccessful attempt to introduce it in the early 1950s (Filmer 1953), before rabbits had been brought under control by other means. In the 1980s and early 1990s, farmers lobbied for its introduction to counter a resurgence of rabbits in Central Otago.

Myxomatosis was deliberately introduced to Australia in 1950 with dramatic results after a hesitant start. In New Zealand, there was some public debate on the need for myxomatosis at that time; but, encouraged by the Australian experience, varying numbers of rabbits (16–500) infected with the virus were released in 1951–3 at 21 sites in the North and South Islands. The weather was cold and wet at the time. This may have limited the numbers of mosquitoes and sandflies (*Austrosimulium* sp.), the most likely native vectors, and the virus died out during the winter of 1952. Further trials in the more promising areas in the summer of 1952–3 likewise failed, again because of the lack of vectors.

Once the relationship between the rabbit, the rabbit flea, and myxomatosis had been described in Britain (for example, Mead-Briggs 1977), the RDC asked the Department of Agriculture to investigate introducing the fleas and the myxoma virus to New Zealand. In 1956, Dr F. N. Ratcliffe and Mr B. V. Fennessy, visiting from Australia, had expressed the opinion that, as myxomatosis could not exterminate rabbits, but only reduce dense populations, which had already been done by other means over most of New Zealand, it 'could not now be expected to make any significant contribution'. So the proposal lapsed for 20 years. In 1976 the APCD asked the TAC to investigate its possible introduction, mainly to reduce rabbits in Central Otago where they were still a problem. The TAC recommended against the introduction at that stage, adding that a full environmental impact assessment would be required to air the views of all interested parties.

By 1981–2 bait-shyness in rabbits was being widely reported from Central Otago. This, combined with a succession of drought years, reduced government funding and farm profitability, strengthened the lobby for myxomatosis. The APDC commissioned its own environmental impact assessment in 1982; it generated 552 submissions. Gibb and Flux (1983) argued against Sobey's (1982) recommendation favouring introduction. The extent of public concern prompted the Minister of Agriculture to request a formal Environmental Impact Report in 1985. The APDC was disappointed but duly commissioned a report (Bamford and Hill 1985). A revised version was submitted to the Commissioner for the Environment in 1987, inviting further public comment. Of 512 submissions received (incorporating an extra 656 petitions), the majority opposed the proposal; 74 per cent expressed the view that myxomatosis would be ineffective, 60 per cent that its introduction was not justified. The Parliamentary Commissioner for the Environment therefore recommended against the introduction 'at this time' (Bilborough 1987); this view was later endorsed by the Government.

The main reasons for the Commissioner's decision were that myxomatosis was not seen as a permanent solution, that rabbits were concentrated in a relatively small area of the South Island where myxomatosis could not be contained, and above all that myxomatosis was not publicly acceptable. Moreover, the Commissioner hoped that the habitat in the drier regions of Otago might be modified to reduce rabbit numbers without recourse to myxomatosis or abandoning pastoral farming; the latter idea was later challenged by a Task Force set up in the wake of the Commissioner's ruling (Rabbit and Land Management Task Force 1988), and at a conference convened by the Ministry of Agriculture and Fisheries in March 1989 (Bell *et al.* 1989).

6.3.5 Control costs

Following the formation of the RDC in 1946 the operating costs of most Boards were met from rabbit rates paid by farmers, plus an equal subsidy from the government. In the early 1960s, when eradication was still the goal, this subsidy was supplemented by special grants to Boards in severely infested districts, such as in Central Otago where burning and overgrazing by sheep and rabbits had reduced the carrying capacity to about one sheep per four hectares (O'Connor 1987). It was argued that runholders in these districts could not afford the high rabbit rates that were necessary, and (probably incorrectly) that if the rabbits were not dealt with there they would spread to surrounding districts. Control was thus seen to be in the national interest and not just of local concern.

The costs of rabbit destruction in Fig. 6.2 are adjusted for inflation using the Consumer Price Index (*NZ Yearbook*) as an index, and expressed in 1983 NZ dollars. Figure 6.2 shows that the annual cost of destruction was sustained at around $18 million from 1948 until 1964. Over this period the number of Boards increased from 100 to 200 and the area covered from about eight to 18 million hectares. Although eradication remained the objective until 1970, steadily declining costs per hectare were inconsistent with it. The basis for measuring the area covered by Boards changed when County Councils began 'acting as Boards' in 1971.

The cost of rabbit destruction was justified by improved farm production until about 1960; financial benefits of up to $33 million annually were claimed in 1964. The Council then urged increased effort in pursuit of the 'last rabbit'. The APDC's Annual Report for 1966 tried to reinforce this objective, warning that 'Any relaxation in the war against the pest would quickly result in their reappearance with disastrous results . . .'. Ironically, a trial had already begun in 1965 demonstrating for the first time that the removal of control (mainly night-shooting) for three years on hill country in the eastern Wairarapa did *not* result in any resurgence in the numbers of rabbits (Gibb *et al.*, 1969). By some quirk, in 1988 the Wairarapa Board was one of the few still reluctant to experiment with reduced control as advocated by the APDC, despite likely cost-saving.

6.3.6 Success to stalemate

The progression from early success to stalemate is a good example of the ultimate ineffectiveness of 'predation' by man on an initially abundant, but rapidly declining, vertebrate pest of agriculture. So long as whole hillsides seemed to move with rabbits, farmers were desperate to be rid of them and there was little difficulty in constituting new boards; the rabbits themselves proved vulnerable to the first furious onslaught. The cry 'To the last rabbit!' spurred boards to greater effort just so long as progress was seen to be made: the problem seemed clear (too many rabbits) and the objective (extermination) straightforward.

The first battles were won astonishingly fast. In March 1948 nearly 20 per cent of the land controlled by Rabbit Boards was classed as heavily infested; five years later, it was only three per cent. As early as 1952–3 the RDC was reporting increased lambing percentages and wool weights; and in 1955, that many farmers had followed up rabbit clearance with improving farming by 'spelling', seeding, top-dressing, growing crops, and clearing scrub. Lighter grazing by sheep reduces overgrazing and the risk of heavy infestation by rabbits, for it is the combination of sheep plus rabbits that is most damaging (Howard 1959).

The very successes of the early years may also have fuelled complacency in successive Ministers of Agriculture, making them reluctant to wield the heavy stick and create new boards without a majority of farmer support. The Council's resolve to kill the 'last rabbit' was thus to some extent sabotaged by its own early successes; but there was also a growing realization among farmers that rabbit numbers were not determined solely by the activities of Rabbit Boards.

Reading the Council's Annual Reports, one detects the gradual change of attitude overtaking the destruction movement. The early years brimmed with 'spectacular progress' (1953), 'continued steady progress' (1954), and 'continuation of the downward trend in rabbit numbers' (1955); but by 1956 the Council observed that boards had (already) 'reduced the pest to a very low level and are finding it increasingly difficult to show further progress'. Some boards were reduced to killing only one rabbit per 40 hectares annually but '. . . further progress must be made if the goal of eradication is ever to be reached' (1956). Then, in 1958, we read, 'It is important that the pest be reduced year by year . . ., the rabbit has amazing powers of recovery' and 'will never be taken cheaply'; and in 1959, 'As the heavy infestations are reduced, the problem of making headway . . . comes into prominence'. Further 'steady progress in eradicating the rabbit' was reported in 1961, while the Council deplored growing complacency in the community.

By 1965, the RDC reported as follows

> the rabbit generally has shown an increase . . . too many Rabbit Boards have departed from a sound killer policy; they have become complacent . . . are quite prepared to live with some rabbits, so long as they are not doing any harm to production. Boards must surely realise that the rabbit never remains static; if complacency sets in rabbits will soon increase.

In 1966, the year after Bart Baker's retirement as Chairman, the Council saw 'very little improvement' and again blamed complacency: 'Any reduction in the war against the pest would quickly result in their reappearance with disastrous results'. By 1968–70 most boards were just 'cropping' the rabbits, which remained at low density except in a few hot spots such as Central Otago.

The unenviable task of officially abandoning the 'last rabbit' policy fell to Chairman G. G. Wilson at a heated meeting of the South Island Rabbit Boards Association in 1971; he was tellingly supported by former Chairman W. R. Kofoed, an earlier supporter of eradication. The change had been signalled in the APDC's Annual Report for 1970, in which the Chairman reported that the Council 'was well aware that with present knowledge and methods the death of the "last rabbit" is only wishful thinking' (see also Gibb 1967; Kofoed 1967) (Fig. 6.21).

6.3.7 Withdrawal of government funding

With the abandonment of the eradication policy in 1971, with its clear sense of purpose, the pest destruction organization began to decline; although this was inevitable anyway as the damage caused by rabbits diminished. In a circular to boards in September 1971 the APDC redefined its objective:

> Boards will be encouraged to make the maximum effort using the most effective known methods. They should particularly ensure that there is no risk of an upsurge in infestation on known rabbit prone country.

In reality many boards continued to pursue a policy of eradication, but within strict financial limitations. The costs of running small boards continued to rise and the APDC encouraged many to amalgamate: from a peak of 208 boards in 1968 the number fell to 105 by 1976 and to 92 by 1986.

From 1975 to 1980 research in the Ministry of Agriculture concentrated on the need to develop cost-effective management reflecting the actual threat of rabbits in different habitats (Williams 1977, 1978). The importance of this work, building on earlier studies in DSIR, was acknowledged in 1979 when the TAC was requested by the Minister of Agriculture to examine 'the methods of operation, objectives and activities of the boards'.

Not surprisingly, the TAC recommended important changes in control policies, among them that government funding should no longer be tied to rabbit rates paid by farmers, but should be given as grants adjusted to the local risk of infestation. The APDC Act was amended in December 1980 to remove the government's commitment to the rate-linked subsidy which had been basic to rabbit destruction since the 1880s. For the years 1981–2 to 1983–4 the government's contribution was pegged at $7 million per year, with the recommendation that more objective ways be found of defining the need for control.

In 1983 yet another Review Committee was charged with evaluating 'the objectives, structure, effectiveness and efficiency of the pest control movement . . .'. It decided that:

(1) most vertebrate pests were at satisfactory levels;
(2) rabbits should not be considered differently from other farm pests;
(3) land should be classified according to its suitability for rabbits;
(4) government funding should be phased out;
(5) land unsuitable for rabbits should not be charged rabbit rates;

(6) some Boards were inefficient;

(7) there should be extensive amalgamations of Boards; and

(8) there should be closer liaison with scientists.

These decisions were milestones in rationalizing rabbit management. Government funding was to be phased out completely by 1992–3 and the amalgamation of boards was accepted as inevitable. An effort was made to adopt a national system for assessing the rabbit proneness of different classes of land; and calls for the introduction of myxomatosis were intensified.

Eventually a separate review of regional government and resource law overtook the reorganization of pest destruction, which became incorporated in a new regional government structure in November 1989. The former autonomy of the APDC and pest destruction boards was removed, and rabbit management had to compete for funds with other demands of the community. Only Central Otago and parts of inland Canterbury and Marlborough were still to receive government support (Rabbit and Land Management Task Force 1988). This was appropriate so long as the only chronic problems remaining were local ones. It remained to be seen whether improved land use would contain rabbit numbers at acceptable levels on all classes of country, without the expedient of myxomatosis.

The reluctant decline of the pest destruction movement in New Zealand was hastened by the inflexible attitude of dominant landholder organizations, and by their inability to adapt to change (Norton and Pech 1988; Williams 1991). The initial policy of delivering the maximum force against rabbits was effective in reducing numbers and generating public support; but the policy became fixed on the unattainable objective of extermination and countenanced no alternative.

Despite the appointment of a Technical Advisory Committee to advise the Minister of Agriculture, rather than the RDC or APDC, on technical and scientific aspects of pest destruction, the Council itself never included an animal ecologist and remained wary of scientific advances in pest management generally. For too long successive Councils saw no place for science in policy formation; the perceived role of scientists was limited to devising better methods of killing rabbits. Howard (1959) and a few others recognized this gulf between science and management, but their advice went unheeded for 20 years until a spate of views in the 1980s forced painful changes that were long overdue.

6.3.8 'Applied research'

A trickle of papers on rabbit control had appeared in the *New Zealand Journal of Agriculture* since 1910, with a deluge of 42 papers between 1912 and 1925. Most of this research was done in the Department of Agriculture and was broadly toxicological, focusing first on strychnine and fumigants.

So long as rabbit destruction was obviously reducing rabbit numbers, the Annual Reports of the RDC ignored research. But by 1958, when Rabbit Boards were beginning to falter, the Council called for more 'applied', as opposed to 'basic', research; in other words, the RDC wanted more research on how to kill rabbits rather than on their population dynamics. The Council suddenly voiced the belief that 'no major further advance is possible without scientists', but science went largely unsupported.

Howard (1959) identified the need to bridge the gap between basic and applied research. As a result, in 1960 the Department of Agriculture appointed a scientist for applied research; while the DSIR continued to concentrate on ecological research. Despite the Agricultural Development Conference's recommendation in 1964–5 that a more scientific attitude to rabbit destruction was required, the Council could detect 'no increase in the amount of applied research' in 1967 or 1969. Yet two years later its official policy shifted from eradication to control, partly as a result of ecological research (Gibb *et al.* 1969) that was not recognized at the time as having 'applied' connotations. This switch in policy was symptomatic of a change in the membership of the APDC which was recruiting a new generation of farmers more receptive to science.

At a Field Day in 1960 DSIR scientists introduced members of the RDC to their current research on rabbits at Kourarau in the Wairarapa. The Chairman of the Council, Bart Baker, made a show of turning away, declaring that the only rabbits he wanted to see were dead ones. This research (Gibb *et al.* 1978) showed how predation might reduce susceptible

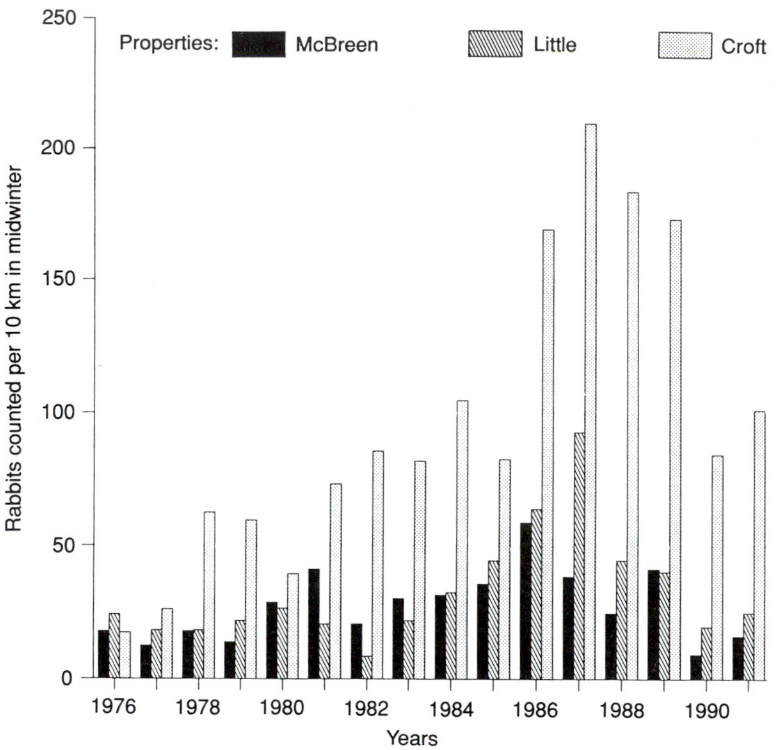

Fig. 6.3 Numbers of rabbits counted per 10 km at night in midwinter, using trail-bikes with spotlights, on three properties in North Canterbury. A standard transect of 25 km was covered on each property. No control was carried out from 1976 to 1988, but the properties had previously been covered by night-shooting (Williams unpublished).

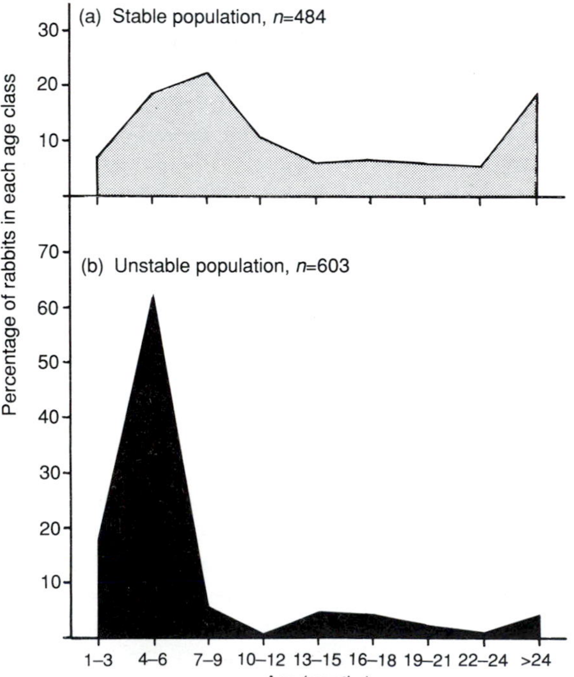

Fig. 6.4 Age composition of rabbits shot in February from a stable population near Wanganui in the North Island and from an unstable population near Kaikoura in the South Island. Note the larger number of juveniles and fewer older rabbits in the unstable population (after Williams and Robson 1985).

rabbit populations. To test this in the wild, the Wairarapa Board collaborated in an experiment whereby the Board undertook to stop routine control on 1200 ha of hill country for three years, while DSIR scientists monitored changes in the numbers of rabbits. The Board supported the trial because it expected that the resulting increase in rabbits would justify its continued operations. The scientists involved doubted whether rabbits would increase at all; they did not (Gibb *et al.* 1969).

Similar trials in later years confirmed that much of the routine work of boards was wasted. Great savings could be made if boards desisted from needless 'control'. The Western Pest Destruction Board, inland of Wanganui, carried out regular night-shooting on some properties but left others alone (Williams and Robson 1985). The shooting reduced rabbit numbers by about 40 per cent on the average, but this was usually made good within a year. Unshot populations contained significantly more older rabbits than the shot populations (see also Gibb *et al.* 1969).

At another site, in North Canterbury, rabbit densities were measured on about 2000 ha of rolling farmland for 13 years without control (Fig. 6.3). Densities remained moderately stable on two properties, but increased on another. On all three properties density increased during a spell of dry years in the mid- to late 1980s. A stable population in this same intensively farmed district contained many more older rabbits and fewer young ones than did an unstable population on open range inland from Kaikoura, a little to the north (Fig. 6.4).

To a large degree boards were willingly duped by the RDC into believing that populations of rabbits would 'explode' if pressure were relaxed. By the mid-1980s control had in fact been withdrawn, experimentally, from some two million hectares of land in the North and South Islands, where previously routine control had been carried out annually; there was no serious resurgence of rabbits (J. Bell and P. Nelson unpublished).

The crucial research that led to the policy of restricting control to where it was actually needed was not accepted by the RDC or APDC until the late 1970s, and then reluctantly; it would never have been commissioned at all under a 'user-pays' regime. Research carried out to help formulate policy was never regarded as a proper concern of scientists; indeed, there may have been some official reluctance to recognize that the costs of rabbit control should be taken into account when formulating agricultural policy. This view was encouraged because, until 1983, government subsidies covered at least 50 per cent of the cost nationally, and up to 80 per cent in the most rabbit-prone districts.

By the mid-1980s research had become focused on four topics: investigating bait-shyness; screening new poisons to supplement 1080 (for example Williams *et al.* 1986); attempting to understand what limits the numbers of rabbits at low density without formal control (Gibb 1981, and unpublished); and defining the suitability for rabbits of different classes of land (Williams 1977; Kerr *et al.* 1987). The overall aim was to determine the real risk posed by rabbits, and to convince farmers and others that rabbits were more susceptible to changes in their habitat than to direct manipulation of their numbers.

6.4 Distribution and habitat

By 1948, when the Rabbit Nuisance Amendment Act 1947 took effect, the rabbit problem had become serious because farms had become run-down and control was weak during the Second World War. Energetic rabbit control during the 1950s and 1960s followed by extensive farm development, and a shift from poisoning to night-shooting which relieved pressure on the predators, resulted in drastically reduced and apparently stable populations of rabbits becoming established over most of New Zealand. Only on the drier 'semi-arid' tussock grasslands of the South Island have rabbits remained, intermittently, at high density.

In New Zealand generally, although rabbits are greatly reduced in numbers, some have survived wherever the habitat is suitable from sea-level to 1000 m a.s.l. On a few off-shore islands, they have either died out or been exterminated; though there are still about 24 islands with rabbits (Table 6.1).

In 1984 the APDC organized a national survey of

rabbit numbers, based on the numbers seen per kilometre from cross-country vehicles with spotlights, as used for night-shooting. Figure 6.1b shows small numbers of rabbits over most of the North Island; but many more east of the main divide in the South Island, especially in Central Otago and in the Mackenzie Basin, North Canterbury, and Marlborough.

Scattered rabbits live in a wide range of habitats (Figs. 6.22–6.25); they are absent only from alpine lands, large blocks of forest or scrub, and from densely settled urban areas. Ideal rabbit habitat has an annual rainfall of less than 1000 mm a year, a sunny aspect, light soil, good drainage, and adequate cover (scrub, logs, or rock outcrops) within easy reach of short herbaceous vegetation and some bare ground. Rabbits especially favour dunes, limestone hills with outcrops, and dry stony riverbeds. They will tolerate rainfall of up to about 2500 mm a year given good drainage, but they avoid cold, dank places. Yet a population of rabbits of domestic origin survives on subantarctic Macquarie I., where it is perennially cloudy, wet and cold (De Lisle 1965). Judging from crude comparison of the numbers killed on the road, rabbits were less common in New Zealand as a whole than in Britain or southern Australia in the 1960s and 1970s (Table 6.2).

Table 6.2 Counts of rabbits and hares in Britain and New Zealand, made from a car by the same observer on tour (Gibb and Flux 1983)

		Britain 1977	New Zealand	
			North Island 1965–66	South Island 1964
Rabbits	Live	396	5	3
	Dead	435	20	19
Hares	Live	27	0	18
	Dead	12	19	24
km travelled		3700	3500	3800

Note: Driving from Perth to Canberra, Australia, in 1971, Mackinnon (1972) counted 694 dead rabbits on 3600 km of road.

6.4.1 Research sites

Over the years research on rabbits in New Zealand has been concentrated in only a few districts (named in Fig. 6.1). Since we refer repeatedly to this research, we summarize here the salient features of these sites and the work done in each.

Sites in the North Island included Hawke's Bay, Wairarapa, Wanganui, and the Orongorongo Valley: the first three have much in common. The predominant land-use is semi-extensive sheep farming with some beef cattle. The site in Hawke's Bay is close to the ranges and at slightly greater elevation than other sites, so its climate is rather cooler and moister than in most of the Wairarapa and near Wanganui; but the overall range of climatic conditions is small (Table 6.3). The topography is also broadly similar: rolling to hilly improved pasture with mainly introduced grasses and clovers, dissected by steep gullies with manuka and kanuka (*Leptospermum scoparium*, *Kunzea ericoides*), gorse (*Ulex europaeus*) and small patches of native bush. The much larger district of the Wairarapa (*c.*2550 km^2) embraces various habitats in addition to sheep pasture; these range from sand-dunes and cliffs, wetlands, horticulture and cropping on the alluvial plain, to rougher hill country with beef cattle and extensive scrublands, and some native bush and pine plantations in the foothills of the ranges to the west and in the eastern hills (Gibb *et al.* 1985).

In each of these districts rabbits were numerous until the mid- to late 1950s, when improved control using the then new poison 1080 in carrot bait, dropped from the air, quickly reduced numbers. By the mid-1960s populations of rabbits were generally scattered, sparse, and stable without the imposition of formal control (Gibb *et al.* 1969). Thereafter the numbers of rabbits in all these districts were 'cropped', but not further reduced, by night-shooting.

Most of the early work on the biology of rabbits in New Zealand was begun when they were still abundant in the eastern foothills of the Ruahine Range in Hawke's Bay (Bull 1953; Tyndale-Biscoe and Williams 1955; Watson 1957). Wodzicki and Taylor (1957) undertook the first trial of 1080 in an enclosure of 8.5 ha at Kourarau in the Wairarapa. When this was finished, Gibb *et al.* (1978) used the same enclosure for a population study. This prompted the first experimental removal of formal control, to test its effectiveness or otherwise, in the eastern Wairarapa (Gibb *et al.*

Table 6.3 Altitude and mean annual temperature and rainfall at research sites (after Wards 1976)

Site	Altitude m a.s.l.	Mean annual temperature (°C)	Mean annual rainfall (mm)
Hawke's Bay	250–550	10.0–12.5	1250
Wairarapa	0–650	10.0–12.5	900–1200
Wanganui	150–450	12.5–15.0	1000–1100
Orongorongo Valley	60–110	11.5–12.0	2500
North Canterbury	150–600	11.0–12.0	600–800
Central Otago	300–1800	7.5–10.0	300–500

1969), and then to the study of the population ecology of rabbits throughout the Wairarapa, drawing on a large sample of rabbits shot by the Wairarapa Pest Destruction Board (Gibb *et al.* 1985). Williams and Robson (1985) undertook more extensive trials on the need for continued control by the Western Pest Destruction Board in the Wanganui district, North Island.

The Orongorongo Valley is narrow, steep, and densely forested on both sides, running 30 km north into the Rimutaka Range from Cook Strait (Campbell 1984). The average annual temperature is about 12.5°C; the average rainfall of 2500 mm is high for rabbits. Severe earthquakes in 1948 and 1855 precipitated landslides from the ridge-tops east of the Orongorongo River; the debris aggraded the main riverbed some years later. This created wide, gravel flats and opened the way for rabbits coming up the valley from the coast. The river now courses through a shifting network of braided channels bordered by river flats averaging 200 m in width. Stands of manuka, kanuka, and tauhinu (*Cassinia leptophylla*) have grown in bends of the river, and ephemeral patches of herbaceous vegetation are alternately established and washed away (Gibb unpublished).

A sparse population of rabbits lives in small groups on the river flats. They do no damage and are never 'controlled'. Their numbers fluctuate in response to periodic floods which sweep away the vegetation on which the rabbits feed. The rabbits live permanently above ground in the scrub by day and feed in the open at night; they breed in shallow 'stops' dug in the river shingle (Gibb 1993). The density of rabbits and the extent of the herbaceous vegetation alongside five kilometres of the river have been measured annually for 18 years (Gibb unpublished). An important feature of the population is that the feral cats (*Felis catus*) and stoats (*Mustela erminea*), which prey on the rabbits, have a choice of other foods in the adjacent forest (Fitzgerald and Karl 1979), so can easily switch from one prey to another.

Research in the South Island has been concentrated in North Canterbury and Central Otago. These two districts are drier than those in the North Island, Central Otago particularly so (Table 6.3). Otherwise the North Canterbury site is not very different from parts of the Wairarapa, being mostly rolling, improved pasture with clover, used for semi-extensive sheep farming and beef cattle; the land is dissected by steep gullies containing scrub and remnant bush. Rabbits were plentiful until reduced by poisoning in the 1950s, leaving only scattered pockets based in the gullies. Bell (1977) described the reproduction of rabbits here, and Williams (unpublished) has since monitored changes in their numbers with and without formal control (Fig. 6.3).

Central Otago is the driest part of New Zealand; though sometimes referred to as 'semi-arid', it does not quite justify this description using Australian criteria. It has a continental climate with an annual rainfall down to 300 mm (more in the ranges), a mean daily minimum temperature of 1.5°C in July and a daily maximum of 22°C in January; severe frosts and droughts are commonplace. Much of the country consists of broad basins 300–600 m a.s.l., surrounded by craggy hills rising to 1800 m, with massive outcrops of schist, and shallow, stony, well-drained soils. Until about 1950 the natural tussock grasslands were severely degraded by repeated cycles of burning followed by overgrazing by merino sheep and rabbits

(O'Connor 1987). The worst-affected ground is now dominated by mats of scabweed (*Raoulia australis*), thyme (*Thymus vulgaris*), stonecrop (*Sedum acre*), and hawkweeds (*Hieracium* spp.), with much bare ground; Mark (1965) called it semi-desert and it still is. Slightly moister ground carries some tussock and introduced grasses, with matagouri (*Discaria toumatou*) and sweetbriar scrub. Grazing by sheep has had to be relaxed in recent years, but rabbits have remained intermittently abundant, though patchy, despite repeated large-scale poisoning. Although Central Otago has been notorious for its rabbits for 100 years, and is sometimes referred to as 'the home of the rabbit' in New Zealand, it is only recently that they have been studied intensively there (for example Fraser 1985, 1988; J. Robertshaw unpublished).

6.5 Predators, parasites, and disease

6.5.1 Predators

The native Australasian harrier (*Circus approximans*) is common over most open country; it rarely kills rabbits heavier than 500 g though it scavenges dead ones, shot, poisoned, or killed on the road. No other bird of prey is capable of killing rabbits in New Zealand, and there are no native carnivores: so it is not surprising that efforts were made to boost the numbers of 'the natural enemy' of the rabbit when damage became acute.

It was expressly forbidden by law to import foxes, which were regarded as a threat to lambs; but numerous ferrets (*Mustela furo*) were introduced from 1882, followed by weasels (*M. nivalis*) and stoats (Wodzicki 1950; King 1990). The Select Committee inquiring into the Rabbit Nuisance in 1876 conclued 'that a grant of money for the purpose of introducing weasels, as a natural check, into the country, would (if the object were attained) be of very great service'. The statement was shrewdly worded: indeed, given the condition in parentheses, it can hardly be faulted. The Committee never actually recommended importing mustelids, which would have been against the advice of two of its informants. Mr William Smith stated that, 'From my own knowledge of weasels and stoats, I believe they would merely go into the bush after the small birds and leave the rabbits alone.' The Hon Mr Robinson, having experienced rabbits in Australia, commented:

> I think I heard mentioned also something about the introduction of ferrets. I think it has never been found that they interfere very much with the increase of rabbits . . . As to the introduction of vermin [i.e., stoats and weasels] to destroy the rabbits, I think there would be a very great objection to that; because they might increase to a very large extent, and destroy other sorts of game which it is desirable to preserve. . . .

Notwithstanding, the Department of Agriculture bred up large numbers of ferrets for release and advertised extensively in the British press for live stoats and weasels for export to New Zealand. In 1884, 1885, and 1886 some 4000 ferrets, 3099 weasels, and 137 stoats were liberated to found local populations.

McCaskill (1970) colourfully described attempts to control rabbits using imported carnivores on the several runs that were later combined to become Molesworth Station (180 476 ha) in inland Marlborough. For example, some runholders bred their own ferrets: they cost Sir N. Campbell £1 a head to breed and feed, and he turned out 200 a year. Stoats imported in 1885 cost £5 a head. Mr William Acton-Adams, runholder of Molesworth, liberated 13 stoats and 141 weasels at Tarndale in 1888, costing in all £800; he preferred stoats at the cost of £3. 5*s*. 0*d*. a head to weasels at £3. 0*s*. 0*d*., believing (probably correctly) that one stoat was worth two weasels for killing rabbits. In 1891 he contracted to buy 500 ferrets a year at 8 shillings each.

Cats were already in the country before mustelids were liberated, and the *Lyttelton Times* reported on 4 November 1867 that they were being released to control rabbits. Mr Acton-Adams brought 200 by wagon from Christchurch to Tarndale in 1885, and 'planned to run a four-horse express with cats' at frequent intervals on this route, despite one load overturning and being drowned in the Acheron River. He turned out 1000 cats a year on Richmond Dale at the cost of one shilling to 1/6*d*. a head.

Table 6.4 Parasites of wild rabbits in New Zealand (after Bull 1953)

Species	Status
Coccidia	
Eimeria steidae	Common in liver of young; sometimes fatal
E. perforans	Widespread, common; may kill
E. pyriformis	Widespread and common
E. irresidus	Uncommon
E. flavescens	Uncertain identification; probably common
E. media	Uncertain identification; probably rare
Trematoda (liver flukes)	
Fasciola hepatica	Local North Island; usual host sheep
Cestoda (tapeworms)	
Cysticercus pisiformis	Widespread, in small numbers
Coenurus serialis	Two records, Wairarapa
Nematoda (roundworms)	
Passalurus ambiguus	Mainly South Island
Trichostrongylus retortaeformis	Widespread and common
T. axei	Single record; usual host sheep
Graphidium strigosum	Widespread South Island and southern North Island; perhaps absent north of North Island
Nematodirus spathiger	Single record Hawke's Bay; usual host sheep
Pentastomida	
Linguatula serrata	Rare, confirmed South Island only; usual host dogs
Ixodoidea (ticks)	
Haemaphysalis bispinosa	Single old record; usual host cattle, local
Acarina (parasitic mites)	
Listrophorus gibbus	Widespread and common
Cheyletiella parasitivorax	Confirmed North Island only
Anoplura (sucking lice)	
Haemodipsus ventricosus	Uncommon; confirmed North Island only
Siphonaptera (fleas)	
Nosopsyllus fasciatus	Single record; usual host rats
Ctenocephalides ?felix	Old record; usual host cats

Similar stories abound in the North Island. One farmer is said to have posted a man with a hatchet beside the chute from the wagon down which the cats were being released, to chop off their tails in case they were recaptured and offered for sale a second time.

Weasels never became abundant, but stoats and ferrets prospered. As predicted by William Smith, the stoats went into the bush and killed many birds, as well as some rabbits (King and Moody 1982). Ferrets were still being bred for rabbit control in the 1920s (Munro 1925). Cats spread into all kinds of habitats, with a varied diet that included rats, rabbits, birds, and lizards (Fitzgerald and Karl 1979).

The release of so many cats and mustelids must have had some effect on the number of rabbits; but despite the faith of some runholders, they failed to hold rabbits at low densities. Vast numbers must have been killed by mistake when trapping rabbits or gassing warrens, and perhaps through eating poisoned carcases. In 1876 William Smith mentioned 16 ferrets caught in rabbit traps in one night, and Mr Waterhouse reported 47 trapped in one month in the Wairarapa.

This selective killing of carnivores presumably continued unabated: in 1955 Wodzicki and Taylor (1957) killed 14 ferrets, 2 stoats, and 2 cats, along with just 43 rabbits, during 200 gin-trap nights at Kourarau. The carnivores may still have suffered even after the use of 1080 became widespread, as they are vulnerable to secondary poisoning from eating poisoned carcases (Marshall 1963). The switch from poisoning to night-shooting, once rabbits were reduced in the 1960s, may have allowed the carnivores to return. Predators (primarily ferrets and cats) are still regarded as potentially important in the natural control of low-density populations of rabbits (Gibb *et al.* 1969, 1978 and unpublished; Fitzgerald and Karl 1979; Newsome *et al.* 1989; Trout and Tittensor 1989; see also Pech *et al.* 1992).

6.5.2 Parasites and disease

The parasite fauna of rabbits in New Zealand (Table 6.4) is similar to that in Britain, whence our rabbits came, except that the rabbit flea *Spilopsyllus cuniculi* and the intestinal tapeworm *Cittotaenia* are absent; presumably they failed to survive the long enforced quarantine on board sailing ships in transit from Britain during the nineteenth century (Bull 1953; summarized by Gibb and Williams 1990). The flea has been introduced to Australia as an additional vector of the myxoma virus (Chapter 7).

Severe coccidiosis of the liver by *Eimeria steidae* (Bull 1958, 1964) has caused heavy mortality among young rabbits 6–11 weeks old, but only where they were numerous (Tyndale-Biscoe and Williams 1955); coccidiosis is probably unimportant at low densities. *Eimeria perforans* can also kill young rabbits but is not known to have caused heavy mortality. With rabbits generally so much scarcer now than before, parasites are even less likely to cause heavy mortality.

Until 1988 myxomatosis was the only disease known to have severely affected rabbit numbers anywhere in recent times (Chapter 7); but it is still absent from New Zealand (see Section 6.3.4). Rabbit viral haemorrhagic disease was first found in wild populations in Europe (Spain) in 1988 (see Section 6.12 and Chapter 3).

6.6 Circadian rhythm and feeding

6.6.1 Activities

The daily routine of rabbits has seldom been adequately described, though properly quantified it can reveal some of the limitations under which rabbits live. Measurements made over a range of population densities in the Kourarau enclosure (Gibb *et al.* 1978) have been used to interpret the behaviour of free-ranging rabbits in the Orongorongo Valley (Gibb 1993).

The rabbits' active 'day' usually begins in late afternoon or early evening; they then stay up all night and return to their daytime resting places, above- or underground, soon after dawn. The principal rest period is from soon after dawn through the morning. Despite Watson's (1954) collection of rabbits at intervals throughout the day and night (to study reingestion), it is still not possible to describe exactly how rabbits spend the time. Using spotlights from a hide at Kourarau we counted the numbers above ground at intervals through the night; but as the rabbits usually paused when the light shone on them it was impossible to judge what they had been doing. This was done more successfully in dimly-lit enclosures in Canberra (Mykytowycz and Rowley 1958), but the results are hardly applicable to free-ranging rabbits in New Zealand. At Kourarau almost all rabbits spent the whole night above ground, dispersed from the warrens; at high density they virtually deserted the warrens at night (Gibb *et al.* 1978).

The number of rabbits active above ground by day depends on how hungry they are, and hence on their

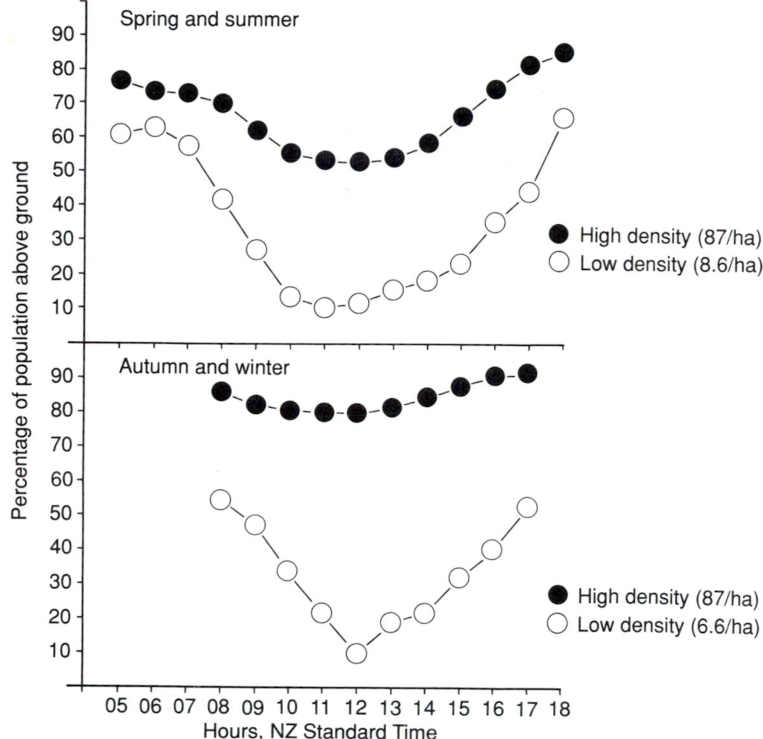

Fig. 6.5 Proportion of the population above ground by day at high and at low density in the Kourarau enclosure (after Gibb *et al*. 1978).

density relative to the food supply (Fig. 6.5). At high densities at Kourarau, more than 60 per cent of the population were above ground through the day in autumn and winter, and more then than in spring and summer; but fewer than 20 per cent were above ground at low density. In Central Otago, the rabbits emerged earlier (relative to sunset) in winter and spring than in summer and autumn (Fraser 1985).

Similarly by day at Kourarau, the rabbits spent longer feeding and engaged in social activities, and much less time inactive, at high as opposed to low density; and they stayed on distant feeding grounds through the middle of the day. At low density recognizably young rabbits spent about as long feeding by day as did the adults; but at high density they spent longer feeding than the adults. So, when food was short, the rabbits virtually sacrificed their morning rest-period in order to spend more time feeding. Rabbits were watched from dawn till dusk at Kourarau; their behaviour was less variable in late afternoon and evening than in the morning.

At Kourarau and in Central Otago (Gibb *et al.* 1978; Fraser 1985), as in other pastoral habitats, rabbits were said to 'emerge' when they first came above ground, usually in late afternoon. In the Orongorongo Valley, however, where the rabbits lived permanently above ground, 'emergence' was taken to be when they first came out of the scrub onto the open riverbed, and could be seen; more adult rabbits were in fact seen in the evening in spring and summer, than in late autumn and winter.

In the Orongorongo Valley the rabbits spent most of the day resting above ground in the scrub. They then emerged at the scrub edge in late afternoon or evening and spent the night feeding on open river flats. They spent more time feeding (never > 50 per cent) at dusk in spring than at other seasons; this was about as long feeding as at low density at dusk at Kourarau, and contrasts with the over 70 per cent of the time feeding at high density. Thus rabbits in the Orongorongo Valley were probably not seriously short of food for maintenance, despite the poor grazing (Gibb 1993).

Rabbits usually remained in their daytime resting places from soon after dawn at least until mid-afternoon. They may spend longer feeding in the morning after a disturbed night; and during the breeding season does often began feeding soon after midday even in fine weather. In very cold districts of the South Island, where severe ground frosts may impede feeding at night and where food may be short, rabbits sometimes feed regularly by day instead of, or as well as, at night (R. J. Pierce unpublished; cf. Fraser 1985).

6.6.2 Reingestion and dung-heaps

The business of feeding and processing food claims much of the rabbits' time. Their continual demand for sustenance in an often harsh environment, and the threat of predation by day, have given rise to the practice of reingestion (Watson 1954), a habit common to all lagomorphs. Watson reviewed earlier work on domestic rabbits, but Southern (1940) was first to report reingestion in wild rabbits.

Watson (1954), a former colleague of H. N. Southern at Oxford, quantified reingestion in wild rabbits in Hawke's Bay (Watson 1954). Adult rabbits feed hard at dusk and dawn and at intervals through the night. The bulk of the food eaten passes through the gut within 48 hours; but some is retained to form soft pellets that accumulate in the rectum 24 hours later, for reingestion the following morning. Young rabbits begin reingesting as soon as they are weaned.

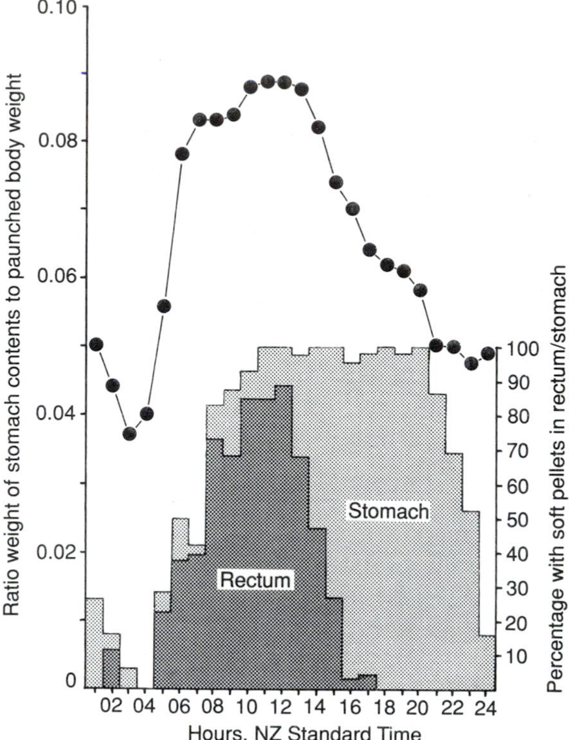

Fig. 6.6 The circadian rhythm of reingestion (after Watson 1954). The upper graph shows the ratio of the weight of the stomach contents to the rabbit's paunched weight; and the stepped histograms show the percentage of the rabbits shot that had soft pellets either in the rectum or the stomach, through the 24 hours.

As Fig. 6.6 shows, reingestion is infrequent during the afternoon; food then remains in the stomach until defaecated as the familiar hard pellets during the following evening. Though some of the ingesta are retained in the caecum for longer, the digestive process is virtually complete by the early hours of the morning of the second day. By this time there are few, if any, soft pellets in the rectum or stomach, and the gross quantity of material in the stomach is at a minimum despite the rabbit's active feeding during the previous evening. The stomach is replenished quickly during an intensive bout of feeding between 0400 and 0600 hours (in January in New Zealand). As Watson pointed out, the quantity of material in the stomach depends on the relative rates of ingestion and defaecation, not just on the amount of food ingested immediately beforehand. The total stomach contents diminish steadily from about midday and through the late afternoon and evening, a period of increasing activity for rabbits, and from then on until after midnight, thereby spanning their intensive feeding at dusk.

6.7 Home range and use of space

6.7.1 Home range at Kourarau

The home ranges of rabbits in the Kourarau enclosure were measured from day and night sightings of tagged animals (Gibb *et al.* 1978). As the enclosure was wholly under improved pasture, the home ranges were always small (and much smaller than the enclosure). Rabbits dispersed after dark, and their night ranges were much larger than their day ranges. As a rule bucks had larger ranges than does, and first-year rabbits of both sexes had larger ranges than older animals; though at night old does had significantly larger ranges than old bucks, presumably because of their pressing need to feed. Increasing population density reduced the size of the day ranges, but not of the night ranges which were exceptionally large at maximum density. Day and night ranges were generally smallest in winter and largest in autumn.

There was little scope for rabbits to shift their ranges very far in the enclosure at Kourarau, and few did. Between successive three-month periods fewer than 40 per cent of rabbits more than three months old shifted the centres of their ranges further than 18 m, and fewer than 5 per cent further than 90 m. Bucks, and especially first-year bucks, were more inclined to shift than does. By their first winter, when most were 9–12 months old and preparing to breed for the first time, first-year bucks and does had shifted on the average slightly less than 50 m from where they were first seen as small young in the previous spring.

Fraser's (1985) measurements of the 'activity ranges' of rabbits in Central Otago are not comparable with ours because he did not know where they had spent the day; nonetheless it is clear that they had small ranges, comparable with those at Kourarau.

6.7.2 Home range in the Orongorongo Valley

In New Zealand, the home ranges of free-ranging rabbits have been measured (by radiotelemetry) only in the Orongorongo Valley (Gibb 1993). Rabbits with transmitters were 'fixed' two to four times during the day, and again at dusk and at night (before midnight). Each rabbit used one of several different resting places by day, up to about 50 m apart. Members of a pair spent the day either close together or alone, usually in thick scrub. Small young rabbits often lay up in cover by day close to their night feeding grounds, instead of returning to the scrub.

A few of the rabbits' home ranges in the Orongorongo Valley are illustrated here to show their variety (Fig. 6.7), using minimum convex polygons. This is a rough and ready method, but more elaborate mapping requires that 'fixes' are distributed at random through the 24-hours, which was not possible. To obtain a better statistical basis for comparing the sizes of the rabbits' home ranges we adapted Hayne's (1949) method to calculate centres of daytime activity (henceforth 'day centres') and then measured the distance of all fixes from this point (see below).

Bucks No. 29 and No. 40 lived around the Field Station and on nearby river flats. They were born respectively in spring 1983 and spring 1984 (or earlier). Buck No. 29 had the distinction of being first trapped some 550 m away to the south, where he occasionally fed at night with a neighbouring group of rabbits. He was often seen with an unmarked female around the Field Station. Though no runt, buck No. 40 was invariably subordinate to No. 29 and was never seen with a mate of his own until November 1989, when he was at least five years old. Buck No. 29 disappeared in December 1990 when at least seven years old, and No. 40 was still alive in November 1992, when at least 8 years old. He probably held the world longevity record for a wild rabbit (cf. Lloyd 1977).

Most rabbits were trapped on the river flat just south of Paua Ridge (Fig. 6.7). Buck No. 4 was first trapped there and fitted with a radio in September 1982. He had been born in spring 1981 and was last seen in July 1989 when about $7\frac{1}{2}$ years old. During his lifetime buck No. 4 consorted with at least five different does, all younger than himself; they included No. 25 and No. 31, mentioned below. He was never seen at points more than about 200 m apart.

When first trapped in October 1984 doe No. 25 was only six weeks old. By the end of November,

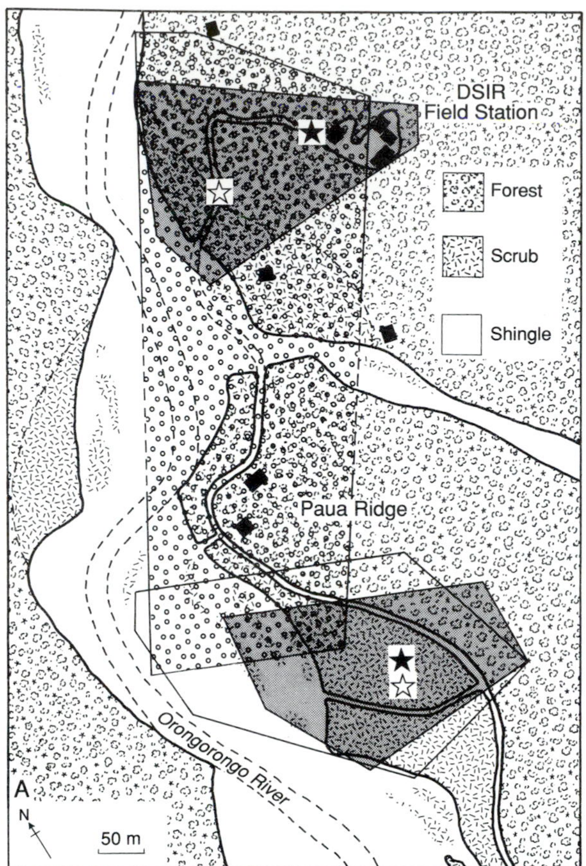

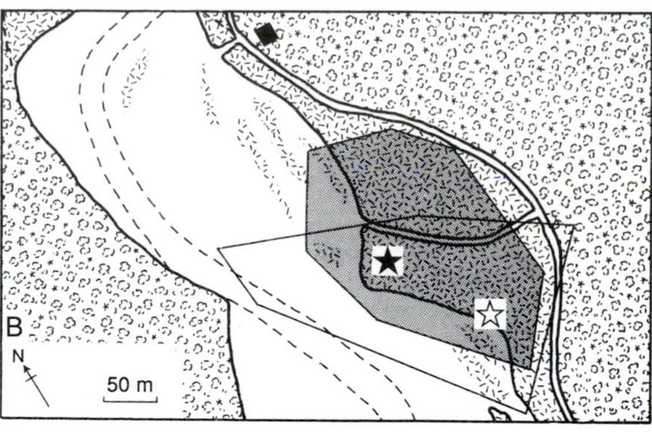

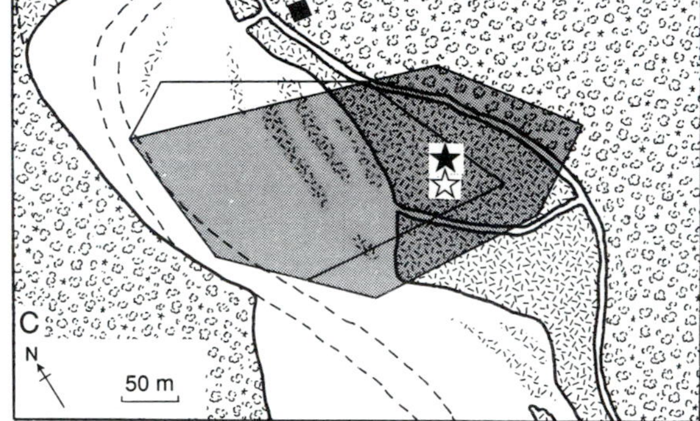

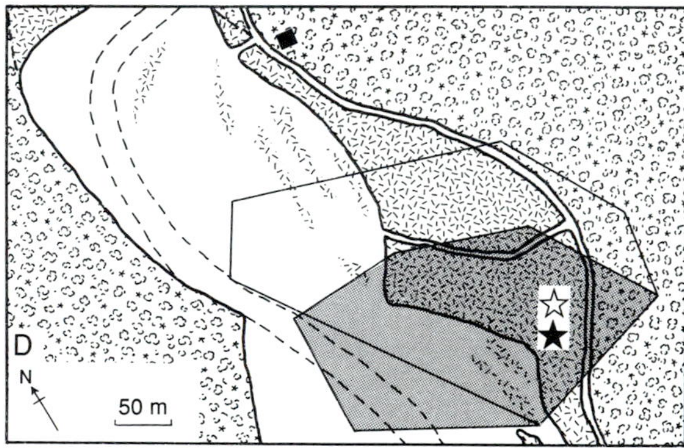

Fig. 6.7 Convex polygons defining the day and night ranges of selected rabbits in the Orongorongo Valley, measured by radiotelemetry. Black stars mark the day centres (see text) of darkly hatched ranges, and white stars those of lightly or unhatched ranges (see also Table 6.6).

Notes: Map A. To the north is the (darker) range of buck No. 40 from September 1986 to January 1987, and the exceptionally large (spotted) range of the dominant buck No. 29 from February to July 1985. The polygon illustrating the range of No. 29 is misleading because his range was constricted (like a figure '8') across the tip of Paua Ridge where he followed the track through the forest. No. 29 spent most of his time north of Paua Ridge, but occasionally fed with a neighbouring group of rabbits to the south.

To the south is the (darkly hatched) range of buck No. 4 from September 1982 to August 1983, and his (unhatched) range from September 1984 to July 1985. No. 4 associated with the does No. 25 and No. 31, also illustrated, and then with another doe (not illustrated) at least until January 1989, when he was over seven years old and living about 100 m north of his earlier range.

Map B. The (darkly hatched) range of doe No. 25 from November 1984 to July 1985, when she switched from buck No. 4 to No. 32 further south; and her (unhatched) range from November 1985 to January 1987, when she was still with No. 32.

Map C. The (darkly hatched) range of doe No. 31 from July 1985 to February 1986 when she was with buck No. 4; and her similar (unhatched) range from March to September 1986 when she was with another buck.

Map D. The (darkly hatched) range of buck No. 32 from June to December 1987, when he associated with doe No. 25; and his (unhatched) range from January to September 1988, when he had transferred to a young doe a little further north. No. 32 was killed by a cat in November 1988, when at least five years old.

when about three months old, she was being courted by buck No. 4 and was probably pregnant when trapped in mid-February 1985. In about March 1985, however, she left (or was deserted by) buck No. 4 and joined up with buck No. 32 (Fig. 6.7B). This pair survived until May 1988; buck No. 32 died in November 1988.

Doe No. 31 was about nine months old when she became the regular mate of buck No. 4 after he left doe No. 25 in March 1985; but by May 1986 she was consorting with another buck as well, and they formed an apparently contented trio. Doe No. 31 survived until December 1988, when she was just over four years old.

Buck No. 32, born in spring of 1983 or earlier, was the long-term mate of doe No. 25 from about March 1985. He accompanied her at least until the end of 1987 and then began consorting with another, younger doe. Buck No. 32 was at least $3\frac{1}{2}$ years old when first given a transmitter in June 1987; he was then still with doe No. 25 and the pair frequented a quite compact home range (Fig. 6.7B). Buck No. 32 was killed by a cat in November 1988, when at least five years old.

6.7.3 Apparent range-size and the number of fixes

Estimates of home-range size measured by minimum convex polygons increase with the number of fixes on which they are based (Table 6.5). Omitting buck No. 29 as exceptional, estimates of home-range size based on day- and night-time fixes continued to expand with the number of fixes exceeding 100. For a given number of fixes (up to about 140), does had ranges one-half to two-thirds as large as those of the bucks.

6.7.4 Extent of ground used by day, at dusk, and at night

Rabbits in the Orongorongo Valley sheltered in scrub by day, emerged on river flats close to the scrub in the evening, and then dispersed to their night feeding grounds as the light faded. To measure the area of ground covered by rabbits during the 24 hours we first calculated the central point of their day-time fixes (i.e., their day centres) and then measured the distance from this point of all fixes made by day, at dusk, and at night, using all the rabbits studied thus far, not just those mentioned here.

Table 6.6 shows that, on the average, adult bucks spent the day within about 33 m of their day centres, and by dusk had moved out to about 70 m from this point; at night they were on average more than 100 m away. Does were slightly less restricted than bucks to a small area by day; but at dusk and at night they had not moved as far as the bucks and spent longer in the scrub. By being in the scrub, does may have been especially vulnerable to predation by cats; but food may have been more readily available there than on open river flats.

The mean distances of rabbits from their day centres are not strictly radii from which the size of the ranges may be calculated because the home ranges were not circular (Fig. 6.7); but if they are treated as radii as an approximation, then the bucks would have had average home ranges of 3.25 ha and does 2.5 ha. This is not grossly larger than the average night range of 2.3 ha embracing 95 per cent of the sightings of marked rabbits at Kourarau, at much greater population density (Gibb *et al.* 1978).

The distance moved by rabbits at night was limited by the size of the enclosure at Kourarau (8.5 ha), and by local topography in the Orongorongo Valley where some rabbits commuted up to 500 m from the scrub to their night feeding grounds. If included in the calculation, this would have increased the estimated size of their home ranges to well over 10 ha. This may not be too much for the rabbits, which covered the distance in about 20 minutes, including generous pauses en route.

6.7.5 The use of space

Though always opportunistic, rabbits remain essentially sedentary for life. Adults rarely shifted ranges completely in the Orongorongo Valley, but they may have to modify the boundaries of their ranges whenever the river changes course; this can happen at any time of year. None of the marked rabbits shifted its range by more than 400 m. Because suitable feeding ground was fragmented, several rabbits had home ranges with two or more foci separated by at least 100 m; such ranges could not be defended effectively at all times, as territories. Sudden disturbances to the rabbits' home ranges in the Orongorongo Valley

Table 6.5 Area of convex polygons describing home ranges as a function of the number of radio fixes; Orongorongo Valley, Wellington (Gibb, unpublished)

	Area of convex polygons (ha) Rabbit No, spring of birth and period of observation					
Bucks	Rabbit No. 29 b. 1983	Rabbit No. 40 b. 1984	Rabbit No. 4 b. 1981	Rabbit No. 32 b. 1983		
No. of fixes	Feb 1985– Jul 1985	Sep 1986– Jan 1987	Sep 1982– Aug 1983	Sep 1984– Jul 1985	Jun 1987– Dec 1987	Jan 1988– Sep 1988
20	3.4	0.6	1.5	2.1	0.9	0.9
40	3.5	3.1	2.2	3.4	2.4	2.4
60	3.7	3.3	2.4	3.8	2.9	3.0
80	4.2	8.4	2.6	4.1	2.9	3.0
100	12.2	3.4	2.7	4.6	3.0	3.0
120		4.5	2.8	4.9	3.1	3.3
140		4.3		4.9	3.1	4.2
160				5.0	3.3	4.2
180				5.0		4.2
200						4.7

Does	Rabbit No. 25 b. 1984		Rabbit No. 31 b. 1984	
No. of fixes	Nov 1984– Jul 1985	Nov 1985– Jan 1987	Jul 1985– Feb 1986	Mar 1986– Sep 1986
20	0.6	0.8	1.4	1.0
40	0.8	1.0	2.8	1.0
60	1.1	1.5	2.9	1.1
80	1.5	1.5	3.0	1.3
100	1.5	1.5	3.4	2.4
120	2.2	1.5	3.4	2.7
140	2.3	1.7	3.4	2.7
160	2.3	2.0	3.4	2.9
180	2.3	2.0		

must have their equivalents on agricultural land where farming practices can be just as unpredictable for them.

However their ranges are measured, the rabbits clearly covered more ground in the Orongorongo Valley than on open pasture at Kourarau, or on chalk downland in England (Cowan 1987). We have no useful information on the dispersal of young from their place of birth in the Orongorongo Valley, because so few survived. Most of the rabbits' home ranges in the Orongorongo Valley included plentiful cover, sheltered grazing for use by day, and more extensive feeding grounds in the open for use at night. This mix was necessary as a defence against predators (Gibb 1993).

In Spain as in Australia, birds of prey are the

Table 6.6 Mean distance from their day centres of marked rabbits by day, at dusk, and at night, in the Orongorongo Valley, Wellington; all seasons pooled (after Gibb 1993)

	Bucks (n = 20)			Does (n = 23)		
	Day	Dusk	Night	Day	Dusk	Night
Number of fixes	688	265	262	714	267	243
Distance from day centre (m)						
Mean	33.26	68.94	101.60	36.86	56.22	89.09
s.d.	26.14	51.45	76.51	25.85	36.44	65.13

Note: Differences between bucks and does are statistically significant ($p < 0.01$).

principal diurnal predators (Delibes and Hiraldo 1981; Chapter 3) from which rabbits must hide. Since they need to feed as well as rest by day, they must also be able to graze under cover. In the Orongorongo Valley the rabbits emerged from the scrub in late afternoon, but waited in the shadows until it was dark, and then made a dash for the night feeding grounds. Their best defence against nocturnal carnivores was to keep out in the open, presumably so as to avoid a surprise attack. Rabbits do this in the Orongorongo Valley despite the generally sparse grazing available on open parts of the riverbed.

By day in the Orongorongo Valley individual rabbits were usually well-dispersed in the scrub, and they tended not to congregate in groups until nightfall. After dark, small groups of rabbits may come together on the night feeding grounds. This larger grouping at night may also assist in defence against prowling carnivores (Burnett and Hosey 1987). The dispersion of rabbits in the Orongorongo Valley seemed consistent with that reported by Daly (1981) at Urana, New South Wales, where the rabbits lived 'in semi-isolated rather than isolated breeding groups'.

6.8 Sociality

6.8.1 Pairs

Until sexually mature, juvenile rabbits live furtively and often alone; but once mature at 3–4 months of age they are liable to form pairs and take their place in a dominance hierarchy during their first breeding season. If still subordinate they may continue to live solitarily or as satellites attatched to established pairs, and continue in this role indefinitely (Gibb 1993). Solitary and satellite rabbits in the Orongorongo Valley were almost always bucks, which outnumbered does in the population. One buck lived solitarily for 2–3 years; while another (No. 40) lived as a satellite with an established pair until first obtaining a doe of his own when five years old.

Members of a pair keep together throughout the year, though the buck is apt to go courting a younger doe towards the end of the breeding season; he may or may not return. Some rabbits pair for life; but as life is often short, the chances of both members surviving for more than a year are slight. Even when both members survived in the Orongorongo Valley, pairs usually broke up inside a year; and as bucks usually lived longer than does, the buck of the pair was, on average, twice as old as the doe. The bucks may have taken fewer risks of predation than the does, by feeding in the open (see Section 6.7.4).

6.8.2 Group-size

At low density the pair is the normal group-size; in the Orongorongo Valley (Gibb 1993) fewer than five per cent of the groups seen comprised three or more rabbits, with a maximum of five. Neighbouring pairs often shared communal feeding grounds at night, and one footloose young buck (No. 29) in the Orongorongo Valley made a habit of visiting a neighbouring group of rabbits 500 m away, where he seemed to be accepted. Another apparently spent the day more than 300 m away from his mate and came looking for her each evening, over a period of several months.

The average group-size in the Orongorongo Valley was small probably because food was sparsely distributed. Most home ranges were too large or too fragmented to be defended effectively. This did not prevent intruders being driven off when and wherever they were found by the resident buck; indeed territorial encounters between neighbouring bucks were surprisingly common, considering their low density.

6.9 Reproduction

From north to south, the reproduction of rabbits in New Zealand has been studied by Watson (1957) in Hawke's Bay, Williams and Robson (1985) near Wanganui, Gibb *et al.* (1985) in the Wairarapa, and Gibb (unpublished) in the Orongorongo Valley, in the North Island; by Bell (1977) and Williams (unpublished) in North Canterbury, and by Fraser (1988) and Robertshaw (unpublished) in Central Otago, in the South Island. There was little variation between sites in most aspects of the rabbits' reproduction, except for Central Otago.

6.9.1 Warrens or stops?

Rabbits breed either in off-shoots of permanent warrens or in isolated breeding 'stops' opening directly to the surface (Lloyd and McCowan 1968). When numerous in New Zealand, until the late 1950s, rabbits commonly lived in large warrens in open country, each harbouring up to about 100 animals. According to Mykytowycz and Gambale (1965), dominant does breed in the warrens, and subordinates less successfully in stops. Yet Watson (1957) found that, where above-ground cover was available in Hawke's Bay, the rabbits lived mostly above ground and dug few holes except for nesting; apparently this was their preferred life-style. Whether rabbits normally spend the day above or under ground probably depends on soil drainage and the likelihood of flooding, and on the ease of digging.

Large concentrations of rabbits in open country were the prime target for early organized control, and were easily dealt with. Increasingly thereafter, most populations have consisted of small groups of rabbits living above ground among rocks or in scrub, without permanent warrens; and they breed mainly in stops even where numerous in Central Otago. This would make it much more difficult to get fleas established if it were decided to introduce them as vectors of the myxoma virus, because: (i) as release points, stops are much harder to find than are open warrens; (ii) when released outside a stop, the fleas might disperse before it was opened up by the doe to feed the young; and (iii) when a stop is found, there is nothing on the surface to show whether or not it contains young, and any disturbance of the ground to find out is likely to make the doe desert. Also, many more stops are dug, and their entrances stopped, than are ever used (Gibb 1993). These points have not been considered in impact assessments for introducing myxomatosis, published thus far.

The stops are dug and the nests built solely by the doe, but the buck watches closely. Many of the stops dug are not used until some weeks or months later, and others are never used. Does normally block the entrance to the stop with spoil whenever they leave, both before and after the young are born; and they unblock the entrance for only about five minutes a night to suckle their young. In New Zealand, ferrets may gain access to stops through the spoil or they may sometimes dig directly into the breeding chamber from overhead (J. Robertshaw personal observation). Stops are liable to become flooded or to collapse in loose soil, and they are sometimes dug at

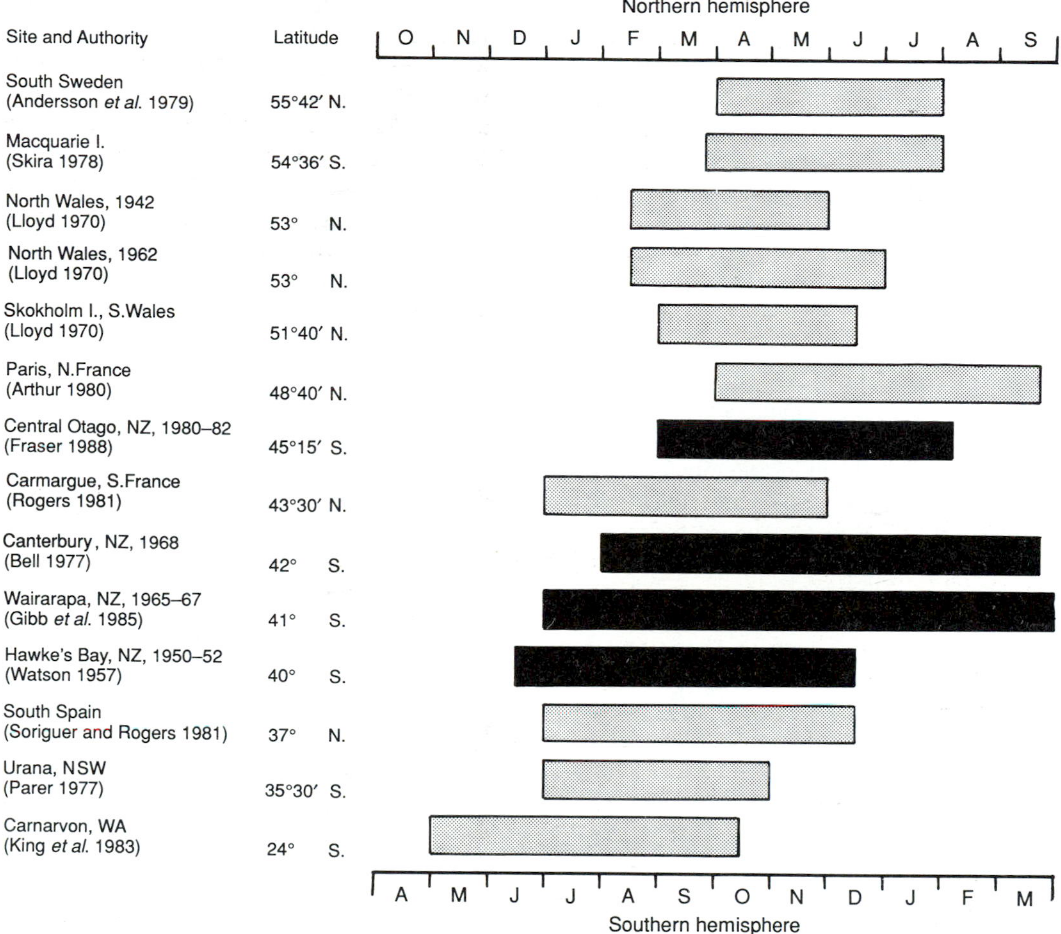

Fig. 6.8 Breeding seasons of rabbits worldwide (>50 per cent of does pregnant), related to latitude; New Zealand sites in black.

some distance from suitable feeding grounds for the young (for example on stony river-flats). R. J. Pierce (personal communication) found some nests above ground among tussock grasses growing on dry, stony riverbeds in a South Island montane basin, where digging would have been laborious; and a few others have been found in rank vegetation on developed farmland in North Canterbury (D. L. Robson personal observation).

6.9.2 Breeding season

Despite some local variation depending upon the habitat, the breeding season of rabbits worldwide tends to commence later the higher the latitude, a trend noticeable across Europe (Chapter 3) and even within New Zealand (Fig. 6.8). Breeding in most of New Zealand is almost continuous, but with a peak in spring; the coincidence with pasture production is obvious in Central Otago. Comparison of the two graphs in Fig. 6.9 suggests that food *per se* is not severely limiting in such districts as the Wairarapa, where the pasture continues growing almost year-round (Radcliffe 1975). There is a marked peak in pasture production from September to November, but usually no marked trough in winter compared with late summer and autumn. The rabbits' breeding likewise peaks in spring, but at present low density it is broadly sustained through the summer when pasture production is minimal; and the decline in April–May–June does not coincide with greatly reduced pasture production.

On unirrigated land in Central Otago (Radcliffe and Cossens 1974) the pasture is at all seasons less productive than in the Wairarapa: peak production in spring is similar to that in mid-winter in the Wairarapa. Yet the modest spring peak in production in Central Otago is enough to produce a surge in breeding: about 85 per cent of the does were pregnant in September and October compared with 95 per cent at this time in the Wairarapa; this level of activity was sustained for $5\frac{1}{2}$ months until February, when pasture production also declined. In Central Otago, without irrigation, almost no rabbits breed in autumn or water (Fraser 1988), and the herbage stops growing from May to August inclusive. It starts growing again in late September instead of in August as in the Wairarapa, which is reflected in the later start to breeding in Otago.

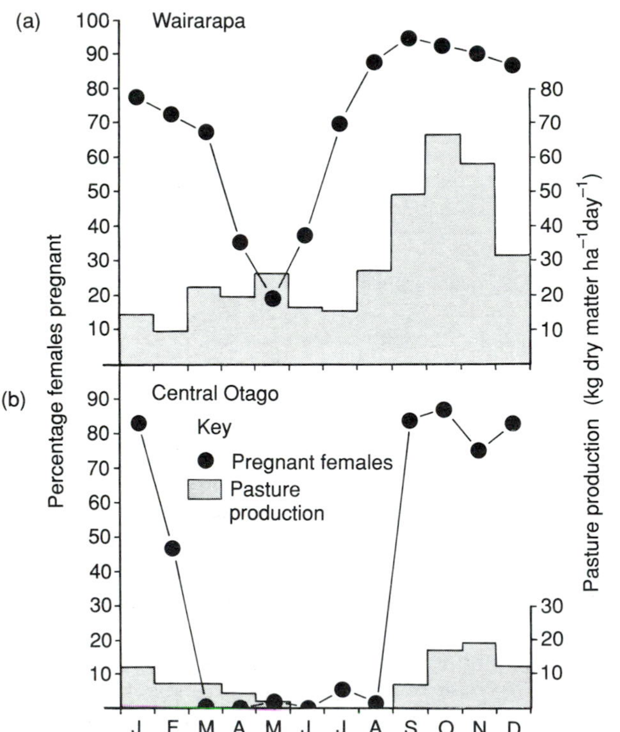

Fig. 6.9 Breeding season of rabbits in the Wairarapa (Gibb *et al.* 1985) and Central Otago (Fraser 1988) related to pasture production (Radcliffe 1975; Radcliffe and Cossens 1974).

Watson's (1957) study of reproduction in Hawke's Bay is the only one covering the high densities then prevailing. His measurements of the breeding season (i.e., percentage of does pregnant) are here multiplied by 30/25 to allow for pregnancies undetected during the first five days of gestation (see Gibb *et al.* 1985; Gilbert *et al.* 1987). This sometimes produces pregnancy rates apparently exceeding 100 per cent, which is impossible; so they are reduced to 100 per cent in Fig. 6.10.

The breeding season of rabbits at high density in Hawke's Bay (Watson 1957) is compared in Fig. 6.10 with that at low density in the Wairarapa (Gibb *et al.* 1985): the main difference is the much smaller proportion of does pregnant in late summer and autumn at high density in Hawke's Bay. This is to some extent compensated for by the apparently slower build-up of breeding in late winter at the lower density in the Wairarapa.

Even in the dense population of rabbits in the early 1950s, Watson (1957) regarded it as 'exceptional in Hawke's Bay to find an adult male without some sperm'. At low density in the Wairarapa (Gibb *et al.* 1985) very few bucks had scrotal testes before they were four months old, compared with 33 per cent of those 5 to 6 months old; thereafter more of the bucks had scrotal testes, increasing to 55 per cent when they were 7 to 9 months old and 62.5 per cent when more than 36 months old. Seasonal variation in the Wairarapa was slight, from 58 per cent with

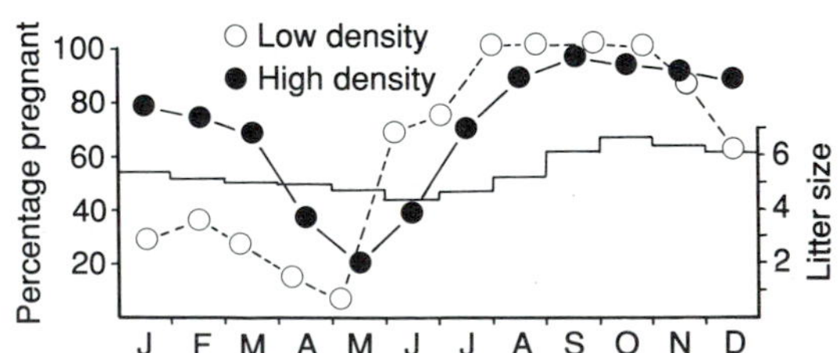

Fig. 6.10 Percentage of does (>3-months-old) pregnant and mean litter-size at birth at low density in the Wairarapa (●——●) and at high density in Hawke's Bay (○ – – ○). On the average litter-size in Hawke's Bay was 0.5 less than in the Wairarapa (Gibb *et al.* 1985; Watson 1957).

Table 6.7 Regional differences in mean litter-sizes at birth and in the reproductive rate (young born per adult doe per year)

Region	Latitude	Mean litter size	Reproductive rate	Authority
Hawke's Bay	40°S	5.03	36–45	Watson, 1957
Wanganui district	40°S	5.23	47.6	Williams and Robson, 1985
Wairarapa	41°S	5.15	44.2–45.9	Gibb *et al*., 1985
North Canterbury	42°S	5.54	42	Bell, 1977
		5.85	37–43	J. M. Williams, unpublished
Central Otago	45°S	5.94	23.1	Fraser, 1988

Note: Watson (1957) and Bell (1977) published lower reproductive rates than those quoted here. They omitted litters born out of the main breeding season because few young survived, and they did not adjust for undetected pregnancies during the first five days of gestation.

scrotal testes in autumn to 67 per cent in spring; but it was more pronounced in Central Otago, where fewer than 20 per cent of the bucks had scrotal testes in autumn and sometimes also in winter (Fraser 1988).

In arid New South Wales, where rain is unpredictable, rabbits breed opportunistically (Wood 1980). The weather is more predictable in New Zealand, but rabbits once bred unseasonably at low density at Kourarau during an unusually warm but moist autumn, when there was plenty of feed. Most does of all ages produced a single extra litter and the young survived well in the virtual absence of predators (Gibb *et al.* 1978).

6.9.3 Litter-size and pre-natal mortality

Regional differences in the rabbits' mean litter-size at birth are slight (Table 6.7), but litter-size varies seasonally: in the Wairarapa it ranged from 4.4 in June to 6.7 in October; in North Canterbury, from 4.7 in June–July to 6.4 in October; and in Central Otago from 5.0 in January–February to 6.5 in October–December, though smaller and less-variable average litter-sizes have been recorded here more recently (J. Robertshaw personal communication). On average through the year in the Wairarapa, young does 4 to 6 months old had litters averaging 4.5 young, and those 10 to 12 months old had the largest litters averaging 6.5 young; progressively older does had slightly smaller litters (Table 6.8).

Table 6.8 Percentage pregnant and mean litter-size of female rabbits of different age in the Wairarapa, all seasons pooled (Gibb *et al*. 1985)

Age (months)	Per cent pregnant	Litter size mean (s.e.)
4–6	48.9	4.53 (0.07)
7–9	64.2	5.49 (0.06)
10–12	87.0	6.46 (0.07)
13–24	85.0	6.18 (0.04)
25–36	79.0	5.80 (0.07)
>36	76.5	

The incidence of pre-natal mortality of embryos is highly variable, affecting both whole litters and single embryos. At times in the Wairarapa, pre-natal mortality was detected in at least 30 per cent of the litters, reducing their average size also by at least 30 per cent; and it was slightly more frequent in litters initially of above average size, reducing them to just below average size. Pre-natal mortality was also more frequent in older rather than in younger does (Fig. 6.11). In the early 1980s Fraser (1988) estimated that about 70 per cent of the litters in Central Otago had some pre-natal loss, affecting on average about 18 per cent of the embryos; about 12 per cent of the litters conceived were lost entirely. More recent estimates are lower than this (J. Robertshaw personal communication).

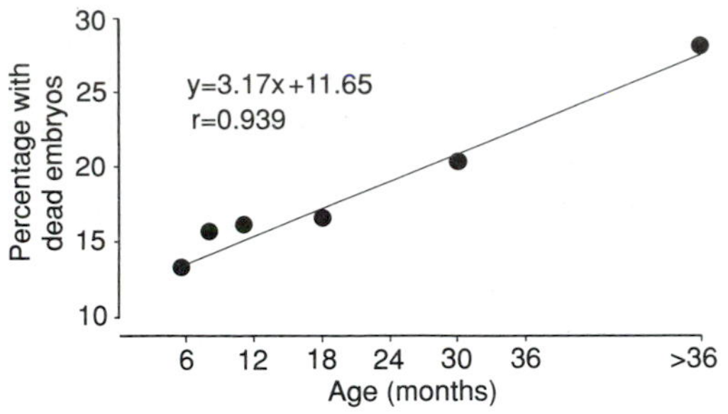

Fig. 6.11 Percentage of does more than five days pregnant having one or more dead embryos in the uterus, related to the does' age in the Wairarapa (Gibb *et al*. 1985).

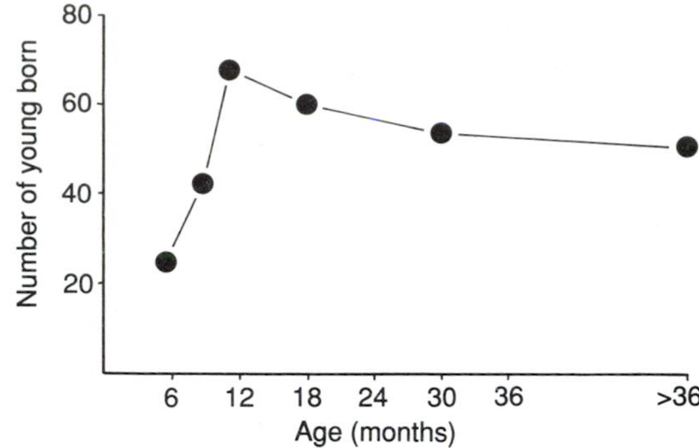

Fig. 6.12 Annual reproductive rate of does according to their age, in the Wairarapa (mean of two years) (Gibb *et al*. 1985).

6.9.4 Reproductive rate

Estimates of the annual reproductive rate of rabbits in New Zealand range from 23.1 young per doe in dense populations in Central Otago to 47.6 young in sparse populations near Wanganui (Table 6.7). The higher estimates are the highest recorded for the species (cf. Gilbert *et al.* 1987); they result more from the exceptionally long breeding season than from large litters.

In the Wairarapa, does were first found pregnant when 3–4 months old. Does aged 10–12 months had the highest pregnancy rates and litter sizes; older does produced progressively fewer young (Fig. 6.12). By contrast, the fertility of bucks continued to increase slowly, at least until they were more than 36 months old. Does do not suddenly become senescent when over 12 months old; perhaps they maximize their reproductive rate when only 10 to 12 months old because their chances of producing another litter fall off very rapidly with age.

6.10 Growth rates and body weights

6.10.1 Growth of young

The growth rates and live body weights of rabbits were recorded at varying population densities at Kourarau (Gibb *et al.* 1978); additional information comes from shot samples of free-ranging rabbits (Tyndale-Biscoe and Williams 1955; Watson 1957; McIlwaine 1962; Bell 1977; Williams and Robson 1985; Gibb *et al.* 1985; Fraser 1988; and J. M. Williams unpublished).

Tyndale-Biscoe and Williams' (1955) growth curve (Fig. 6.13) is used here as an optimum for comparing the growth of retarded young. Young rabbits weigh 30–35 g at birth and 200–250 g on leaving the nest three weeks later. They begin grazing immediately they leave the nest, but may suckle for another week or so if the mother is not again pregnant.

As a rule young born early in the spring, when the

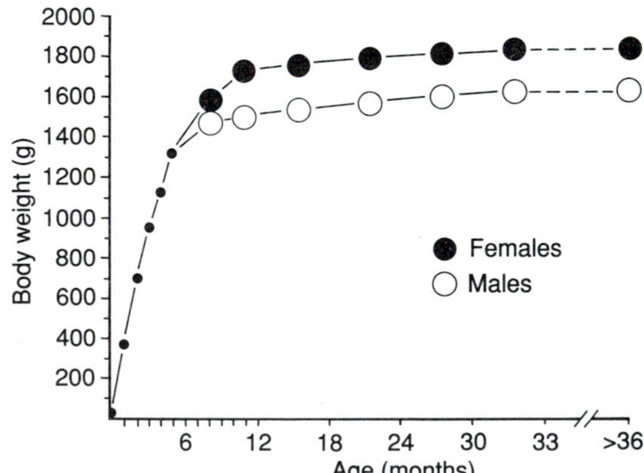

Fig. 6.13 Mean live body weight of rabbits in Hawke's Bay and the Wairarapa, pooled. Sexes of rabbits less than 6-months-old were not separated. Maximum s.e. of means 6.5 g (Gibb *et al*. 1978, 1985).

pasture is still lush, grow at near the optimum rate; but the growth rate of young born later in the summer declines as feeding conditions deteriorate. The summers of 1959–60 and 1965–6 in the Kourarau enclosure illustrate this (Gibb *et al.* 1978). In each summer the density of rabbits began at a moderate level and finished at peak (Table 6.9); a drought in January 1960 reduced the growth of late-born young, born in late October or November 1959, to 0.35 of the optimum. This rate applied only to young which survived long enough to be weighed twice—others died earlier. The pasture quickly responded to rain at the end of January and the growth of the young increased to 0.91 of the optimum.

The pasture was in good condition at the start of the 1965–6 breeding season and the young grew well until the end of January. The problem came in February when growth rates slumped to 0.33 of the optimum. This was not triggered by a drought, but came simply when the rabbits' increasing demand for food exceeded its supply in late summer. Harsh times persisted through the autumn and winter, and the population crashed. Young whose growth was retarded late in the season, remained skeletally small for life and never quite achieved normal body weight for age.

6.10.2 Body weights

Watson (1957) found no difference in the paunched weights of bucks and does. Live weights varied seasonally with the proportion of does pregnant, which itself varied with population density. In the Kourarau enclosure, bucks and does were of similar live weight for age until 4 to 6 months old. Thereafter, throughout one population cycle subjected to unrestricted predation, bucks were on the average 3.2 per cent heavier than the does in summer, autumn, and winter; but when most does were pregnant or lactating in spring, they became on average 7.6 per cent heavier than the bucks.

At much lower density in the Wairarapa, when more than 50 per cent of the does were pregnant in all months except April, May, and June, the does were heavier than the bucks by an amount that varied with the proportion of does pregnant (Gibb *et al.* 1985). The live weights of rabbits were not recorded in Central Otago, where the length of breeding season was much shorter.

The mean paunched weights of rabbits in different districts are given in Table 6.10. The generally heavier rabbits from North Canterbury are believed to be descendants of original liberations of large domestic stock. The rabbits of domestic origin that were eradicated from Motunau I., off the North Canterbury coast, were still heavier than any recorded elsewhere (Cox *et al.* 1967).

In the wild, the mean live weight of both sexes continues to increase slowly until they are at least 3 to 4 years old, which is the limit to which rabbits can be aged.

6.11 Population dynamics

6.11.1 Hawke's Bay

Tyndale-Biscoe and Williams (1955) first addressed the population dynamics of wild rabbits in New Zealand; as with Watson's (1957) study of reproduction, theirs is the only one dealing with the dense populations then prevailing. The two studies were done together, so are complementary.

The survival rate of young and old rabbits was calculated from the recovery of marked animals. For their first 100 days out of the nest, the survival rate of the young per 30-day month was about 0.65, and that of rabbits more than 120 days old about 0.87 (Fig. 6.14). These rates were presumably higher than they would have been if the population had not been increasing since the previous year's trapping. The sudden transition from the lower juvenile to the higher adult survival rate at about 120 days of age (i.e., when the young became mature) was statistically significant.

Tyndale-Biscoe and Williams identified coccidiosis and predation by cats, mustelids, and harriers as the most likely causes of death of the young. They pointed out that young rabbits were most severely infected with coccidia when 40 to 80 days old, which was when most young died. Infection declined to very low levels by the time the young were 200 days

Table 6.9 Growth rates of young born early, mid, and late in the season during two summers (after Gibb *et al*. 1978)

Year and range of density	Months when young weighed	No. young weighed	Growth rate cf. 'normal'[3]
Summer 1959–60	Sep–Dec	57	1.02
(46–125 rabbits per ha)	Jan early[1]	28	0.65
	Jan late[2]	22	0.35
	Feb–Apr	39	0.91
Summer 1965–66	Nov–Dec	27	0.84
(35–200 rabbits per ha)	Jan	18	0.78
	Feb–Mar	20	0.33

[1] Young born before end September.
[2] Young born October–November.
[3] For 'normal' growth rate see Table 23 in Gibb *et al*. (1978).

Table 6.10 Mean carcase weight and length of adult rabbits (>6 months old) in three districts (after Gibb and Williams 1990)

Season		Western North Island[1]		North Canterbury[1]		Central Otago[2]	
		Bucks	Does	Bucks	Does	Bucks	Does
				Carcase weight (g)			
Spring	Mean	1344	1371	1560	1583	1453	1441
	N	3792	3820	3384	2916	296	294
	s.e.	2.1	2.1	2.2	2.4	10.2	10.7
Summer	Mean	1370	1360	1521	1536	1551	1546
	N	1138	972	1795	1468	235	166
	s.e.	3.9	4.2	3.1	3.4	10.9	12.6
Autumn	Mean	1353	1325	1506	1477	1415	1358
	N	2335	2067	2908	2342	267	254
	s.e.	2.7	2.9	2.4	2.7	11.7	10.3
Winter	Mean	1366	1388	1528	1530	1312	1239
	N	2570	2344	1520	1162	444	390
	s.e.	2.6	2.7	3.3	3.8	9.7	10.0
				Length (mm)—tip of nose to tail			
All year	Mean	484	490	492	494		
	N	9565	7937	966	822		
	s.e.	1.9	2.2	0.7	0.9		

[1] Carcase weight includes kidneys and associated fat.
[2] Carcase weight excludes kidneys and associated fat.

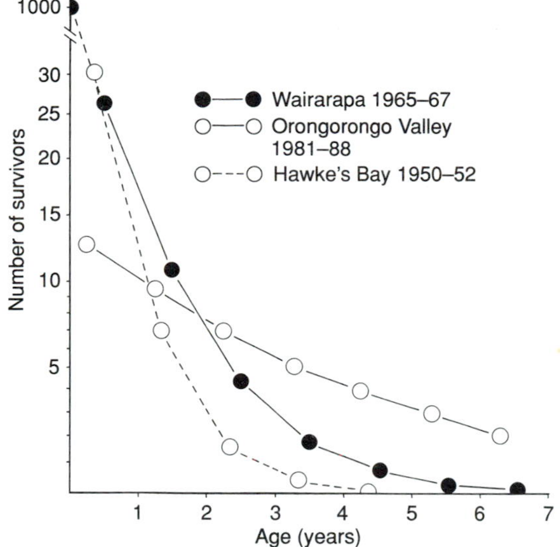

Fig. 6.14 Estimated survival of rabbits from birth in the Wairarapa, assuming an annual reproductive rate of 45 young per adult doe, an adult survival rate of 0.41 from age 6 months and a stable population density (after Gibb *et al.* 1985); with comparable estimates from a dense population in Hawke's Bay (Tyndale-Biscoe and Williams 1955) and a sparse population in the Orongorongo Valley (Gibb unpublished).

old; but there was no sudden drop in infection rates when the rabbits were about 120 days old to account for their improved survival then. Southern (1940) likewise concluded that most young rabbits on a warren in Oxfordshire disappeared before they were four weeks old as a result of predation by cats and dogs. Tyndale-Biscoe and Williams commented that the low adult survival rate 'suggests that most rabbits only survive for one breeding season', which agreed with Southern's (1940) observations in the absence of formal control. Later on, we compare Tyndale-Biscoe and Williams' estimates with more recent ones, but first we consider the population at Kourarau.

6.11.2 Kourarau

Over the ten years of the Kourarau study (Gibb *et al.* 1978) the population of rabbits in the enclosure went through two cycles of abundance (Fig. 6.15), the first with unrestrained predation by cats, ferrets, and harriers, and the second, rising even higher, with most cats and ferrets excluded. The second peak alone was followed by mass starvation of the young.

The ultimate limit to the density of rabbits at Kourarau was set by the supply of food. The rabbits' social behaviour was probably more important in deciding which rabbits survived, rather than how many. After the first peak outright starvation among full-grown rabbits was forestalled by predation by cats and ferrets, which found it easier to catch rabbits when they risked predation to get enough food. When cats and ferrets were excluded during the second cycle, the population of rabbits temporarily overshot the carrying capacity of the land; all of

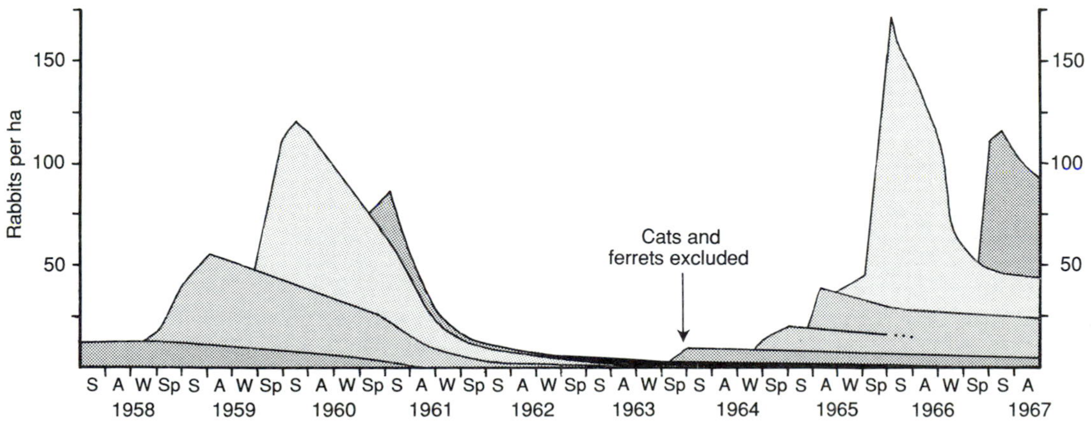

Fig. 6.15 Population density of rabbits in the Kourarau enclosure, with approximate age composition. The long decline through 1960–63 was reversed with the exclusion of carnivores in late 1963 (Gibb *et al.* 1978).

them lost weight and many (especially bucks) died of starvation (Gibb *et al.* 1978).

The reasons for the $3\frac{1}{2}$-year decline in the population between the two peaks were not at first obvious. The intensity of predation, measured by the frequency with which cats and ferrets were seen, continued to increase for some months after the rabbits peaked in the summer of 1959–60. This resulted in a typically delayed predator–prey oscillation at the end of which the numbers of both predators and prey returned to low levels (Fig. 6.16). Very few young had survived for long after the 1960 breeding season, and none at all in 1961–2, 1962–3, or the first half of the 1963–4 season; and the total population crashed from over 1000 in January 1960 to 13 (including only two does) by September 1963.

Two-thirds of the way through the 1963–4 breeding season we switched on an electric fence around the enclosure, designed to keep out the cats, having previously trapped and removed the ferrets. Within 2–3 weeks numbers of young rabbits were surviving above ground for the first time for $3\frac{1}{2}$ years. This emphatic result confirmed the key role of cats and ferrets in the early mortality of young rabbits, as in Australia (Richardson and Wood 1982) and North Canterbury (D. L. Robson personal communication), and in the subsequent decline of the population.

Evidently the carnivores were able to reduce the numbers of rabbits so low because at least some of them survived, subsisting mainly on other foods but killing the odd rabbit when they could. As with Pearson's (1966) cats preying on *Microtus*, the existence of an alternative supply of food allowed the cats to exaggerate and prolong the rabbits' decline. Chitty (1967) described the similarly long decline of voles and lemmings after reaching peak density as their 'most puzzling feature'; he dismissed predation as a possible cause, at that time preferring genetically-induced behavioural changes in the prey as an explanation (Krebs 1978).

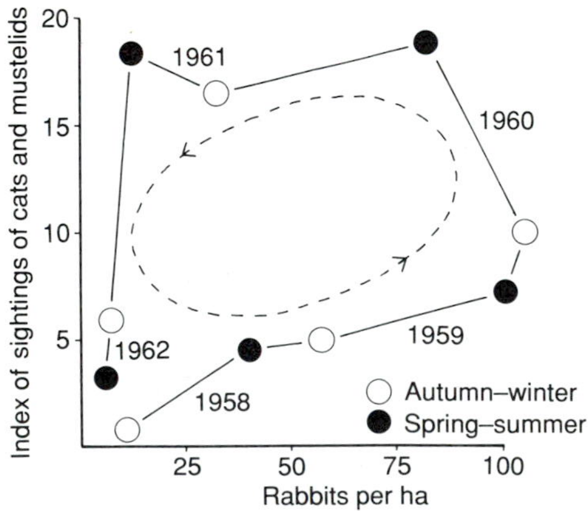

Fig. 6.16 Relationship between the density of rabbits and the number of carnivores seen in the Kourarau enclosure (Gibb *et al.* 1978). Changes in the number of carnivores lagged behind changes in the density of rabbits, producing a 'delayed' predator–prey cycle.

At relatively low density, as in the breeding season of 1964–5 (Fig. 6.17), young born late in the season survived just as well as those born earlier. In the breeding season of 1965–6, however, at much higher density, young born early in the season survived better than those born mid-season, and those born late in the season survived worst of all. Conditions improved somewhat in autumn, but deteriorated again in the winter of 1966. During the ensuing crash, the three-year-old rabbits survived better than the two-year-olds, and two-year-olds better than rabbits in their first year; and among these first-year rabbits, the older ones born early in the 1965 season survived better than younger ones born later.

In the Kourarau enclosure several facets of the rabbits' biology changed in ways conducive to density-dependent control (Gibb *et al.* 1978; Garson 1981).

1. With increasing density before cats and ferrets were removed, the breeding season was shortened and the increase in numbers correspondingly reduced, while mortality between breeding seasons increased.
2. The quantity and quality of the pasture deteriorated, the rabbits spent longer above ground and feeding, and less time inactive, and they dispersed further from the warrens; they also took less notice of passing predators.
3. With increasing density there were fewer burrow entrances per rabbit, which probably facilitated predation (Parer 1977); the main warren became overcrowded and surplus rabbits had to open up new ground.
4. The size of the defended territories contracted, and the rabbits had to use communal feeding

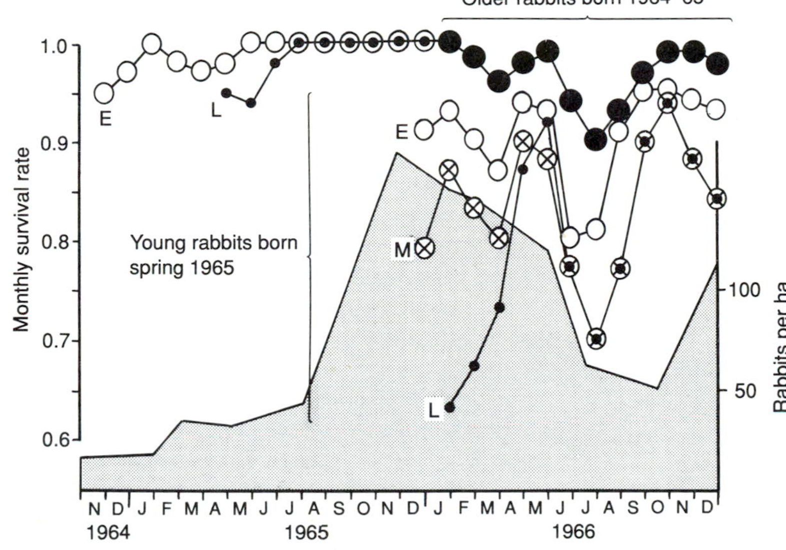

Fig. 6.17 Monthly survival rates of young rabbits born early (E), mid (M) and late (L) in the 1964–65 and 1965–66 breeding seasons, compared with older rabbits in the Kourarau enclosure. The graph shows the better survival of young born early in the season at high density in 1966, but not at low density in 1965; and the better survival of older rabbits than of young ones during the crash in 1966 (Gibb *et al.* 1978).

grounds far from the relative safety of the warrens.

5. With increasing density, young rabbits became more heavily infected with coccidia and they lost weight; their mortality (from predation) increased.
6. Until removed in late 1963, cats and ferrets were seen more often and ate more rabbits, as density increased.
7. These trends were spread over the whole range of rabbit densities, not just at extremes.

The enclosed condition of the Kourarau population (with restricted dispersal and the removal of predators) led to widely fluctuating extremes in numbers, as in voles (Krebs 1971). The population nearly died out in mid-1963 and was saved only by excluding the predators; and numbers rose so far above carrying capacity in 1965–6 that many rabbits starved.

6.11.3 Wairarapa

Marked short-term fluctuations in the numbers of rabbits are typical of enclosed populations (Krebs 1971), whereas sparse free-ranging populations on the New Zealand mainland are usually stable for years. Density-dependent predation seemed likely to explain this stability in eastern Wairarapa (Gibb *et al.* 1969); but a more extensive study of rabbit populations in the whole district (2535 km^2) was required to examine the effects of control operations by the Rabbit Board. This lasted for two years and involved the autopsy and ageing of more than 17 000 rabbits (Gibb *et al.* 1985).

Tyndale-Biscoe and Williams (1955) and Watson (1957) established that young rabbits born early in the season may themselves breed later in the same season, but that their young rarely survive for long. Using the Wairarapa data, Neil Gilbert (in Gibb *et al.* 1985) estimated that the survival rate of young conceived between December and April, after the main breeding season, was less than 0.01 from mid-pregnancy to 6 months of age, compared with about 0.1 for young conceived between May and November (Fig. 6.18). The mean annual survival rate of rabbits over 6 months old was 0.41; but the proportion of bucks shot increased with the rabbits' age, so they evidently survived on average longer than does (Table 6.11). Thus, in the Wairarapa at large the adult survival rate was higher, and the juvenile rate lower, than estimated by Tyndale-Biscoe and Williams (1955) at higher density in Hawke's Bay.

A crude index of population density was allocated to each property in the Wairarapa, based on the

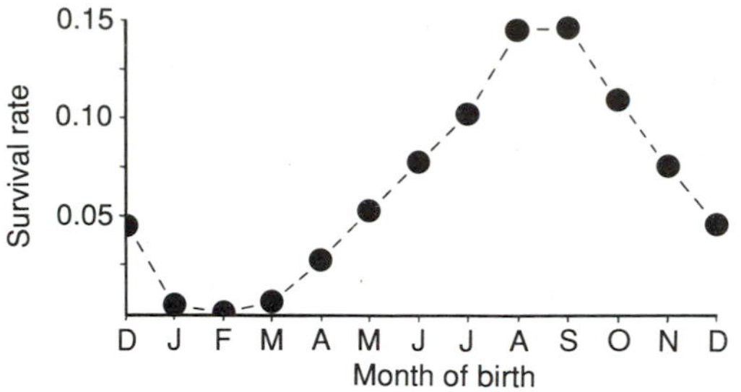

Fig. 6.18 Approximate survival rate of young rabbits from birth to 6 months of age, according to the month of birth in the Wairarapa (Gibb *et al*. 1985).

number of rabbits shot per square kilometre each year; the properties were then placed in groups of low, medium, or high density (Gibb *et al.* 1985). Table 6.12 shows that samples shot on low-density properties contained fewer first-year rabbits, and more over three years old, than did those from high-density properties. In other words, mortality factors were different on properties with varying densities of rabbits. This was evident even over the range of generally low densities in the Wanganui area (Williams and Robson 1985), as well as in the Wairarapa as a whole compared with higher densities in Hawke's Bay in the early 1950s; and it was most pronounced in the Orongorongo Valley (see below).

Gilbert (in Gibb *et al.* 1985) fitted sine curves to the age composition of rabbits shot through the year in the Wairarapa. Despite the high reproductive rate of 45 young per female per year, Fig. 6.19 shows that the maximum population at the end of the breeding season in March was only twice the minimum in September.

There were also significant local differences in the age composition of rabbits shot in the Wairarapa, apparently related to differences in rainfall, but pos-

Table 6.11 Sex ratio of rabbits shot in the Wairarapa, by age (Gibb *et al*. 1985)

Age (months)	Number of rabbits shot	Per cent bucks
1–12	9607	48.3
13–24	5080	54.2
25–36	1343	60.3
>36	1028	65.4

Table 6.12 Age composition of rabbits shot in districts of increasing population density in the Wairarapa (Gibb *et al*. 1985)

Age (months)	Percentage age composition Population density		
	Low	Medium	High
1–12	52.2	59.5	59.6
13–24	30.6	26.9	28.6
25–36	9.4	8.4	7.1
>36	7.8	5.2	4.7

Key to population density:
Low: 1–4 rabbits shot per km^2 per year
Medium: 5–9 rabbits shot per km^2 per year
High: >9 rabbits shot per km^2 per year

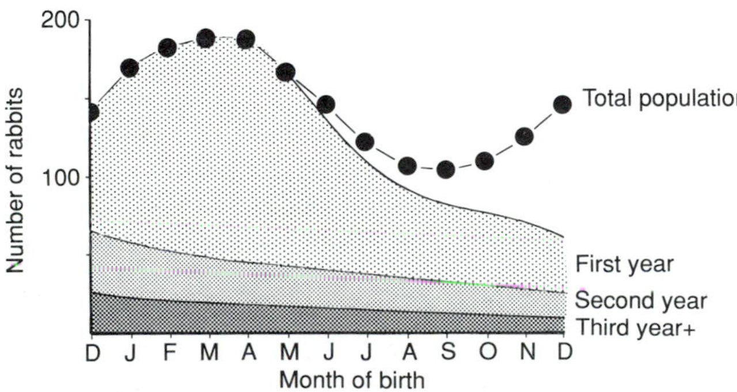

Fig. 6.19 Sine curves showing seasonal changes in the age composition and total population (scaled to an arbitrary minimum of 100) of rabbits in the Wairarapa (Gibb *et al*. 1985).

sibly also a function of differences in the intensity of shooting (Gibb *et al.* 1969, Williams and Robson 1985). Although it is often necessary to pool samples collected from different localities and at different times of year to emphasize broad trends, the regional population has to be seen as a kaleidoscope of local populations, each liable to change asynchronously with the rest (Andrewartha and Birch 1984, Gibb *et al.* 1985; see also Gilbert *et al.* 1987).

6.11.4 Orongorongo Valley

The Orongorongo Valley supports a sparse population of rabbits living on stony river flats hedged in by dense forest on both sides of the valley. The valley closes in and rabbits peter out 3–4 km upstream of the study area, but similarly sparse populations exist for 10 km downstream to the coast.

Changes in the distribution and amount of rabbit sign along 5 km of river flats were measured in late summer every year but one from 1971 to 1990 (Gibb unpublished). Although the gross amount of sign changed considerably over the years, it did so gradually and apparently in response to changes in the extent of plant cover, and perhaps also to the number of cats (Fig. 6.20).

Preliminary results suggest that the breeding season of rabbits in the Orongorongo Valley starts 3–4 weeks later in the spring than in the Wairarapa (Gibb *et al.* 1985); that all breeding takes place in stops, many of which become flooded or collapse; that some adult does probably do not breed continuously all through the breeding season; and that young out of the nest survive very poorly, partly (at any rate) as a result of heavy predation by feral cats and stoats. Against this, the annual survival rate of

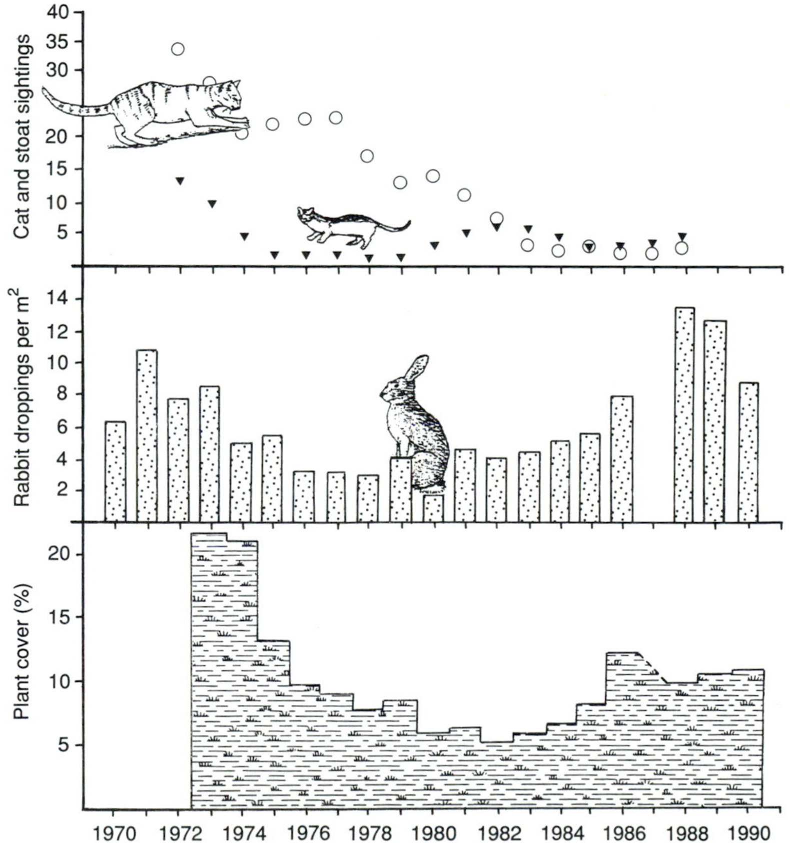

Fig. 6.20 Relationship between the density of rabbit 'sign' (mainly droppings) and the extent of plant cover, measured in late summer for 15 years, with indices of sightings of feral cats (after Fitzgerald 1988) and stoats in the Orongorongo Valley (Gibb unpublished).

adult bucks is about 0.75–0.80, and that of does about 0.55, which is much higher than the average of 0.41 in the Wairarapa.

6.11.5 Population regulation

We agree with Gilbert *et al.* (1987) that there is no such thing as *the* population dynamics of the rabbit (see also Gibb *et al.* 1985): different populations are regulated differently and the factors involved are complex. We also agree with Myers (Chapter 5) in rejecting single-factor explanations as unrealistic.

Andrewartha and Birch (1984) chose the rabbit in Australia to illustrate their view on population regulation. Impressed with the instability of Australian ecosystems, they claimed that population regulation is an illusion of scale due to supposedly asynchronous fluctuations of numerous small, local populations. Up to a point we agree, of course, that a regional population is a mix of local populations (Gibb *et al.* 1985); but the stability of regional populations in most of New Zealand cannot be accounted for merely by the supposed asynchrony in the fluctuations of local populations, because local populations are themselves equally stable (Gibb *et al.* 1969; Gibb unpublished).

Natural populations of rabbits are usually subject to heavy predation (Delibes and Hiraldo 1981), and many of the predators involved, including feral cats (Catling 1988; Fitzgerald and Karl 1979; Fitzgerald 1988) and mustelids (King and Moody 1982), concentrate on young rabbits, which are available only in the rabbits' breeding season.

In Central Otago, for instance, the rabbits' breeding season is short (Fraser 1988) and there may be too few other prey to support all the predators in the long hiatus between breeding seasons, causing them to decline. This, along with the otherwise favourable habitat, may help to explain why rabbits remain more abundant there than elsewhere in New Zealand. Similarly overseas, rabbits have remained abundant on the island of Skokholm (*c.* 100 ha) off the Pembrokeshire coast of Wales, where their only predators are occasional birds of prey and gulls; and in the Brecklands of East Anglia where predatory 'vermin' are suppressed by strict gamekeeping (for example, Tapper *et al.* 1982, J. A. Gibb personal observation). In semi-arid regions of Australia, too, the rabbits' breeding season is as short as in Central Otago (Chapter 5); and this, with periodic droughts, may destabilize populations of predators and their prey (Newsome *et al.* 1989).

Though obviously substantial, the mortality of nestlings in the Orongorongo Valley is hard to measure because it depends on the erratic incidence of floods and heavy rain, which vary from year to year. The mortality of young after leaving the nest is also hard to measure. Young rabbits are the cats' and probably the stoats' preferred prey during the breeding season; and they will presumably go on feeding on them, rather than on other prey such as rats, so long as it pays them to do so. But the point at which they switch from hunting young rabbits to other prey must also vary from year to year, if only because the numbers of other prey fluctuate. Consequently we expect considerable variation between years in the mortality of young rabbits: conditions for their survival may be relatively favourable one year and unfavourable the next, more or less regardless of their density.

Over a period of 18 years in the Orongorongo Valley the population of rabbits has changed with the extent of herbaceous vegetation, but not greatly compared with the possibilities suggested by their high reproductive rate. Thus the amount of rabbit sign over five kilometres of river flats in late summer varied from one year to the next within the limits of +45 to −40 per cent; and the number of adult rabbits counted for eight years at the start of the breeding season on 2.5 kilometres of river flats varied similarly between +45 and −20 per cent. In North Canterbury, too, spotlight counts in mid-winter ranged from 2.2 to 6.0 rabbits per kilometre over 13 years (Williams unpublished).

Rabbits have a high reproductive rate and, in the Orongorongo Valley, an unusually high annual adult survival rate of about 0.70 (Gibb unpublished); consequently even fewer than usual of the young can expect to join the breeding population in the following year. Small fluctuations in the mortality of first-year rabbits would have a big effect on the size of the next year's breeding population.

In the Wairarapa, for instance, a static population needs only about 2.7 per cent of the young born to compensate for a year's adult mortality; and in the Orongorongo Valley only about 1.7 per cent are needed. In both districts this implies an average daily loss of 8.0–8.2 young from the total production

Fig. 6.21 The late J. Sharon Watson, having supposedly shot the elusive 'last rabbit' in New Zealand, in 1957. (Photo by S. N. Beatus.)

of 50 pairs; a daily loss of 7.6–7.7 young would result in a doubling of the breeding population a year later.

Thus remarkable precision in the mortality of young rabbits is required to account for the observed relative stability in numbers between years. This could hardly be achieved by random fluctuations in the mortality of the young, even allowing some degree of density-dependence. Instead, we suggest that once rabbit populations have been substantially reduced, probably by accidental deaths and predation, their size may be finally adjusted by the adults' own spacing behaviour—modified by the availability of food.

The spacing behaviour of animals (which need not be 'territorial' *sensu stricto*) is often seen as a means of preventing over-crowding (for example, Wynne-Edwards 1962; Mykytowycz and Dudzinski 1972). If subordinate rabbits are made especially vulnerable to predation or are forced to occupy second-rate habitat, 'then behaviour can be indicted as a . . . regulating factor' (Myers, Chapter 5). Moreover, if resources such as food are scarce, dominant rabbits may resist intrusion more forcibly, and if resources are abundant they may relax. But in either case we suppose that spacing behaviour can succeed in matching numbers to the available resources (in the Orongorongo Valley, safe feeding grounds) only if extrinsic mortality has first reduced the population to a manageable size. This early mortality may or may not be density-dependent.

6.12 Developments from 1988 to 1991

Following the New Zealand government's decision in 1987 not to introduce myxomatosis, $16.4 million (plus regional funds) were set aside for a five-year Rabbit and Land Management Programme (R&LMP). This was to expand research on the biology and control of rabbits on the 'semi-arid' tussock grasslands of South Canterbury and Otago. Much has already been done and knowledge of the rabbit in this harsh environment will be greatly increased by the end of the five-year term; but we have had to impose an end-of-1991 deadline for unpublished material to be included in this chapter.

It is at least becoming recognized that the heavy infestation of rabbits on these lands and the recent spread of introduced weeds, for example *Hieracium* spp., are symptoms of 150 years of overgrazing by rabbits and sheep, as Howard (1959) perceived. The worst affected land is now of no productive value as pasture; other uses such as plantation forestry are being considered. It may still be necessary first to reduce rabbit numbers. Consequently there are renewed calls for introducing myxomatosis to reduce rabbits long enough (up to 15 years?) to provide a useful 'window of opportunity for change'. It is by no means clear, however, just what changes are feasible, and acceptable. There would certainly be daunting social as well as financial and ecological difficulties in attempting to replace pastoralism as the dominant land use; and it is agreed that government subsidies for farming and pest control must ultimately be withdrawn.

Runholders still see the rabbit as the prime culprit and ardently wish to introduce myxomatosis immediately. The R&LMP has probably not softened their stance appreciably. Towards the end of 1991 the NZ Federated Farmers sought authority once again to introduce the rabbit flea and the myxoma virus. Public submissions on the proposal were invited

Fig. 6.22 One of the worst rabbit areas in Central Otago, South Island, 1985, illustrating the extreme patchiness of over-grazing. (Photo by J. E. C. Flux.)

(over the Christmas/New Year holiday period!). At last, in June 1993, the government announced that myxomatosis was still not to be introduced.

There now remains the option that rabbit (or viral) haemorrhagic disease (RHD) may be introduced at some later date. This disease was first reported in domestic rabbits in China in 1984 and has since been reported from Korea and Mexico, and from Europe in 1988 (Xu and Chen 1989). It has killed large numbers of domestic rabbits in Italy, for instance, and has been found in wild populations in Spain (Chapter 3). RHD is highly contagious in all breeds of *Oryctolagus*; hares are also susceptible (Westbury 1989). It has an incubation period of 20–48 hours and a mortality rate approaching 100 per cent in severely affected animals. No intermediate host is involved and infected rabbits usually die in 2–3 days. It affects mainly rabbits at least two months old. Young rabbits are immune and survivors of sub-acute strains are resistant. Vaccines to protect domestic stock are effective and available. Further research on the possible use of RHD to control wild rabbits is being funded jointly by Australia and New Zealand, at the Australian Animal Health Laboratory in Geelong, Victoria.

Other research, led by Dr C. H. Tyndale-Biscoe in the CSIRO Division of Wildlife and Ecology in Canberra, aims to use a benign strain of the myxoma virus to carry a gene that would sterilize infected rabbits. This might be used to control rabbits in New Zealand; but there are still serious problems and research is at an early stage.

Finally, one objection to introducing myxomatosis to New Zealand has been that it could not be contained where it was actually needed; indeed the European rabbit flea might eventually be distributed (with the virus) everywhere **except** where it was needed, i.e., in the 'semi-arid' lands that are likely to remain inhospitable to it. Two species of fleas natural to arid parts of Spain have now been identified. If introduced to New Zealand as vectors, they might conceivably make it possible to restrict myxomatosis to 'semi-arid' parts of the South Island; on the other hand they might well spread more widely if relieved of competition with *Spilopsyllus cuniculi*.

(a)

(b)

Fig. 6.23 Over-grazed land in Central Otago, 1985, being invaded by (a) native scabweed (*Raoulia* sp.), and (b) introduced hawkweed (*Hieracium* sp.), both members of the Astetaceae. (Photos by J. E. C. Flux.)

Fig. 6.24 Hill pasture severely damaged by rabbits in the foreground. Earnscleugh Station, Central Otago, 1985. Annual rainfall *c*.400 mm.

Fig. 6.25 Hill country in the Eastern Wairarapa district about 100 km ENE of Wellington, North Island, 1958. These hills were formerly overrun by rabbits. Annual rainfall *c*. 1000 mm.

6.13 Summary

The first rabbits introduced to the New Zealand region (34–38°S) were marooned on subantarctic islands of the Southern Ocean in the mid-eighteenth century, along with pigs and goats as a source of food for castaways. By about 1830 they were being distributed throughout New Zealand for sport. Early introductions were of domestic breeds of French origin ('silver greys'), many of which failed to establish in the wild. Wild-type rabbits followed from about 1850, and spread quickly as land was opened up for grazing. They were regarded as pests by 1870 and reached peak abundance by 1890. Vast numbers were killed and industries based on their products sprang up.

A new era of control began with legislation in 1947. This established a Rabbit Destruction Council with local Rabbit Boards pursuing a 'killer policy', whereby: (i) rabbit rates paid by farmers were subsidized £ for £ by the government; (ii) all work was done by trained employees of the boards; (iii) rabbits and their products were 'decommercialized'; and (iv) farmers were required to cooperate. This policy was dramatically successful, aided in time by aerially-dropped baits and the poison sodium monofluoroacetate (1080) replacing strychnine.

By the mid-1960s rabbits no longer threatened agricultural production. The earlier policy of extermination was replaced in 1971 by one of control, the costs of which were to be matched by increased production. This unpopular change was reinforced by the discovery that sparse populations of rabbits rarely increased at all when control was relaxed. Improved farming methods were held responsible.

Since the early 1970s sparse populations have remained stable with little active control, except in the dry tussock grasslands of Central Otago and Canterbury where a resurgence of rabbits began in the mid-1980s. This coincided with a run of dry years with over-grazing by rabbits and sheep; and it led to renewed calls by farmers for myxomatosis, so far resisted.

The size of a rabbit's home range depends in part on the distribution of cover and food. When they were common, many rabbits lived in large warrens surrounded by good grazing; they then had home ranges of less than one hectare. Large warrens are now rare, even in Central Otago, and most rabbits live in small groups in scattered burrows or above ground in thick scrub; and most can still feed close to their daytime refuges. Where cover is at a distance from suitable feeding grounds, however, their ranges may exceed 10 ha in size and individual rabbits may travel 500 m or more betwen the cover and the feeding grounds. Young rabbits may disperse several kilometres, but once they have settled down they are likely to remain sedentary for life.

Most of New Zealand has a temperate climate; rain is well distributed and grass grows for most of the year. Consequently, on improved pasture, more than 50 per cent of the adult does are pregnant for at least nine months of the year. On average, each doe then produces *c.*45 young per year, and young born in spring may themselves breed in late summer or autumn. In Central Otago and probably in other harsh environments, each doe may produce as few as 25 young per year. Average litter-size varies seasonally from about 4.5 to 6.5.

The survival of young rabbits is extremely variable. In some districts many young die in flooded stops, or are killed in the nest by ferrets. Cats and stoats kill many of the young that do leave the nest. The survival of adults is equally variable; usually fewer than 50 per cent survive for more than a year; though at very low density up to 80 per cent may do so. Some rabbits may then live to over seven years of age. Conversely, dense (and unstable) populations contain many young and few old rabbits.

Wild populations on the mainland are usually stable for years on end, locally and regionally. Food shortage and ultimately starvation set the theoretical upper limit to density. However, predation and social intolerance, probably modified by food shortage, seem responsible for density-dependent mortality, which usually results in densities much lower than would be set by food shortage alone.

Acknowledgements

It is a pleasure to acknowledge assistance from many quarters. Brian T. Robertson provided information about the early spread of rabbits in New Zealand. Dr John Flux has responded equably to a barrage of questions that taxed his encylopaedic knowledge of lagomorphs; he has advised on the layout of the chapter and has provided excellent photographs. The late Jim Bell, John Robertshaw, Don Robson, and Don Ross, of the Ministry of Agriculture and Fisheries, produced unpublished material and commented on the manuscript. Peter C. Nelson helped with the maps in Fig. 6.1. Dr Michael Fitzgerald has studied cats and predation with the first author for 25 years; he has obviously influenced our views. Finally, the Editors scrutinized the manuscript and made many valuable suggestions; they have jollied us along almost beyond the point of exhaustion, but with remarkably good humour. We thank them all.

References

Andersson, M., Dahlbäck, M., and Meurling, P. (1979). Biology of the wild rabbit, *Oryctolagus cuniculus*, in southern Sweden. I. Breeding season. *Swedish Wildlife Research 'Viltrevy'*, **11**, 103–27.

Andrewartha, H. G. and Birch, L. C. (1984). *The ecological web. More on the distribution and abundance of animals.* University of Chicago Press.

Arthur, C. P. (1980). Démographie du lapin de garenne *Oryctolagus cuniculus* (L.) 1758 en région parisienne. *Bulletin mensuel office national de la chasse. Numéro spécial, scientifique et technique,* Décembre 1980, 127–62.

Bamford, J. and Hill, J. (1985). Environmental impact report on a proposal to introduce myxomatosis as another means of rabbit control in New Zealand. Bamford Associates, Nelson, New Zealand.

Beaglehole, J. C. (ed.) (1967). *The journals of Captain James Cook*. Hakluyt Society, Cambridge University Press.

Bell, J. (1977). Breeding season and fertility of the wild rabbit, *Oryctolagus cuniculus* (L.) in North Canterbury, New Zealand. *Proceedings of the New Zealand Ecological Society*, **24**, 79–83.

Bell, J., Ross, W. D., and Batcheler, C. L. (compilers) (1989). Surveillance and research needs for integrated rabbit and land management in the semi-arid regions of New Zealand. Ministry of Agriculture and Fisheries, Lincoln.

Bilborough, N. (ed.) (1987). Investigation of the proposal to introduce myxomatosis for rabbit control. Office of the Parliamentary Commission for the Environment, Wellington.

Bull, P. C. (1953). Parasites of the wild rabbit, *Oryctolagus cuniculus* (L.) in New Zealand. *New Zealand Journal of Science and Technology*, **B34**, 341–72.

Bull, P. C. (1958). Incidence of coccidia (Sporozoa) in wild rabbits, *Oryctolagus cuniculus* (L.) in Hawke's Bay, New Zealand. *New Zealand Journal of Science*, **1**, 289–329.

Bull, P. C. (1964). Ecology of helminth parasites of the wild rabbit *Oryctolagus cuniculus* (L.) in New Zealand. *New Zealand DSIR Bulletin*, **158**.

Burnett, L. and Hosey, G. R. (1987). Frequency of vigilance behaviour and group size in rabbits (*Oryctolagus cuniculus*). *Journal of Zoology, London*, **212**, 367–8.

Campbell, D. J. (1984). The vascular flora of the DSIR study area, lower Orongorongo Valley, Wellington, New Zealand. *New Zealand Journal of Botany*, **22**, 223–70.

Catling, P. C. (1988). Similarities and contrasts in the diets of foxes, *Vulpes vulpes*, and cats, *Felis catus*, relative to fluctuating prey populations and drought. *Australian Wildlife Research*, **15**, 307–17.

Chitty, D. (1967). The natural selection of self-regulatory behaviour in animal populations. *Proceedings of the Ecological Society of Australia*, **2**, 51–78.

Cowan, D. P. (1987). Aspects of the social organisation of the European wild rabbit (*Oryctolagus cuniculus*). *Ethology*, **75**, 197–210.

Cox, J. E., Taylor, R. H., and Mason, R. (1967). Motunau Island, Canterbury, New Zealand. An ecological survey. *New Zealand DSIR Research Bulletin*, **178**.

Daly, J. C. (1981). Social organisation and genetic structure in a rabbit population. In *Proceedings of the World Lagomorph Conference* (1979) (ed. K. Myers and C. D. MacInnes), pp. 90–7. University of Guelph, Ontario.

Delibes, M. and Hiraldo, F. (1981). The rabbit as prey in the Iberian mediterranean ecosystem. In *Proceedings of the World Lagomorph Conference* (1979) (ed. K. Myers and C. D. McInnes), pp. 614–22. University of Guelph, Ontario.

De Lisle, J. F. (1965). The climate of the Auckland Islands, Campbell Island and Macquarie Island. *Proceedings of the New Zealand Ecological Society*, **12**, 37–44.

Fennessy, B. V. (1958). Control of the European rabbit in New Zealand. *Wildlife Survey Section Technical Paper 1*. CSIRO, Melbourne.

Filmer, J. F. (1953). Disappointing tests of myxomatosis as rabbit control. *New Zealand Journal of Agriculture*, **87**, 402–4.

Fitzgerald, B. M. (1988). Diet of domestic cats and their impact on prey populations. In *The domestic cat: the biology of its behaviour* (ed. D. C. Turner and P. Bateson), pp. 123–44. Cambridge University Press.

Fitzgerald, B. M. and Karl, B. J. (1979). Foods of feral house cats (*Felis catus* L.) in forest of the Orongorongo Valley, Wellington. *New Zealand Journal of Zoology*, **6**, 107–26.

Flux, J. E. C. (1967). Hare numbers and diet in an alpine basin in New Zealand. *Proceedings of the New Zealand Ecological Society*, **14**, 27–33.

Flux, J. E. C. and Fullagar, P. J. (1983). World distribution of the rabbit, *Oryctolagus cuniculus*. *Acta Zoologica Fennica*, **174**, 75–7.

Fraser, K. W. (1985). Biology of the rabbit (*Oryctolagus cuniculus* (L.)) in Central Otago, New Zealand, with emphasis on behaviour and its relevance to poison control operations. Unpublished Ph.D. thesis. University of Canterbury, New Zealand.

Fraser, K. W. (1988). Reproductive biology of rabbits, *Oryctolagus cuniculus* (L.), in Central Otago, New Zealand. *New Zealand Journal of Ecology*, **11**, 79–88.

Garson, P. J. 1981. Social organisation and reproduction in the rabbit: a review. In *Proceedings of the World Lagomorph Conference* (1979) (ed. K. Myers and C. D. MacInnes), pp. 256–70. University of Guelph, Ontario.

Gibb, J. A. (1967). What is efficient rabbit destruction? *Tussock Grasslands and Mountain Lands Institute 'Review'*, **12**, 9–14.

Gibb, J. A. (1981). Limits to population density in the rabbit. In *Proceedings of the World Lagomorph Conference* (1979) (ed. K. Myers and C. D. MacInnes), pp. 654–63. University of Guelph, Ontario.

Gibb, J. A. (1993). Sociality, time and space in a sparse population of rabbits (*Oryctolagus cuniculus*). *Journal of Zoology, London*, **229**, 581–607.

Gibb, J. A. and Flux, J. E. C. (1983). Why New Zealand should not use myxomatosis in rabbit control operations. *Search*, **14**, 41–3.

Gibb, J. A. and Williams, J. M. (1990). European rabbit. In *The handbook of New Zealand mammals* (ed. C. M. King), pp. 138–60. Oxford University Press.

Gibb, J. A., Ward, G. D., and Ward, C. P. (1969). An experiment in the control of a sparse population of wild rabbits (*Oryctolagus cuniculus* (L.)) in New Zealand. *New Zealand Journal of Science*, **12**, 509–34.

Gibb, J. A., Ward, C. P., and Ward, G. D. (1978). Natural control of a population of rabbits, *Oryctolagus cuniculus* (L.), for ten years in the Kourarau enclosure. *New Zealand DSIR Bulletin*, **223**.

Gibb, J. A., White, A. J., and Ward, C. P. (1985). Population ecology of rabbits in the Wairarapa, New Zealand. *New Zealand Journal of Ecology*, **8**, 55–82.

Gilbert, N., Myers, K., Cooke, B. D., Dunsmore, J. D., Fullagar, P. J., Gibb, J. A. *et al.* (1987). Comparative dynamics of Australasian rabbit populations. *Australian Wildlife Research*, **14**, 491–503.

Hayne, D. W. (1949). Calculation of size of home range. *Journal of Mammalogy*, **30**, 1–8.

Hodgkinson, S. (Chairman) (1876). Report of the Rabbit Nuisance Committee. *Appendix to the Journals of the House of Representatives, 1876*, 1–5.

Howard, W. E. (1959). The rabbit problem in New Zealand. *New Zealand DSIR Information Series*, **16**.

Kerr, I. G. C., Williams, J. M., Ross, W. D., and Pollard, J. M. (1987). The classification of land according to the degree of rabbit infestation in Central Otago. *Proceedings of the New Zealand Grassland Association*, **48**, 65–70.

King, C. M. (ed.) (1990). *The handbook of New Zealand mammals*. Oxford University Press.

King, C. M. and Moody, J. E. (1982). The biology of the stoat (*Mustela erminea*) in the National Parks of New Zealand. *New Zealand Journal of Zoology*, **9**, 49–144.

King, D. R., Wheeler, S. H., and Schmidt, G. L. (1983). Population fluctuations and reproduction of rabbits in a pastoral area on the coast north of Carnarvon, W.A. *Australian Wildlife Research*, **10**, 97–104.

Kofoed, W. R. (1967). A defence of the rabbit control policy. *Tussock Grasslands and Mountain Lands Institute 'Review'*, **13**, 8–10.

Krebs, C. J. (1971). Genetic and behavioural studies of fluctuating vole populations. *Proceedings of the Advanced Study Institute on 'Dynamics of Numbers in Populations'*, Oosterbeek, 1970, pp. 243–56. Pudoc, Wageningen.

Krebs, C. J. (1978). A review of the 'Chitty Hypothesis' of population regulation. *Canadian Journal of Zoology*, **56**, 2463–80.

Lloyd, H. G. (1970). Variation and adaptation in reproductive performance. In *Variation in mammalian populations*. Symposia of the Zoological Society of London (No. 26) (ed. R. J. Berry and H. N. Southern), pp. 165–82. Academic Press, New York.

Lloyd, H. G. (1977). Rabbit *Oryctolagus cuniculus*. In *The handbook of British mammals* (2nd edn.) (ed. G. B. Corbet and H. N. Southern), pp. 130–9. Blackwell Scientific Publications, Oxford.

Lloyd, H. G. and McCowan, D. (1968). Some observations on the breeding burrows of the wild rabbit *Oryctolagus cuniculus* on the island of Skokholm. *Journal of Zoology, London*, **156**, 540–9.

McCaskill, L. W. (1970). *Molesworth*. Reed, Wellington.

McIlwaine, C. P. (1962). Reproduction and body weights of the wild rabbit, *Oryctolagus cuniculus* (L.) in Hawke's Bay, New Zealand. *New Zealand Journal of Science*, **5**, 325–41.

McIntosh, I. G. (1958). 1080 poison: outstanding animal pest destroyer. *New Zealand Journal of Agriculture*, **97**, 361.

MacKinnon, M. (1972). Slaughter on the roads. *Animals*, **14**, 111.

McLean, W. H. (1966). *Rabbits galore*. Reed, Wellington.

Maitland, J. P. (1885). Particulars of runs abandoned in Otago during the years 1877–84. *Appendix to the Journals of the House of Representatives*, 1885, C.-9.

Mark, A. F. (1965). Vegetation and mountain climate. In *Central Otago* (ed. R. G. Lister and R. P. Hargreaves). New Zealand Geographical Society.

Marshall, W. H. (1963). The ecology of mustelids in New Zealand. *New Zealand DSIR Bulletin Information Series*, **38**.

Mead-Briggs, A. R. (1977). The European rabbit, the European rabbit flea and myxomatosis. *Advances in Applied Biology*, **2**, 183–261.

Munro, H. (1925). The rabbit pest. *New Zealand Journal of Agriculture*, **15**, 206–9.

Mykytowycz, R. and Dudzinski, M. L. (1972). Aggressive and protective behaviour of adult rabbits, *Oryctolagus cuniculus* (L.) towards juveniles. *Behaviour*, **43**, 96–120.

Mykytowycz, R. and Gambale, S. (1965). A study of the inter-warren activities and dispersal of wild rabbits, *Oryctolagus cuniculus* (L.), living in a 45-ac paddock. *CSIRO Wildlife Research*, **10**, 111–23.

Mykytowycz, R. and Rowley, I. (1958). Continuous observations of the activity of the wild rabbit, *Oryctolagus cuniculus* (L.) during 24-hour periods. *CSIRO Wildlife Research*, **3**, 26–31.

Newsome, A. E., Parer, I., and Catling, P. C. (1989). Prolonged prey suppression by carnivores—predator-removal experiments. *Oecologia*, **78**, 458–67.

Norton, G. A. and Pech, R. P. (1988). Vertebrate pest management in Australia. *Project Report No. 5. Division of Wildlife and Ecology*, CSIRO, Australia.

O'Connor, K. F. (1987). The sustainability of pastoralism. *Proceedings of the 1987 Hill and High Country Seminar, Tussock Grasslands and Mountain Lands Institute*. Lincoln College, New Zealand.

Parer, I. (1977). The population ecology of the wild rabbit, *Oryctolagus cuniculus* (L.), in a Mediterranean-type climate in New South Wales. *Australian Wildlife Research*, **4**, 171–205.

Pearson, O. P. (1966). The prey of carnivores during one cycle of mouse abundance. *Journal of Animal Ecology*, **35**, 217–33.

Pech, R. P., Sinclair, A. R. E., Newsome, A. E., and Catling, P. C. (1992). Limits to predator regulation of rabbits in Australia: evidence from predator-removal experiments. *Oecologia*, **89**, 102–12.

Rabbit and Land Management Task Force (1988). The report of the Rabbit and Land Management Task Force to the Rt Hon. C. J. Moyle, Minister of Agriculture. Mosgiel, New Zealand.

Radcliffe, J. E. (1975). Seasonal distribution of pasture production in New Zealand. VII. Masterton (Wairarapa) and Maraekakaho (Hawke's Bay). *New Zealand Journal of Experimental Agriculture*, **3**, 259–65.

Radcliffe, J. E. and Cossens, G. G. (1974). Seasonal distribution of pasture production in New Zealand. III. Central Otago. *New Zealand Journal of Experimental Agriculture*, **2**, 349–58.

Richardson, B. J. and Wood, D. H. (1982). Experimental ecological studies on a subalpine rabbit population. I. Mortality factors acting on emergent kittens. *Australian Wildlife Research*, **9**, 443–50.

Rogers, P. M. (1981). The wild rabbit in the Camargue, Southern France. In *Proceedings of the World Lagomorph Conference* (1979) (ed. K. Myers and C. D. MacInnes), pp. 587–99. University of Guelph, Ontario.

Skira, I. J. (1978). Reproduction of the rabbit, *Oryctolagus cuniculus* (L.), on Macquarie Island, Subantarctic. *Australian Wildlife Research*, **5**, 317–26.

Sobey, W. R. (1982). Should New Zealand use rabbit fleas and myxomatosis to assist with rabbit control? *Search*, **13**, 71–2.

Soriguer, R. C. and Rogers, P. M. (1981). The European wild rabbit in Mediterranean Spain. In *Proceedings of the World Lagomorph Conference* (1979) (ed. K. Myers and C. D. MacInnes), pp. 600–13. University of Guelph, Ontario.

Southern, H. N. (1940). The ecology and population dynamics of the wild rabbit (*Oryctolagus cuniculus*). *Annals of Applied Biology*, **27**, 509–26.

Tapper, S. C., Green, R. E., and Rands, M. R. W. (1982). Effects of mammalian predators on partridge populations. *Mammal Review*, **12**, 159–67.

Thomson, G. M. (1922). *The naturalization of animals and plants in New Zealand*. Cambridge University Press.

Trout, R. C. and Tittensor, A. M. (1989). Can predators regulate wild rabbit *Oryctolagus cuniculus* population density in England and Wales? *Mammal Review*, **19**, 153–73.

Tyndale-Biscoe, C. H. and Williams, R. M. (1955). A study of natural mortality in a wild population of the rabbit, *Oryctolagus cuniculus* (L.). *New Zealand Journal of Science and Technology*, **B36**, 561–80.

Wards, I. (ed.) (1976). *New Zealand Atlas*. Government Printer, Wellington.

Watson, J. S. (1954). Reingestion in the wild rabbit, *Oryctolagus cuniculus* (L.). *Proceedings of the Zoological Society of London*, **124**, 615–24.

Watson, J. S. (1957). Reproduction of the wild rabbit, *Oryctolagus cuniculus* (L.) in Hawke's Bay, New Zealand. *New Zealand Journal of Science and Technology*, **B38**, 389–99.

Westbury, H. A. (1989). Viral haemorrhagic disease of rabbits. Unpublished report of a study tour funded by the Committee of Nature Conservation Ministers of Australia. Australian Animal Health Laboratory, Geelong, Victoria.

Williams, J. M. (1977). A possible basis for economic rabbit control. *Proceedings of the Ecological Society of New Zealand*, **24**, 72–82.

Williams, J. M. (1978). A basis for economic rabbit control in New Zealand. *Working papers, Australian Vertebrate Pest Control Conference*, 110–19.

Williams, J. M. and Robson, D. L. (1985). Rabbit ecology and management in the Western Pest Destruction Board: a report and management recommendations based on studies between 1978 and 1983. New Zealand MAF Research Division, Christchurch, New Zealand.

Williams, J. M., Bell, J., Ross, W. D., and Broad, T. M. (1986). Rabbit (*Oryctolagus cuniculus*) control with a single application of 50 ppm brodifacoum cereal baits. *New Zealand Journal of Ecology*, **9**, 123–36.

Wodzicki, K. (1950). Introduced mammals of New Zealand. *New Zealand DSIR Bulletin*, **98**.

Wodzicki, K. and Taylor, R. H. (1957). An experiment in rabbit control: Kourarau poison trials. *New Zealand Journal of Science and Technology*, **B38**, 389–99.

Wood, D. H. (1980). The demography of a rabbit population in an arid region of New South Wales, Australia. *Journal of Animal Ecology*, **49**, 55–79.

Wynne-Edwards, V. C. (1962). *Animal dispersion in relation to social behaviour*. Hafner, New York.

Xu, Z. J. and Chen, W. X. (1989). Viral haemorrhagic disease in rabbits: a review. *Veterinary Research Communications*, **13**, 205–12.

7

Myxomatosis

Frank Fenner and John Ross

7.1 Introduction

Myxomatosis, caused by a virus of a South American rabbit, was introduced into wild European rabbit populations in Australia and Europe in the 1950s. It had an immediate and catastrophic effect on rabbit numbers, and it has continued to be an important factor in the ecology of the rabbit.

The study of myxomatosis in the Australian wild rabbit was the principal preoccupation of the Wildlife Survey Section of CSIRO from its formation in 1949 until it was transformed into the CSIRO Division of Wildlife Research in 1960. This work has been described in detail in the book *Myxomatosis* (Fenner and Ratcliffe 1965). Myxomatosis has continued to be a major part of the work of the CSIRO Division (now Wildlife and Ecology), and has recently (1992) been expanded by the establishment of a Cooperative Research Centre, which aims to enhance the performance of myxoma virus for rabbit control by introducing a sterilizing factor into the genome of the virus by genetic engineering.

Myxomatosis has been the most important new disease of wildlife in Europe during this century, having a major impact in France and Britain in particular. Scientific studies of the disease in France have been summarized by Fenner and Ratcliffe (1965), in a two-volume book, *La myxomatose* (Joubert *et al.* 1972, 1973), and more recently by Joubert *et al.* (1982). No book has examined in detail the course of myxomatosis in Britain, but summaries of its effects on rabbits have been published in Fenner and Ratcliffe (1965), and Ross (1982).

The aims of this chapter are: (*a*) to provide zoologists with a succinct account of myxomatosis as a component of the environment of rabbits in Australia and Europe since 1950; (*b*) to provide enough information on myxomatosis to render references to it, made elsewhere in the book, intelligible; and (*c*) to bring up to date the account of changes in disease, virus, and host in the twenty-nine years since the publication of the book by Fenner and Ratcliffe (1965).

7.2 History

Myxomatosis was first recognized in 1896 as a lethal new disease amongst laboratory (European) rabbits in South America (Sanarelli 1898). Subsequently it caused sporadic lethal infections in domestic and laboratory rabbits in Brazil, and in 1930 severe outbreaks of myxomatosis occurred in domestic (European) rabbit breeding establishments in California (Kessel *et al.* 1931).

7.2.1 Introduction into Australia

Myxomatosis was first suggested as a means of controlling pest populations of wild rabbits in Australia in 1918, by Dr H. de Beaurepaire Aragão, who had worked on the disease at the Oswaldo Cruz Institute in Brazil. However, objections were raised by Sir Harry Allen, Professor of Pathology in the

University of Melbourne, who thought that its importation should be forbidden unless 'information of a reassuring character' could be supplied about 'the relationships of the disease' (presumably its species specificity). Further, the response of the Institute of Science and Industry in Australia, although at variance with the views of the Stock-owner's Association, whose members were anxious to get rid of rabbits, effectively prevented the development of Aragão's ideas:

> The trade in rabbits both fresh and frozen, either for local food or for export, has grown to be one of great importance and popular sentiment here is opposed to the extermination of the rabbit by the use of some virulent organism. . . .

Some years later, Dr H. R. Seddon, then Director of Veterinary Research in the Department of Agriculture, New South Wales, sought and obtained a specimen of myxoma virus from Dr Aragão. Inconclusive experiments on its capacity to spread from infected rabbits in one burrow to rabbits in another burrow led to abandonment of experiments with the virus (White 1929).

In 1934, unaware of this episode, Dr (later Dame) Jean Macnamara, a Melbourne paediatrician, wrote to Mr Stanley Bruce (later Lord Bruce), then High Commissioner for Australia in London, after seeing cases of the disease in the laboratories of Dr Richard E. Shope, at the laboratories of the Rockefeller Institute of Medical Research in Princeton, New Jersey. Referred to Australia, the Health Department determined that the virus should be kept in strict quarantine. The Council for Scientific and Industrial Research (CSIR) in Australia commissioned Sir Charles Martin to undertake investigations in England, where quarantine restrictions against myxomatosis were not in force, to determine whether it had potential for rabbit control. After a series of well-designed experiments, Martin (1936) reported favourably, and Dr L. B. Bull of the Division of Animal Health of CSIR was asked to investigate the virus further. He demonstrated that it did not infect domestic animals or common Australian native animals (Bull and Dickinson 1937), and carried out experiments on the mechanical transmission of the virus by local mosquitoes and by the native stickfast flea (*Echidnophaga myrmecobii*) (Bull and Mules 1944). After some difficulty in obtaining suitable experimental sites because of objections by quarantine authorities, Bull and Mules (1944) carried out field trials on Wardang Island, in South Australia, and later in the semi-arid interior of that state. The results were successful where stickfast fleas were present, but in their absence the disease failed to spread from one warren to another. In summing up the work that terminated in 1943, Bull and Mules concluded:

> . . . that myxomatosis cannot be used to control rabbit populations under most natural conditions in Australia with any promise of success. Nevertheless, it seems possible that in some parts of Australia under special conditions, including the presence of insect vectors in abundance and the absence of predatory animals, the disease could be used with some promise of temporary control of a rabbit population.

At this juncture the CSIR considered that if further trials with myxomatosis were to be done, they might properly be undertaken by the State authorities, who had the responsibility for vermin control within their territories; but the Director-General of Health refused to release the virus from quarantine restrictions, and work on myxomatosis was discontinued for a second time.

A very serious rabbit situation developed in Australia in the years immediately following the Second World War. In 1949 the Commonwealth Scientific and Industrial Research Organization (CSIRO—the successor to CSIR) established a Wildlife Survey Section under the leadership of F. N. Ratcliffe, which in addition to its responsibilities for the study of the Australian native fauna undertook to explore the possibilities of dealing in a scientific fashion with the control of Australia's major animal pest, the rabbit. Experiments with myxomatosis were undertaken again almost immediately. Stimulated by the insistence of Dame Jean Macnamara, in a newspaper controversy, that the early trials had been carried out in dry country in which there were few mosquitoes, and that adequate experiments in well-watered country had still to be done, and following Bull's suggestion that release should be attempted where there were abundant vectors, trials were conducted in several sites in the Murray Valley in the period May–December 1950. The initial results were similar to those obtained by Bull and Mules in their South Australian liberations; the disease remained localized to the warren and

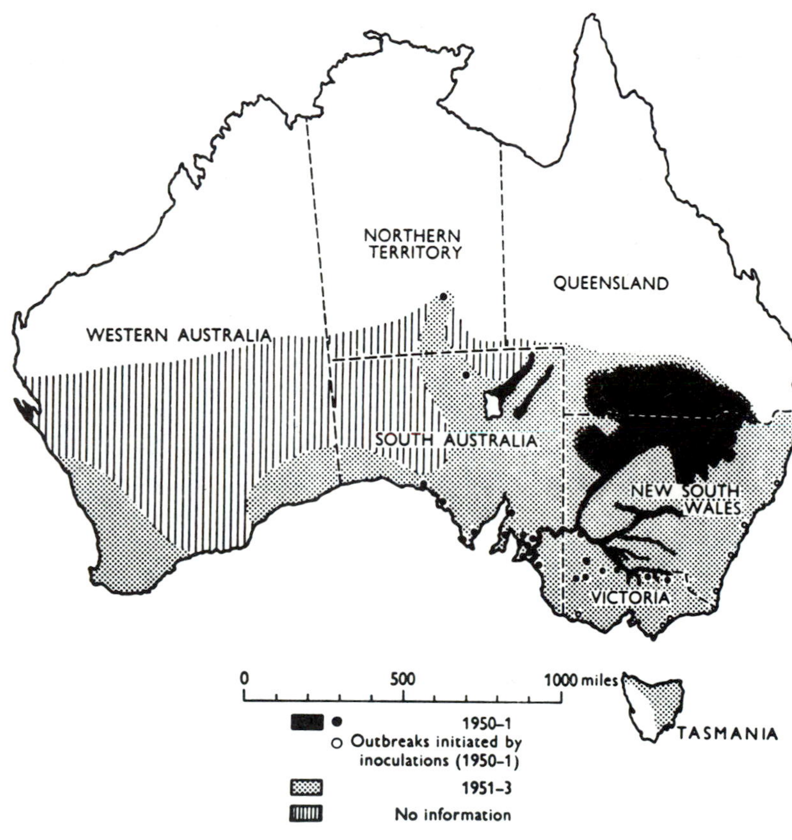

Fig. 7.1 The spread of myxomatosis in Australia during the three years following its escape, late in 1950, from a test site in northern Victoria. The major spread early in 1951 (dense black) marks out the Murray–Darling–Murrumbidgee river system of south-eastern Australia. Inoculation campaigns were widely undertaken each summer during the 1950s in all the rabbit infested agricultural and pastoral lands of southern Australia. Aided by natural spread, they rendered myxomatosis co-extensive with the rabbit by 1954. (From Fenner and Ratcliffe 1965.)

apparently died out after a few generations (Myers 1954).

However, in December 1950 myxomatosis flared up near one of the trial sites, and was reported almost simultaneously at several points along the river nearby; and in the following few weeks it spread with dramatic speed over south-eastern Australia (Ratcliffe *et al.* 1952). The extent of the outbreak in the summer of 1950–1 is difficult to appreciate now, when rabbit numbers are so much smaller than they were then. Over a period of a few months the disease spread along the waterways of the Murray–Darling system (Fig. 7.1). The mortality in the very dense rabbit populations then present in the riverfront country was enormous; where counts were made the case-fatality rate exceeded 99 per cent and in many localized areas all rabbits were killed. Myxomatosis has never since disappeared, and is now firmly established as an enzootic disease of the Australian wild rabbit, with epizootics developing periodically in association with local and seasonal vector activity.

With some reluctance, the quarantine restrictions imposed by the Director-General of Health were cancelled in 1951, when it was apparent that they could no longer apply. However, because of public anxiety about the use of fibroma virus vaccine for the protection of domestic rabbits raised for commercial purposes (see section 7.12), the Director-General of Health in 1962 re-applied restrictions upon scientific laboratories studying fibroma and myxoma viruses, but not upon the distribution of myxoma virus for rabbit control.

7.2.2 Introduction into Europe

Following Martin's (1936) experiments in Cambridge, several unsuccessful attempts were made to introduce myxomatosis into wild rabbit populations in Europe, using virus supplied by Martin, on the island of Skokholm off the coast of Wales, in 1936–8; on Vejrø Island, Denmark, in 1936–8; and in Dufeke Estate in Skaane, Sweden, in 1938 (Fenner and Ratcliffe 1965).

The dramatic effect of myxomatosis on rabbits in Australia in 1951–2 made headlines in the press in Europe. In January 1952, a scientist at the Pasteur Institute in Paris wrote to Ratcliffe of CSIRO requesting ampoules of the virus and instructions as to its distribution, because

> Les lapins constituant dans certain parties de la France, un véritable fléau, l'Institut Pasteur désirerait entreprende la lutte biologique contre ce Rongeur et assayer de développer une épidemie de myxomatosis en Sologne. . . .

Ratcliffe passed the letter to one of us (F.F.), who sent ampoules of freeze-dried virus and instructions for its use to the Pasteur Institute. The receipt of this material was acknowledged, but before it could be used in the field French officials became aware of the local quarantine implications of the use of myxoma virus and no further work with it was undertaken by the Institut Pasteur until after the virus had been introduced by Dr Delille (see below).

The publicity given to the effectiveness of myxomatosis for rabbit destruction was not lost on members of the French public. In mid-1952 Dr P. F. Armand Delille, a paediatrician who was concerned with the numbers of rabbits in his 270-ha walled estate at Maillebois, obtained a strain of myxoma virus from a friend in the Laboratorie de Bacteriologie, Lausanne, Switzerland. He inoculated two rabbits with this virus on 14 June 1952, and by the end of August virtually all rabbits on the estate were dead and outbreaks of myxomatosis had occurred in villages several kilometres from Maillebois (Delille 1953). The first official identification of myxomatosis in France was made by workers of the Institut Pasteur, from material from a wild rabbit taken at Rambouillet in October 1952 (Jacotot and Vallée 1953). Details of the spread through France were given in Thompson (1954) and is summarized in Chapter 3.

From France, the disease spread during 1953 to Belgium (Devos *et al.* 1963), Holland (Lankamp 1953), western German (Muller-Using 1953), Luxemburg (Fioretti 1985) and Spain (Sanchez *et al.* 1954). By the end of 1954, it had reached Switzerland (Bouvier 1967), Czechoslovakia (Blazek and Kral 1954) and Ireland (Fioretti 1985). During 1955, myxomatosis is reported to have reached Berlin (Lubke 1968), but did not spread to the whole of eastern Germany until 1961 (Kotsche 1964). In 1955 the disease also reached Austria (Fioretti 1985), Italy (Corazzola and Zanin 1970) and Poland (Steffen 1971). Scandinavia apparently remained free of myxomatosis until its presence was confirmed in Denmark late in 1960 (Basse 1961) and in southern Sweden in August 1961 (Borg 1962). Myxomatosis has also been reported in Portugal, Yugoslavia, and Rumania, but no dates can be found for the first cases seen.

At about the same time as Delille introduced myxomatosis into France, Shanks *et al.* (1955) made an attempt to establish the disease in the rabbit-infested Heisker Islands in the Hebrides, Scotland, using virus obtained from Dr Weston Hurst of ICI. Although it spread, causing considerable mortality in 1953, the virus did not persist. Permanent establishment of myxomatosis in Britain dates from autumn 1953. Deliberate carriage, by an English resident, of an infected rabbit from France, and its release near Edenbridge, Kent, resulted in the first observed cases of myxomatosis (Armour and Thompson 1955). As soon as it was recognized, attempts were made to contain or eradicate the disease, but these were unsuccessful. The disease spread slowly at first, but by the end of 1955 few parts of Britain were unaffected.

7.3 The virus and its natural hosts

Myxoma virus particles were first recognized in stained smears by Aragão (1927), who called attention to the close resemblance between them and the causative agents of smallpox, molluscum contagiosum, and fowlpox. His judgement was correct, for all these viruses are now accepted as being members of the family *Poxviridae*, subfamily *Chordopoxvirinae*, although each belongs to a different genus. Myxoma virus is the type species of the genus *Leporipoxvirus*, which comprises three closely-related viruses found

Table 7.1 Types of clinical disease produced by viruses of the genus *Leporipoxvirus* (family *Poxviridae*) in their natural hosts (leporids and squirrels) and in the European rabbit (*Oryctolagus cuniculus*)[a]

Virus	Natural host	Location	Clinical signs in *Oryctolagus cuniculus*	Eponym
Rabbit fibroma	*Sylvilagus floridanus*	Eastern United States	Localized, benign fibroma	Shope's fibroma
Brazilian myxoma	*Sylvilagus brasiliensis*	South America	Generalized, lethal disease, gross external signs	Aragão's fibroma
Californian myxoma	*Sylvilagus bachmani*	Western United States	Generalized lethal disease, few external signs	Marshall–Regnery fibroma
Hare fibroma	*Lepus europaeus*	Europe	Localized, benign fibroma	—
Squirrel fibroma	*Sciurus carolinensis*	Eastern United States	Multiple fibromas	—
Western squirrel[b] fibroma	*Sciurus griseus griseus*	Western United States	Localized benign fibroma	—

[a] Data from Fenner and Ratcliffe (1965).
[b] Regnery (1975).

in *Sylvilagus* rabbits in the Americas, two viruses of North American squirrels and the more distantly related hare fibroma virus, whose natural host is *Lepus europaeus* (Table 7.1). The eponyms used in Table 7.1 indicate the nature of the lesions caused by leporiviruses in their natural hosts; all are fibromas, i.e. they produce localized benign connective tissue tumours in the skin.

Other than as an experimental infection of laboratory rabbits (for example Hurst 1937), myxomatosis was unknown outside the Americas until 1950, and in that continent it was recognized only intermittently as a sporadic 'spontaneous' disease of domestic or laboratory European rabbits (Sanarelli 1898; Aragão 1927; Kessel *et al.* 1934). Clearly there was some other reservoir of the virus in wild animals of the regions where these outbreaks occurred. Aragão (1943) demonstrated that the natural host of myxoma virus in South America was the tapeti, *Sylvilagus brasiliensis* (Fig. 7.2), and later Marshall and Regnery (1960) recovered a different strain of myxoma virus from the brush rabbit of California, *Sylvilagus bachmani*. Little more work has been carried out on Aragão's fibroma, but Regnery and his colleagues have elucidated the natural history of the disease in California, culminating in the description of a natural epizootic among brush rabbits (Regnery and Miller 1972). The California strain of myxoma virus has never been released among wild *Oryctolagus cuniculus*.

In its natural hosts (*S. brasiliensis* in South and Central America, *S. bachmani* in California), myxoma virus produces a small fibroma, often located at the base of the ear (Fig. 7.3; Regnery and Miller 1972). There is usually no evidence of generalized disease, although in a juvenile *S. brasiliensis* inoculated intradermally, Aragão (personal communication, quoted in Fenner and Ratcliffe 1965) found that generalized lesions were produced, including mild blepharoconjunctivitis.

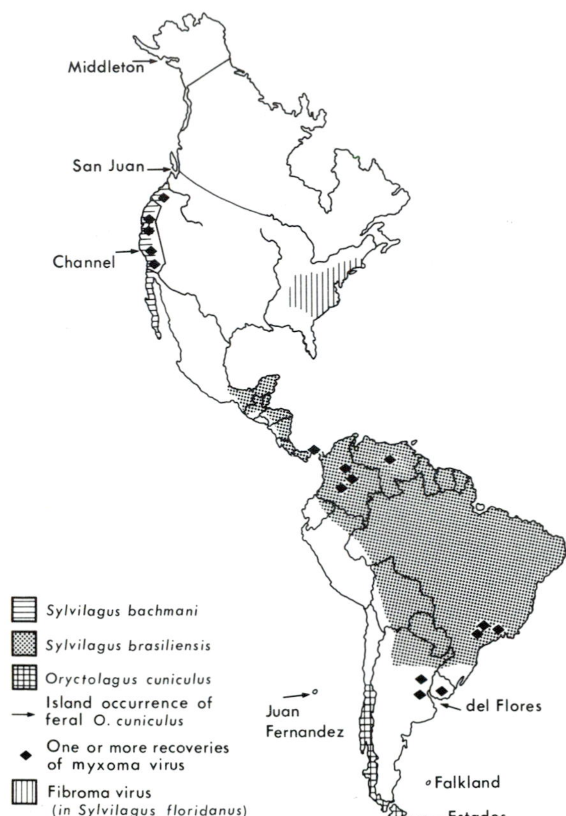

Fig. 7.2 Map of the Americas, showing the distribution of leporids significant in the natural history of myxomatosis and locations where myxoma and fibroma viruses have been isolated. (From Marshall (1961), updated with data from P. J. Fullagar, personal communication 1978 and from D. C. Regnery, personal communication 1992). (1) *Sylvilagus brasiliensis* is the natural host of Aragão's fibroma (South American myxoma virus) in South and Central America. (2) *Sylvilagus bachmani* is the natural host in the Marshall–Regnery fibroma (Californian myxoma virus) in California. (3) *Sylvilagus floridanus* occurs in eastern USA; it is the natural host of Shope's fibroma virus throughout its range. (4) Introduced *Oryctolagus cuniculus* (now deliberately and enzootically infected with South American myxoma virus) occurs in many parts of Chile. (5) *Oryctolagus cuniculus* has also been introduced into a large number of offshore islands, from Alaska to Terra del Fuego, but has not persisted on all of these.

Fig. 7.3 Marshall–Regnery fibroma in *Sylvilagus bachmani*. (Courtesy D. C. Regnery.) Aragão's fibroma produces a similar lesion in its natural host *Sylvilagus brasiliensis*; good photographs are not available but see Aragão (1943).

7.4 Myxomatosis in *Oryctolagus cuniculus*

In contrast to the trivial symptomatology of myxoma virus infection in its natural hosts, myxomatosis in *Oryctolagus cuniculus*, produced by mosquito transfer of either the Californian or the South American strain of virus from the appropriate species of *Sylvilagus*, or by intradermal inoculation, is a severe generalized disease that is almost always fatal. The California virus produces a rapidly lethal disease with only slight development of skin lesions. It will not be discussed further here, since all the later studies of the ecology of myxomatosis in *Oryctolagus* have been made with Brazilian strains.

Two Brazilian strains, and the variants that developed after their establishment in populations of wild European rabbits in Australia and Europe, have been investigated in detail (Fenner and Marshall 1957). One (designated 'Standard Laboratory' or 'Moses' strain) was originally recovered from a naturally-infected laboratory rabbit in Rio de Janeiro (Moses 1911) and was used by Shope in the Rockefeller Institute studies in the 1930s, by Hurst and Martin in England, and by Bull and Mules in Australia, before being successfully released in Australia in 1950. It had been maintained by intermittent passage in laboratory rabbits for nearly 40 years before its natural spread in Australia. The other strain ('Lausanne') was recovered in 1949 (Bouvier 1954) and had had very few laboratory passages before it was used to establish the disease in wild rabbits in France in 1952. It subsequently spread all over Europe. Both strains were almost always lethal for genetically unselected laboratory and wild rabbits, although 'Lausanne' proved to be more virulent when tested in genetically resistant rabbits (see Section 7.9). However, the diseases produced by the two strains differed in clinical signs. The passaged 'Standard Laboratory' strain (Fig. 7.4B) and all its field derivatives of lower virulence (Fig. 7.4C, 7.4D) produced

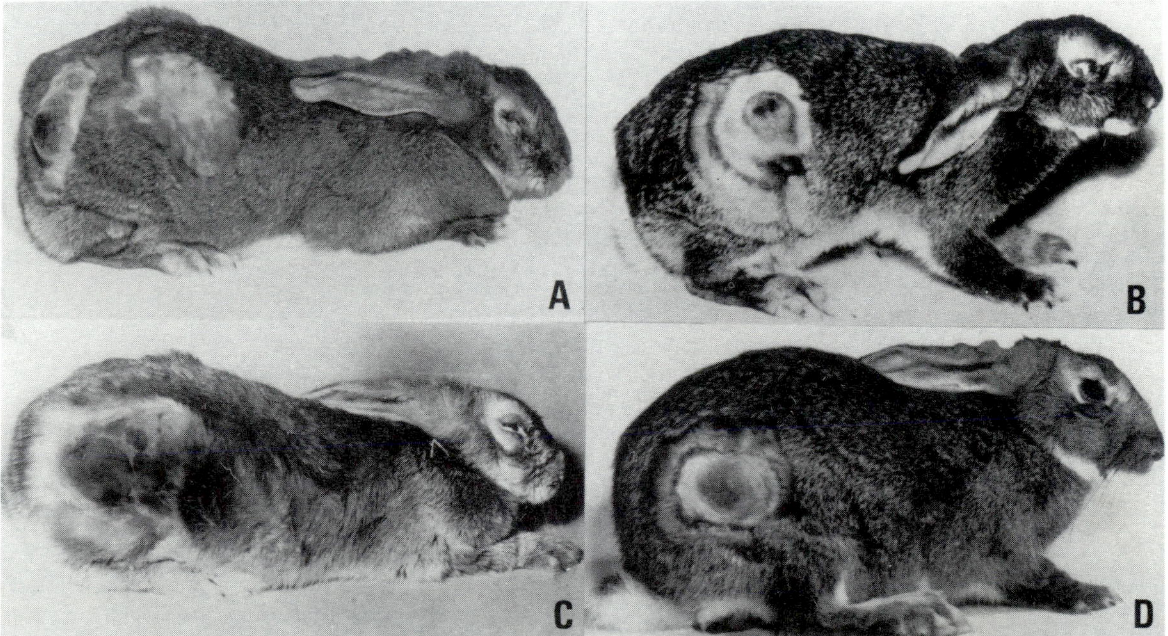

Fig. 7.4 Normal (genetically unselected) laboratory rabbits infected with various strains of myxoma virus. A. Lausanne strain, 10 days after infection. B. Standard Laboratory (Moses) strain, 10 days after infection. C. Australian field strain of grade III virulence, 21 days after infection. D. Australian field strain of grade V virulence, 16 days after infection.

relatively flat skin lesions; 'Lausanne' (Fig. 7.4A) and most of its field derivatives produced protuberant lesions.

The pathogenesis of myxomatosis in *Oryctolagus* after natural or experimental infection follows the same pattern as that found in other generalized poxvirus infections (Fenner and Woodroofe 1953). Sequential replication of the virus at the inoculation site and the regional lymph node is followed by cell-associated viremia and then generalization throughout the body (Strayer and Sell 1983), especially in the skin, in which areas of diffusely thickened dermis and subcutis later progress to produce nodular tumours. Virulent strains of myxoma virus almost invariably kill genetically unselected *Oryctolagus*, usually within 9–13 days after intradermal infection with a small dose of virus.

Some years after its establishment in wild rabbits in Australia and Europe, most strains recovered from the field were attenuated, although they still killed between 50 per cent and 95 per cent of rabbits tested under standard laboratory conditions. The clinical signs in animals that survived for more than 14 days were modified by the influence of the developing immune response. Virulence measured in terms of lethality was found to be inversely correlated with survival time, which enabled tests for the virulence of field strains to be developed. Although there is probably a continuous spectrum of virulence, for analytical purposes Fenner and Marshall (1957) established a series of 'virulence grades' (I, II, III (later IIIA and IIIB), IV, V—see Tables 7.3 and 7.4) in order to follow secular changes in the virulence of field strains.

Recently other variants have been found. Strains of myxoma virus causing fewer and smaller skin lesions, but massive pulmonary oedema, were found from about 1980 in domestic and wild rabbits in France (Brun *et al.* 1981*b*; see Section 7.9.2). Another variant, which has been called malignant rabbit fibroma virus, was isolated from tumours induced by an uncloned stock of 'Shope's fibroma virus' (Strayer *et al.* 1983). This virus produces a disease which is different from both Shope's fibroma and myxomatosis in its clinical and pathological features. Using cloned DNA probes from Shope's fibroma virus and myxoma virus, Block *et al.* (1985) showed that it was a recombinant between fibroma and myxoma viruses, the major part of its DNA being derived from myxoma virus. It seems likely that it arose inadvertently during the passage of fibroma virus in a laboratory in which myxoma virus was also being used.

7.4.1 Shope's fibroma

In contrast to the extreme virulence for *Oryctolagus* of natural American strains of myxoma virus (Aragão's and Marshall–Regnery fibromas, Table 7.1), the antigenically related Shope's fibroma virus (natural host, *Sylvilagus floridanus*) produces only a localized fibroma in mature *Oryctolagus*, although larger tumours and a severe disease may occur in newborn rabbits (Duran-Reynals 1940). In contrast to the fibromas in *S. floridanus*, which often persist for some months (Kilham and Fisher 1954; Kilham and Dalmat 1955), those in *Oryctolagus* regress rapidly, and are poor sources of virus for mosquito vectors.

7.5 Mechanisms of transmission

Even in the absence of vectors, myxomatosis can spread between rabbits caged together, infection probably occurring by the respiratory route; this seems to have become the normal mode of transmission for certain strains of myxoma virus recently isolated in France (see Section 7.4). However, as Aragão (1943) showed, much the most important mechanism of transmission is by arthropod vectors. At different times a wide variety of arthropods have been implicated (Fenner and Ratcliffe 1965).

7.5.1 Mosquito transmission

In Australia, early field investigations (Myers *et al.* 1954) demonstrated the importance of mosquitoes as vectors. The mechanism of mosquito transmission was studied in detail by Fenner and Day (Fenner *et al.* 1952, 1956; Day *et al.* 1956), and the earlier findings of Aragão (1943) and Bull and Mules (1944) that it was mechanical were confirmed. The mosquito behaved like a 'flying pin', virions in skin lesions but

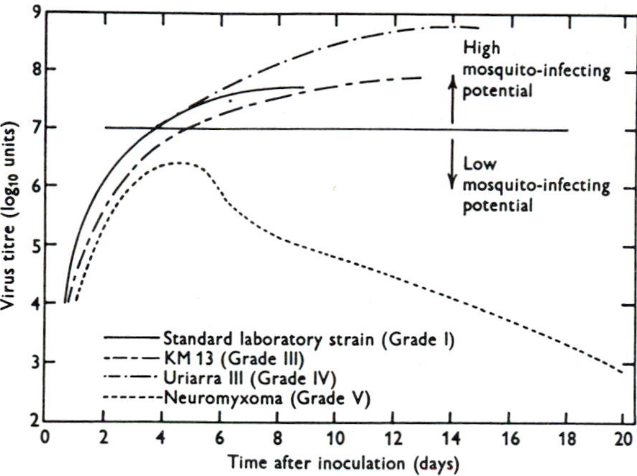

Fig. 7.5 Correlation between titre of myxoma virus in skin lesions in *Oryctolagus cuniculus* and the suitability of the lesion as a source of virus for mechanical transmission by mosquitoes. Except for a few grade V strains from which some infected rabbits rapidly recovered, all field strains were highly infectious for mosquitoes. However the rapid death produced by highly virulent strains removed the infectious source in a few days, whereas moderately virulent strains produced lesions that were infectious for two weeks or occasionally much longer. With more attenuated strains, the lesions healed within ten days. The laboratory variant, neuromyxoma (Hurst 1937) was so attenuated and replicated so poorly in the skin that mosquito transmission was rarely achieved. (From Fenner and Ratcliffe 1965.)

not in the blood serving as the source of infection. Thus any arthropod that probes or bites through a lesion of an infected rabbit and then probes or bites a susceptible rabbit is a potential vector. Tests with strains of differing virulence, and at different periods of the disease, showed that the efficiency of transmission was correlated with the titre of virus in the skin (Fig. 7.5)—a fact of major importance in determining which strains of virus were best transmitted in the field. Mosquito transmission of the very attenuated 'neuromyxoma' strain, a laboratory variant recovered by Hurst (1937), was found to be very inefficient (Fenner *et al.* 1956).

Shope's fibroma virus is even more difficult than neuromyxoma for mosquitoes to transmit from tumours in *Oryctolagus* (Day *et al.* 1956; Dalmat 1959; Dalmat and Stanton 1959), although readily transmitted from fibromas in *S. floridanus* (Kilham and Dalmat 1955).

Intensive study of vectors in the field revealed the importance of *Culex annulirostris* and to a lesser extent *Anopheles annulipes* in epizootic spread in most parts of Australia where myxomatosis became established, but also showed that a wide variety of other winged and, less commonly, wingless arthropod vectors also played a role in particular places and at particulat times (Fenner and Ratcliffe 1965).

Recent studies with marked female mosquitoes have shown that *Culex annulirostris* and *Anopheles annulipes* disperse over much longer distances than was previously thought (M. O'Donnell, G. Berry, J. Bryan, and T. Carvan, personal communication). The estimated average flight range for *C. annulirostris* was 6.8 km; for *An. annulipes* 3.2 km. Calculations suggested that 50 per cent of the marked *C. annulirostris* travelled 4.8 km from the release point and 10 per cent, 15.0 km or more. Rarely, with particularly favourable meteorological conditions, mosquitoes carrying myxoma virus probably moved over much longer distances [for example the Woody Island episode (Fenner and Ratcliffe 1965)].

Andrewes *et al.* (1956) showed that myxoma virus could persist for long periods (up to 220 days) on the mouthparts of hibernating mosquitoes, and mosquitoes are probably responsible for most outbreaks of the disease in commercial rabbitries in Europe. However, compared with fleas (see below) they are of minor importance in the transmission of myxomatosis in wild rabbit populations in Britain (Muirhead-Thomson 1956; Service 1971).

7.5.2 Flea transmission

Fleas are efficient vectors of myxomatosis in *Oryctolagus*. Bull and Mules (1944) regarded the native stickfast flea (*Echidnophaga myrmecobii*) as particularly important in the early field trials in arid areas, but later work showed that mosquitoes were much more important vectors over most of Australia. This led to the expectation of a similar situation in continental Europe and in Britain. However, Rothschild (1953) suggested that the European rabbit flea (*Spilopsyllus cuniculi*) was perfectly adapted to function as the principal vector in Britain. This was confirmed by Lockley (1954) and Muirhead-Thomson (1956) who found that rabbit fleas could, like mosquitoes, act as mechanical vectors. Myxoma virus can persist for over 100 days on flea mouthparts

(Chapple and Lewis 1965) and fleas can survive in the cold for even longer periods without feeding on a host (Allan 1956). It is now agreed that rabbit fleas are the main vectors in Britain (Andrewes *et al.* 1959); this is consistent with the slower spread of the disease and less marked seasonal prevalence in Britain, as compared with Australia. However, although rabbit fleas are present and feeding on rabbits throughout the year in sufficient numbers to ensure effective transmission of myxomatosis, there is considerable variation in the numbers of fleas on individual rabbits and the rate at which they transfer between rabbits. It has been suggested that these variations influence the seasonal prevalence of myxomatosis in Britain (Ross *et al.* 1989).

Believing that fleas leave rabbits only when they die, Fenner and Marshall (1957) and Andrewes *et al.* (1959) speculated that flea transmission might be associated with no selection for attenuated strains, or perhaps a much slower selection than had been found with mosquito transmission. However, Mead-Briggs (1964) showed that there was considerable interchange of fleas between healthy rabbits. By 1962, selection for attenuated strains had resulted in moderately virulent strains being the most common in Britain (Fenner and Chapple 1965) and in France (Joubert *et al.* 1972), just as they were in Australia by 1958–9 (Marshall and Fenner 1960). Mead-Briggs and Vaughan (1975) showed that moderately virulent strains of virus were transmitted by fleas more effectively than either fully virulent or highly attenuated strains. Nevertheless, changes in the virulence of field strains of myxoma virus in Britain have been different from those in Australia, due, at least in part, to the different insect vectors (see Section 7.9.2).

The rabbit flea was not introduced into Australia (or New Zealand) with the rabbit in 1859, or subsequently. The belief that flea transmission might be associated with a slower selection for attenuated strains than had been found with mosquito transmission, as discussed above, and the possibility that fleas might ensure continuing transmission in the absence of mosquitoes, led to a sustained campaign to breed and release *Spilopsyllus cuniculi* in Australia (see Section 7.6.3).

7.5.3 Recurrent infectivity of rabbits

An epidemiologically important possibility that has been suggested by Williams *et al.* (1972*a*), but cannot yet be regarded as proven, is that field strains of myxoma virus may persist in a latent form in rabbits that have recovered from acute myxomatosis, and that in some of these animals disease may recur, with lesions which are infectious. If so, it would obviously provide an important mechanism whereby myxoma virus could persist over a period of several months in a rabbit population in the absence of either frank disease or persistence on the mouthparts of arthropod vectors. Other Australian workers (Marshall, personal communication, 1986; Sobey, personal communication, 1986) have failed to find any evidence of recurrent infectivity in experiments designed to explore this possibility further. Furthermore, such a pattern would be unusual among generalized poxvirus infections, in which recovery is accompanied by elimination of the virus.

However, in a study of myxomatosis in Britain in the 1970s (Ross *et al.* 1989), five rabbits (2 per cent) were seen with symptoms on two or more occasions separated by at least three months. A further thirteen rabbits (6 per cent) which had detectable antibodies to myxoma virus were later seen with symptoms. In the majority of these cases only mild symptoms were recorded; they could have been the results of re-infection rather than recrudescent activity of latent virus.

7.6 Myxomatosis as a method of rabbit control in Australia

The events of the summer of 1950–1 transformed rabbit control efforts in Australia, which thereafter were directed towards making maximum use of myxomatosis as a weapon for biological control. In parallel with the field investigations by CSIRO, virological aspects of myxomatosis were studied by members of the Department of Microbiology of the John Curtin School of Medical Research, Australian National University (ANU), and a long-term programme was undertaken to monitor the virulence of viruses spreading in the field and the factors affecting rabbit resistance (see Sections 7.9 and 7.10).

7.6.1 Establishment: 1951–1954

Two features of the initial spread of myxomatosis in the summer of 1950–1 stand out; the extent of the epizootic, which occurred patchily over an area measuring nearly 1600 km from south to north and 1760 km from east to west, and the dominating importance of the Murray–Darling river system, due, as it was later learnt, to the breeding there in summer of the most important vectors in Australia, the mosquitoes *Culex annulirostris* and *Anopheles annulipes*.

The disease must have persisted through the winter of 1951 in isolated areas throughout south-eastern Australia, for independently of inoculation campaigns (see below), there were outbreaks in many areas during the following summer and they extended many miles back from the rivers. During the third summer season (1952–3) the disease spread throughout the whole area, wherever rabbits lived, and it has remained enzootic ever since, extending out from over-wintering foci and from groups of inoculated rabbits in an erratic fashion that depended upon both vector activity and the size and age structure of the rabbit populations.

7.6.2 Inoculation campaigns

State rabbit control authorities promoted the spread of myxomatosis by inoculating wild rabbits with virulent virus and releasing them again in the locality from which they were obtained, to act as infectious foci. Myxoma virus for these inoculation campaigns was initially produced by the ANU team, and then undertaken commercially by the Commonwealth Serum Laboratories, and in some states by State authorities, notably the Veterinary Research Institute at Glenfield, New South Wales. At different times, three strains of virus have been used. These appeared to be equally (and very highly) virulent when first tested in genetically fully susceptible rabbits, but were shown to differ in virulence when tested in genetically resistant rabbits. They are the Standard Laboratory Strain, an early field derivative called Glenfield (Douglas 1962), and the Lausanne strain; their virulence increases in that order (see Table 7.6).

Inoculation campaigns were pursued with varying enthusiasm in different states. In some, large-scale inoculation and release of wild rabbits were conducted by rabbit control officers; in others the responsibility was left with landholders. The most determined efforts were made in Victoria, where, owing to Dame Jean Macnamara's representations, Mr G. W. Douglas was appointed to a senior position in the state public service especially to pursue rabbit control activities. The Victorian effort was outstanding among the state activities, and valuable work was also carried out by state officials in South Australia and Western Australia, especially after the introduction of the European rabbit flea (see Section 7.6.4). In Victoria, not only was careful research done to determine the optimum time to carry out inoculations (in relation to vector activities), but facilities were established, and continued to be used, to monitor changes in virus virulence and host resistance (see Tables 7.4 and 7.6). Some idea of the scale of these inoculation campaigns can be gauged from the fact that over 250 000 wild rabbits was inoculated and released by inspectors of the Victorian Department of Lands and Survey during the first 12 years of the campaign, and private landholders supplemented this activity.

The need for the repeated introduction of highly virulent virus was apparent from surveys of the virulence of field strains (see Section 7.9) which revealed the dominance after 1953 of viruses that killed 90 per cent rather than 99 per cent of genetically susceptible rabbits. But field experiments carried out in 1955 (Fenner *et al.* 1957) suggested that the selective advantage of these less virulent strains was such that inoculation would have to be on too large a scale to be practical, if it was hoped to establish virulent strains permanently in situations where there was direct competition with attenuated strains. Inoculation was wasted effort if vector mosquitoes were not abundant (and it is difficult to predict their local abundance accurately), and the field experiments and subsequent experience showed that inoculation was of limited value if field strains were circulating at the time of the inoculation campaign. Nevertheless, as Sobey (1960) emphasized, introductions of virulent virus represented the only way in which we could hope to increase the effectiveness of the disease as a control agent. Depending upon their local experience, rabbit control authorities in different states, after the first five years, attached varying importance to the continuation of inoculation, although virus has always been made available for use by landholders. In Victoria, where Dame Jean Macnamara's influence persisted (Zwar 1984) and where extensive laboratory facilities and a field organization were set up to promote the use of myxomatosis, widespread inoculation was carried out under central direction until 1983. Since then, inoculation has been carried out in districts where rabbits were numerous, on a more selective and concentrated basis. Other states had ceased widespread inoculation campaigns by the early 1960s, although in some states virus was introduced with fleas throughout the 1970s (see below).

7.6.3 Release of myxoma virus-contaminated fleas

By the mid-1960s Australian scientists were looking for methods of maintaining the effectiveness of myxomatosis as a method of rabbit control, and of extending its use to areas where spread by mosquitoes was relatively ineffective. It was decided that two approaches should be combined; the use of the rabbit flea as a vector, and the use of the more virulent Lausanne strain of myxoma virus. Under quarantine, fleas from Britain were introduced into rabbits housed in an artificial warren in November 1966 (Sobey and Menzies 1969). They survived and multiplied and in January 1968, following the demonstration that they would not remain on marsupials, quarantine restrictions were lifted. Fleas bred in the laboratory were then introduced into warrens of wild rabbits in three properties in New South Wales, some uninfected and some rendered infectious by probing through the skin lesions of rabbits infected with the Lausanne strain (Sobey and Conolly 1971). Within two breeding seasons the fleas had spread to other warrens, and myxomatosis due to the Lausanne strain of virus had prevented the expected summer build-up in the rabbit populations.

After the early difficulties in the breeding of rabbit fleas in the laboratory had been solved by the demonstration of the intimate connection between the life-cycle of the flea and that of the rabbit (Mead-Briggs and Rudge, 1960; Mead-Briggs and Vaughan 1969; Rothschild and Ford 1964, 1966), Sobey *et al.* (1977) developed methods for breeding large numbers of fleas in the laboratory. Methods for rendering them infectious were also simplified; fleas were shaken up with a concentrated suspension of myxoma virus (1 ml per 5000 fleas) and allowed to dry on a filter paper (Sobey and Conolly 1975). Such insects were effective in introducing virulent strains of virus into free-living populations of rabbits (Parer *et al.* 1981). Over the next decade this method of introducing myxoma virus was tried, with some success, by rabbit control authorities in New South Wales, Queensland, Victoria, and South Australia.

Several observers noted changes in the epidemiology of myxomatosis in areas where fleas had been established. Apart from the acute outbreaks associated with the introduction with the fleas of the Lausanne strain of virus, outbreaks in Victoria due to enzootic attenuated strains started in early spring rather than mid-summer, and caused a high mortality in young kittens, except in a year when the breeding season was much later than usual (Shepherd and Edmonds 1978). In a controlled trial in South Australia, Cooke (1983) found that in areas where fleas had been introduced there was a heavy mortality in young rabbits, so that they formed a much lower proportion of summer populations than in areas where fleas were not present. In Western Australia, also, there were epidemics of myxomatosis in winter after the introduction of fleas, which caused greater reductions in rabbit numbers than did the mosquito-transmitted summer outbreaks experienced earlier (King and Wheeler 1985; King *et al.* 1985).

The flea has become established in Australian rabbit populations living in a range of climatic conditions (Williams and Parer 1971; Shepherd and Edmonds 1979; King *et al.* 1985); except in arid areas in South Australia and Western Australia, possibly because of the failure of rabbits to breed during periods of prolonged drought, or because of the low humidity during hot summers causing high mortality among the fleas (Cooke 1984).

7.6.4 Myxomatosis as an enzootic disease

Although the situation differed considerably in different parts of Australia, myxomatosis was, in general, accepted as part of the ecological scene from about 1960. It had brought, and continues to bring, incalculable benefit to Australia, converting a rabbit pest problem that was uncontrollable in 1950 to one that remains challenging, but is manageable. Outbreaks of myxomatosis continue, and combined with

Table 7.2 Long-term effects of myxomatosis on rabbit numbers at four CSIRO study sites in south-eastern Australia[a]

	Before myxomatosis (1950)	After myxomatosis (1970)
Rutherglen, Vic., 13 acres		
Number of warrens	15	0
Warrens per acre	1.15	0
Number of rabbits	1000	3
Rabbits per acre	77	0.23
Active burrows	850	4
Coreen, NSW, 17 acres		
Number of warrens	26	0
Warrens per acre	1.5	0
Number of rabbits	400	2
Rabbits per acre	24	0.12
Active burrows	600	2
Balldale, NSW, 17 acres		
Number of warrens	150	5
Warrens per acre	8.8	0.3
Number of rabbits	300	20
Rabbits per acre	18	1.18
Active burrows	1250	17
Urana, NSW, 340 acres[b]		
Number of rabbits	5000	11
Rabbits per acre	167	0.37

[a] From Myers (1970).
[b] Site of studies on virus virulence and genetic resistance.

the greater effectiveness of predation in small rabbit populations and on sick rabbits, the disease continues to exert a substantial influence on rabbit numbers, in spite of the changes in virus virulence and rabbit resistance outlined below (Sections 7.9 and 7.10). Accurate data are impossible to obtain, but the best estimate is that rabbit populations in most open agricultural areas in southern Australia now fluctuate around 1–10 per cent of pre-myxomatosis levels. However, there have been significant increases in numbers in some wooded, sandy, and rocky environments where farming activities are restricted, and in many favourable arid and semi-arid environments plagues reminiscent of pre-myxomatosis times now occur (K. Myers, personal communication 1991). In four sites where CSIRO conducted long-term field studies, Myers (1970) recorded very substantial reductions in the rabbit populations (Table 7.2), which appear to have been maintained since then.

In an attempt to determine the effects of myxomatosis in each of the four years 1978–81, Parer *et al.* (1985) introduced a highly attenuated 'immunizing' strain of myxoma virus into two populations of rabbits at Urana, in which *Spilopsyllus cuniculi* was well-established. Within two years, one population had increased by a factor of 8 and the other by a factor of 12, suggesting that the field strains of myxoma virus were still playing an important part in suppressing these rabbit populations.

7.7 Effects of myxomatosis on rabbit numbers in Britain

Despite severe rabbit damage to agriculture and horticulture, estimated to cost 40–50 million pounds per year (Thompson and Worden 1956), and despite earlier suggestions that myxomatosis should be introduced into Britain, official policy when the disease did appear in Kent in September 1953 was to attempt eradication of the virus (Chapter 4). Rabbit-proof fences were erected round the areas of the first outbreaks, and all the rabbits enclosed were destroyed. These efforts were unsuccessful, and the disease soon extended beyond the fences. Within two years it had spread over most of Britain (Armour and Thompson 1955), aided by the deliberate use of infected rabbits to spread myxomatosis, despite a Section in the Pests Act 1954 which made this illegal.

It was estimated that over 99 per cent of rabbits were killed as the disease spread over the country (Hudson *et al.* 1955; Brown *et al.* 1956). Rabbit populations remained very low generally for some years, but were subject to further sporadic outbreaks of myxomatosis where rabbit numbers had increased locally (Lloyd 1970). By 1959, Andrewes *et al.* (1959) noted an increase in damage to crops, indicating an increase in rabbit numbers. During the 1960s, rabbit populations increased only slowly over the country as a whole, because of regular outbreaks of myxomatosis and the influence of predators (Lloyd 1970), although there were large local fluctuations, with rabbits disappearing from some properties and appearing on others.

More recently, a number of surveys giving some measure of distribution and/or abundance of rabbits in Britain have shown continuing increases in both (Trout *et al.* 1986). Lloyd (1981) estimated the overwinter rabbit population in 1979 to have reached 20 per cent of the pre-myxomatosis population. It is likely that rabbit numbers will rise further, especially if the degree of resistance to myxomatosis continues to increase without a compensating increase in the virulence of field strains of virus.

Studies of the epidemiology of myxomatosis illustrate the changes which have allowed rabbit numbers to recover. In the original 1953 outbreak in Kent, Armour and Thompson (1955) found that virtually all rabbits were infected with the fully virulent virus strain involved and died from the disease. In 1962, an outbreak in Yorkshire, caused by moderately virulent virus strains (Chapple and Lewis 1964), resulted in the death of about 90 per cent of the population. Although almost all rabbits appeared to have been infected, the disease spread much more slowly than in the original Kent outbreak, probably because of the lower density of rabbits in 1962. During outbreaks investigated in the 1970s (Ross *et al.* 1989), the spread of the disease was slow, incomplete (in that it failed to reach all areas with rabbits), and inefficient (since some susceptible rabbits within affected areas escaped infection). On average, only 25–27 per cent of rabbits present during the (prolonged) outbreaks were known to have been infected, and 47–69 per cent of these died. Thus, in rabbit populations with relatively high proportions of immune rabbits (particularly adults), and in which considerable genetic resistance to the disease had developed (see section 7.10), only 12–19 per cent of the total rabbit population was known to have died directly or indirectly from myxomatosis, despite some increase in the virulence of field strains of virus.

Trout *et al.* (1993) described an experiment in which the prevalence of myxomatosis in a rabbit population fell after reducing the numbers of rabbit fleas for two years. Flea numbers were then increased by systematic introductions for two years. The four-year cycle was then repeated. Changes in disease prevalence followed changes in flea numbers. Rabbit numbers increased two- to three-fold during flea reduction and decreased during the intervening periods, suggesting that myxomatosis continues to have a role in determining rabbit populations numbers.

7.8 Myxomatosis in France

The rabbit occupied a unique place in France, because wild rabbits were highly valued as a major game animal and as a source of skins and hair for hat-making, while there was also an extensive domestic rabbit industry involving both large breeding farms and backyard production (Chapter 3; Fig. 7.6). It was therefore predictable that Delille's action in introducing myxoma virus should meet with a mixed reception; foresters and some farmers were pleased, but domestic rabbit breeders, the hunting industry and other farmers, who rented land for hunting, were infuriated (Fenner and Ratcliffe 1965).

In May 1953 myxomatosis was made a notifiable disease in France and efforts were made to limit its spread. However, myxomatosis spread throughout the country, and by the end of 1954 it was estimated that 90 per cent of the wild rabbits in France had been killed by myxomatosis. Extreme measures were taken to lessen the effects; attempts were made to create 'sanitary barriers' by immunizing large numbers of wild rabbits with fibroma virus (see Section 7.12) and then releasing them, and attempts were made to introduce *Sylvilagus floridanus* from the USA and even to import genetically resistant *Oryctolagus cuniculus* from Australia.

A good indication of the impact of the disease on the wild rabbit population can be obtained from records of the numbers of rabbits shot during the hunting season. Very few rabbits were shot during the season following the first outbreaks of myxomatosis, but there were small increases in the following two seasons (Giban 1956) and further increases in succeeding years. Giban (1959) pointed out, however, that variations in the timing of the major outbreaks of myxomatosis in relation to the hunting season caused large fluctuations in the numbers of rabbits shot.

More recent hunting records (Arthur *et al.* 1980) showed that numbers of rabbits shot in the Sologne region (expressed as percentages of the mean number shot per season before myxomatosis) increased from 1.5 per cent in 1953–8 to 13.5 per cent in 1967–72 and to 17.6 per cent in 1972–7. Waechter (1983) recorded that, in Alsace, rabbits were numerous only in pockets, concentrated particularly in the valley of the Rhine, where rabbit browsing prevented regeneration of elm trees (Schortanner 1976).

National surveys in France in 1977 and 1978 (Arthur *et al.* 1980) showed that rabbit density varied from region to region due to environmental factors, such as type of soil, habitat and agricultural practice, and was also affected by hunting pressure and predation. Rabbits were most abundant in north-western France and much less common in the south and west, where they often lived only in isolated pockets. The epidemiology of myxomatosis in France is similar to that in Britain, although mosquito transmission is more important. Major outbreaks are most

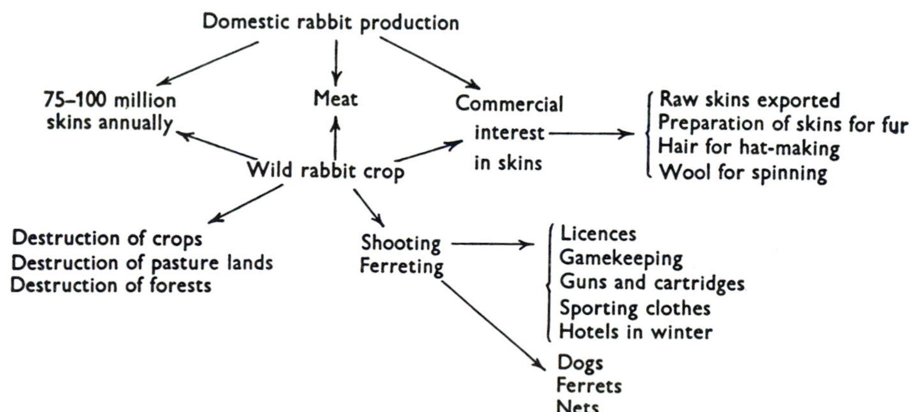

Fig. 7.6 The place of the rabbit in the domestic economy of France. (From Fenner and Ratcliffe 1965).

likely in late summer–autumn. Recent surveys of epizootics in three areas of France (Arthur and Louzis 1988) revealed that an estimated 30–60 per cent of the populations were infected during the annual epidemics, of which 54–69 per cent died, giving an overall mortality of 22–36 per cent, somewhat higher than the mortality found in the British study (see Section 7.7).

7.8.1 Measures taken against myxomatosis

Outbreaks of myxomatosis among domestic rabbits in France were very common in 1953 and 1954, and continued in subsequent years. Vaccination with fibroma virus was used on a large scale (tens of millions of doses annually); after 1970 attenuated strains of myxoma virus were also used for immunization (see Section 7.12.2). More recently, infection of domestic rabbits has decreased as the numbers of wild rabbits and the extent of myxomatosis among them has diminished, but protection by insect screening and vaccination is still recommended.

Because of the value of wild rabbits as game animals, measures to protect them against myxomatosis were still being taken in the 1980s. Joubert *et al.* (1982) describe methods for limiting the reservoir of myxoma virus, for controlling vectors, and protecting wild rabbits by vaccination, for which they recommend vaccination first with fibroma virus, followed 6–8 weeks later by attenuated myxoma virus, both given by dermojet (see Section 7.12).

7.9 Changes in virus virulence

When it was clear that myxomatosis had become established in the wild rabbit populations of Australia and Europe, the immediate concern of scientists in Australia was whether the virulence of the virus would attenuate.

7.9.1 Observations in Australia

The first indications that the severity of the disease might be changing came from field observations at Lake Urana, in the riverine plain between the Murray and Murrumbidgee Rivers, and laboratory tests of viruses isolated from outbreaks there. The results of the first major study (Myers *et al.* 1954) are illustrated in Fig. 7.7. By inoculating about 100 local rabbits with virulent standard laboratory strain myxoma virus, and then releasing them, myxomatosis was established in early summer in a very dense rabbit population which had not previouisly been infected (transect count, 5000 rabbits). Owing to the presence of a large population of efficient mosquito vectors, the disease soon spread to uninoculated rabbits and produced a catastrophic fall in rabbit numbers; the case-fatality rate was calculated to be 99·8 per cent. Regular observations were maintained throughout the winter, and isolated cases of myxomatosis were observed almost every month, so there must have been some new infections throughout the winter and spring. The surviving recovered and uninfected rabbits bred in the spring, and by early summer the population had built up from its very low post-epizootic level, by breeding and migration, to reach a total of about 500, as estimated by transect counts. Another outbreak began independently of inoculation in October 1952, and in late summer serological tests on a sample of the survivors suggested that all rabbits in the area had been infected. However, on this occasion a substantial number had recovered; the case-fatality rate among previously uninfected rabbits was calculated to be only 90 per cent. Furthermore, viruses obtained during the second epizootic produced quite different signs in laboratory rabbits, with a much more prolonged course and a case-fatality rate of about 90 per cent. One such virus provided the prototype strain of grade III virulence (Fig. 7.4C).

This result provided conclusive evidence of rapid attenuation of the virus, albeit to a strain that still caused a very high mortality. Investigations were therefore undertaken systematically to determine the virulence of field strains over several succeeding years, using the survival times of rabbits infected with small doses of virus as a way of allocating strains to one of five (later six) 'virulence grades'. The results obtained by the ANU team over the period 1950–8 (Marshall and Fenner 1960; Table 7.3) are unequivocal. Myxomatosis had survived in the Australian rabbit population because natural selec-

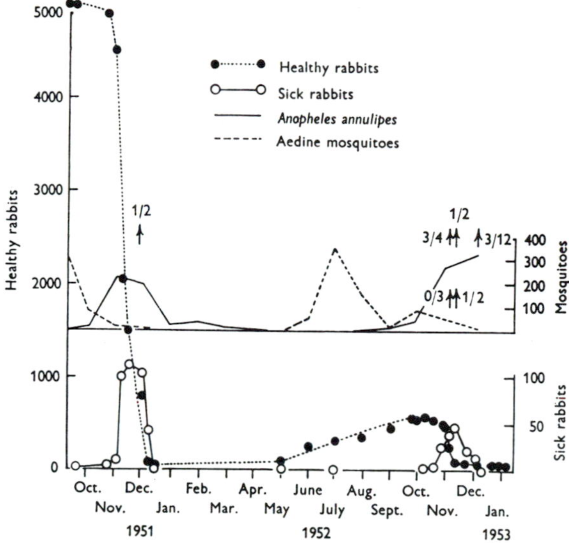

Fig. 7.7 Numbers of healthy and sick rabbits, and adult mosquitoes, before and after the first and second epizootics of myxomatosis at Lake Urana (dotted and solid lines indicate rabbit numbers found on standardized transect counts). The case-fatality rates were calculated as 99.8 per cent in the first outbreak and 90 per cent in the second outbreak. Vertical arrows indicate occasions on which myxoma virus was isolated from batches of mosquitoes. The virus was transmitted mechanically by *Anopheles annulipes* mosquitoes (solid line). Aedine mosquitoes (dashed line), although a pest to man, failed to spread myxomatosis because they rarely bit rabbits. (From Myers *et al*. 1954.)

tion had operated at the level of transmission to select for viruses of somewhat reduced virulence, which therefore did not eradicate the only host animal, *Oryctolagus cuniculus*, and with it the virus. Evidence since 1959 shows that viruses of grade III virulence have remained the dominant strains (Dunsmore *et al.* 1971; Edmonds *et al.* 1975; Fenner 1983; see Table 7.9). The results suggest that over the decade 1970–80 there may have been a slight increase in the levels of virulence of viral strains recovered from a region of Victoria subject to annual outbreaks, but not from other parts of that state, where outbreaks were less frequent (see Section 7.11).

A field experiment designed to assess the value of inoculation campaigns with the virulent Lausanne strain of myxoma virus (Fenner *et al.* 1957) provided an opportunity to compare the efficiency of transmission of virulent and moderately attenuated (field) strains of myxoma virus. Although inoculation of Lausanne virus was carried out on a large scale, the resulting epizootic, while causing a high initial mortality (due to Lausanne virus), was dominated in its later stages by the enzootic grade III field strain, because of the selective advantage for transmission afforded by the longer period of infectivity of rabbits infected with this strain. Similar results were reported by Shepherd and Edmonds (1977) in the presence of rabbit fleas; the Lausanne strain was found to persist for only ten weeks in the presence of field strains of the virus.

7.9.2 Observations in Europe

For different reasons, deliberate inoculation of virulent virus was forbidden in both Britain and France, and the evolutionary changes since introduction of the Lausanne strain in 1952 are therefore relatively uncomplicated. Furthermore, transmission in Britain is due very largely to rabbit fleas (Andrewes *et al.* 1959); in France, both fleas and mosquitoes play a role in transmission.

The absence of any reintroductions of virulent virus might be expected to speed up the appearance and dominance of attenuated strains of myxoma virus in Europe, as compared with Australia, but in fact the reverse was observed (Table 7.4). Nine years after its introduction, over 20 per cent of strains recovered from wild rabbits in Britain were of grade I or grade II virulence compared with only 6 per cent in Australia. In France, ten years after the introduction of myxomatosis, 30 per cent of field strains were of grade I or grade II virulence (Joubert *et al.* 1972). Thus, both in Britain and in France, the initial selection for viral strains of reduced virulence was somewhat slower than in Australia. Although no major survey of the virulence of field strains of virus was carried out between 1962 and 1975, the results of tests of small numbers of strains in the intervening years (Ross 1982) indicated that the attenuated strains did not become increasingly common in Britain, in contrast to Australia (Fenner and Woodroofe 1965) and France (Joubert *et al.* 1972). This could well have been due to the difference in selection pressures exerted by flea transmission in Britain and mosquito transmission in Australia and France.

Table 7.3 The virulence of strains of myxoma virus recovered from the field in Australia between 1951 and 1981, calculated on the basis of survival times (expressed as percentages)

Virulence grade	I	II	III	IV	V	Number of
Case fatality rate (%)	>99	95–99	70–95	50–70	<50	samples
Mean survival time (days)	≤13	14–16	17–28	29–50	—	
1950–51[a]	>99[d]					1
1952–55[a]	13.3	20.0	53.3	13.3	0.0	60
1955–58[a]	0.7	5.3	54.6	24.1	15.5	432
1959–63[b]	1.7	11.1	60.6	21.8	4.7	449
1964–66[b]	0.7	0.3	63.7	34.0	1.3	306
1967–69[b]	0.0	0.0	62.4	35.8	1.7	229
1970–74[b]	0.6	4.6	74.1	20.7	0.0	174
1975–81[c]	1.9	3.3	67.0	27.8	0.0	212

[a] Data from Marshall and Fenner (1960).
[b] Data from Edmonds *et al*. (1975).
[c] Data from J. W. Edmonds and R. C. H. Shepherd (personal communication, 1982).
[d] Although only one strain was tested, the very high mortality rates in the initial outbreaks justify this extrapolation.

Table 7.4 The virulence of strains of myxoma virus recovered from the field in Britain and France between 1953 and 1981, calculated on the basis of survival times (expressed as percentages)

Virulence grade	I	II	III	IV	V	Number of
Case fatality rate (%)	>99	95–99	70–95	50–70	<50	samples
Mean survival time (days)	≤13	14–16	17–28	29–50	—	
Britain						
1953	>99[a]	—	—	—	—	
1962[b]	4.1	17.6	63.6	14.0	0.9	222
1962[b]	1.6	25.8	66.4	5.5	0.8	128
1981[c]	0.0	35.8	62.6	1.6	0.0	123
France						
1953[d]	>99[a]	—	—	—	-	
1962[d]	11	19.3	55.4	13.5	0.8	
1968[d]	2.0	4.1	35.1	58.8	4.3	

[a] Based on very high mortality rates in the initial outbreaks.
[b] From Fenner and Chapple (1965).
[c] From Ross and Sanders (1987).
[d] From Joubert *et al*. (1972).

Subsequent changes in the virulence of field strains followed different patterns in the three countries. In Australia, reduction in virulence continued until the late 1960s, by which time over 30 per cent of strains tested were of grade IV virulence and there were few strains of grade II virulence. In Britain, strains of grade IV virulence were never common, and there were significant increases in virulence between 1962 and 1975 and again between 1975 and 1981 (Ross and Sanders 1987). In France, the virulence of field strains was further reduced between 1962 and 1968, by which time strains of grade IV virulence made up nearly 60 per cent of the total. No more recent information is available.

Several factors could have contributed towards these differences. (a) The different insect vectors may have exerted different selection pressures (see Section 7.5). (b) Increases in virulence of field strains are associated with increases in the degree of genetic resistance in rabbits, and rabbits in Britain may have recently developed a greater degree of resistance than Australian wild rabbits (see Section 7.10.2), leading to a further increase in virulence in Britain. (c) The 'Standard Laboratory' strain of myxoma virus which was released in Australia had been passaged many times in the laboratory, whereas the 'Lausanne' strain which initiated the French and British outbreaks had been passaged only a few times; this may have influenced subsequent changes in the virulence of field strains (Ross 1982).

An interesting development was noted in France in the late 1970s. On intensive rabbit farms Brun *et al.* (1981a) detected a number of strains which produced much less-pronounced skin lesions but were associated with severe respiratory signs. The appearance of these strains was at first thought to be connected with the widespread use of live attenuated myxoma virus (SG33) as a vaccine (see Section 7.12.2). However they were shown to be distinct from SG33, and were also found in rabbit farms not using vaccine. Such 'respiratory' strains were shown serologically to be authentic myxoma virus, but were readily transmissible by close contact in the absence of insect vectors and did not revert on passage to 'classical' myxoma strains. Joubert *et al.* (1982) found that these 'respiratory' strains ranged from very virulent to attenuated and that infected rabbits frequently contracted 'snuffles' (pasteurellosis) due to secondary bacterial infection. The latter complication was also observed in rabbits infected with Australian attenuated strains (Fenner, unpublished observations); rabbits infected with virulent strains died before 'snuffles' became apparent. It is probably due in part to an immunosuppressive effect of the viral infection, observed by Strayer *et al.* (1983) in rabbits infected with a myxoma-rabbit fibroma recombinant virus (Block *et al.* 1985). Recent studies of recombinants between fibroma virus and fragments of myxoma virus DNA demonstrated that the immunosuppressive effect was due in part to a gene which enabled the myxoma virus and the recombinant to replicate in lymphocytes (which fibroma virus fails to do) and thus suppress immune function (Heard *et al.* 1990).

7.10 Changes in rabbit resistance

The factors that determine whether rabbits die of myxomatosis, or how long they live after being infected, can be considered under four headings: acquired immunity, genetic resistance, 'paternal immunity', and non-specific factors, such as the effects of ambient temperature on the disease process and predator activity.

7.10.1 Acquired immunity

In myxomatosis, as in other generalized viral infections, there is an immune response that can be first detected by *in vitro* tests about seven days after infection (Fenner and Woodroofe 1953), and reaches peak levels (in animals that survive) by about 28 days. Antibodies persist for prolonged periods, and an altered response to reinfection, amounting to absolute immunity for many months, lasts for life.

Immune does transfer antibody to their kittens, and at birth the antibody titres in mother and young are the same. It is usually difficult to detect antibodies in the serum of kittens by *in vitro* tests for more than 4–5 weeks, but passively immune kittens respond to infection differently from normal kittens for at least eight weeks (Fenner and Marshall 1954). This is particularly important because non-immune

Table 7.5 Case-fatality rates and severity of myxomatosis in groups of non-immune wild rabbits from Lake Urana, after successive epizootics of myxomatosis. Challenge infection with virus of grade III virulence[a]

Number of epizootics to which population had been exposed	Case-fatality rate (%)	Clinical signs in challenged rabbits (%)		
		Severe (including fatal)	Moderate	Mild
0	90	93	5	2
2	88	95	5	0
3	80	93	5	2
4	50	61	26	12
5	53	75	14	11
7	30	54	16	30

[a] Data from Marshall and Fenner (1958), Marshall and Douglas (1961).

kittens are much more susceptible to myxoma virus, even to attenuated strains, than are adult rabbits (Fenner and Marshall 1954). Passive immunity can completely protect newborn rabbits from infection; in older kittens it can convert a severe or lethal infection into a mild disease which gives prolonged immunity to reinfection. The effect of acquired immunity is enhanced by both genetic resistance of the host (and any non-specific factors that enhance resistance) and the replacement of highly virulent strains (which may overwhelm host resistance) by attentuated strains of virus.

While rabbit numbers were high and population turnover rapid, acquired immunity, active or passive, was probably not very important ecologically. When population pressure was lower and rabbits commonly survived for more than one myxomatosis season, active immunity decreased the proportion of susceptible rabbits. Passive immunity might also then provide some protection for kittens borne by immune does, although experiments by Sobey and Conolly (1975) suggested that this had little effect on kitten survival.

7.10.2 Genetic resistance

Initially, the nearly universal mortality provided no opportunity for rabbits to be selected for resistance to myxomatosis. When the virulence of the dominant strains fell, so that about 10 per cent of animals recovered (even though many males were left sterile: Sobey and Turnbull 1956), there was an opportunity for selection to favour more resistant rabbits. In Australia, tests carried out on wild rabbits that had not been previously infected with myxomatosis showed that such selection was effective and rapid (Table 7.5; Marshall and Fenner 1958; Marshall and Douglas 1961). On theoretical grounds, Rendel (1971) predicted that the average level of resistance would continue to rise rapidly, but this does not appear to have happened in Australia (G. W. Douglas, personal communication 1979; Table 7.6).

Laboratory breeding experiments from rabbits that had recovered from infection with either virulent or attenuated strains of virus showed that under these conditions also, resistance was selected for quite rapidly (heritability, $h^2 = 0.35–0.45$) (Fig. 7.8; Sobey 1969). However, as in regions of frequent epizootics in the field (see Table 7.6, 'Mallee wild rabbits'), there was no increase in resistance in a continuation of these breeding experiments with domestic rabbits between 1969 and 1974 (Sobey and Conolly 1986). In both situations there seems to have been an initial quite rapid selection for increased resistance, followed by a longer period characterized by a relatively stable host-virus balance.

The development of genetic resistance in Britain was apparently very much slower than in Australia. Vaughan and Vaughan (1968) found what they interpreted as the first signs of developing resistance in

Table 7.6 Case-fatality rates of non-immune rabbits challenged with strains of myxoma virus of grade I virulence

	Case fatality rate (%)		
	s.l.s.[a]	Glenfield strain[b]	Lausanne strain[c]
Selection experiments[d]			
grade 0	99		
grade 4	88		
grade 6.5	79		
Gippsland wild rabbits[e]			
1961–66	94	98	
1967–71	90	99	
1972–75	85	98	*c.* 100
1975–81[f]	79	95	*c.* 100
Mallee wild rabbits[e]			
1961–66	68	98	
1967–71	66	94	
1972–75	67	96	*c.* 100
1976–81[f]	60	91	98
1984[g]			88

[a] Standard laboratory strain: virus used for introduction of myxomatosis into Australia (1950).
[b] Glenfield strain: a field isolate of high virulence recovered in New South Wales in February 1951.
[c] Lausanne strain: virus used for introduction of myxomatosis into France (1952).
[d] From Sobey (1969).
[e] G. W. Douglas, personal communication, 1978. Gippsland is a part of Victoria in which epizootics of myxomatosis recurred at intervals of several years. The Mallee is a part of Victoria that had experienced annual outbreaks of myxomatosis since 1951.
[f] J. W. Edmonds and R. C. H. Shepherd, personal communication, 1982.
[g] From Williams *et al.* (1990).

the extended survival times of some rabbits from two areas, when injected with a moderately virulent (grade III) strain, although case-fatality rates were unaffected. The first definite evidence for increased resistance in Britain was found in 1970 (Ross and Sanders 1977). The case-fatality rate of rabbits from one site, challenged with the same moderately virulent strain as that used by Vaughan and Vaughan (1968), was significantly lower than in previous years (Table 7.7). Similar resistance to this strain was also found in rabbits from other widely separated areas (Ross and Sanders 1984). Rabbits from one of these areas (Wiltshire, Table 7.7) have shown significant resistance to a fully virulent grade I strain, which was derived from the Lausanne strain, to which rabbits in Australia appear to have little enhanced resistance (see Table 7.6). Thus, although direct comparative tests have not been carried out, resistance to myxomatosis in wild rabbits in Britain may have developed beyond the level reached in Australia. Since only a small number of wild rabbits were originally introduced into Australia, the genetic variability of Australian wild rabbits may be relatively small, thus limiting the extent of selection for resistance to myxomatosis.

The failure to detect the development of resistance in Britain until 15–17 years after the disease first appeared is difficult to explain. Regular annual tests for resistance were restricted to rabbits from two areas (and later one area), and it is possible that

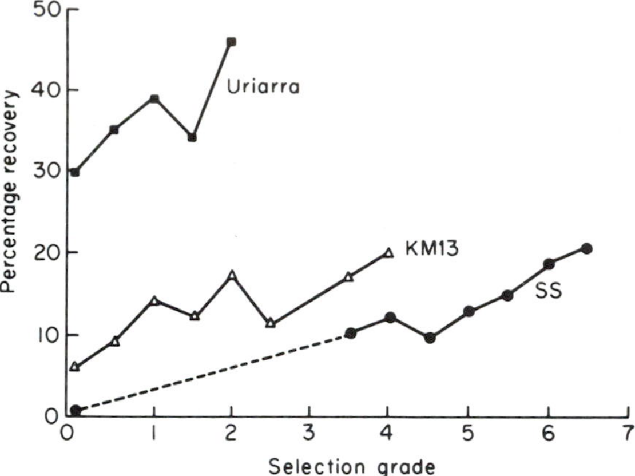

Fig. 7.8 Changes in the genetic resistance of laboratory rabbits to myxoma virus during breeding from animals that had recovered from infection with various strains of virus. 'Selection grade' indicates the parentage of rabbits under test: 0.5, one parent had recovered from myxomatosis; 1, both parents had recovered; 2, both parents and both grandparents had recovered, etc. For the initial selection two attenuated strains were used: Uriarra (grade IV virulence) from selection grade 0 to 2, ■——■; KM13 (grade III virulence) from selection grade 0 to grade 4, Δ——Δ. At selection grade 3.5 a selection line was begun with virulent myxoma virus, SS (= Standard Laboratory Strain) ●——●, a strain that killed almost all genetically unselected rabbits. (Data from Sobey 1969.)

resistance developed more quickly elsewhere, although the main area studied had higher than average rabbit densities, and experienced annual outbreaks of myxomatosis at least from 1960—factors likely to favour selection for resistance. The greater persistence of virulent strains of virus in Britain compared with Australia (see Section 7.9) may have limited the opportunity for selection. In addition, major outbreaks of myxomatosis in Australia, being mosquito-borne, were confined mainly to the summer months, and the high summer temperatures experienced in many parts of that country are known to alleviate the effects of the disease (see Section 7.10.4) and so would have favoured selection for resistance.

In France, the only evidence of genetic resistance is from one experiment in which small numbers of rabbits from three sites were infected with the Lausanne strain (Galaup, 1988). Rabbits from two of the sites showed little resistance (1/7 and 1/4 survived) but all seven rabbits from the third site survived infection.

7.10.3 'Paternal immunity'

In the continued breeding experiments, the poor reproductive performance of does selected for resistance to myxomatosis led to the introduction of matings between selected sires and unselected does. In their analysis of the data from these experiments, Sobey and Conolly (1986) observed that bucks mating within seven months of infection with myxoma virus sometimes conferred partial resistance on progeny born to the mated doe in the following seven months, even if the biological father was not immune. They speculated that the semen of recovered bucks contained some factor that enhanced survival of the doe's progeny, when they were infected with myxoma virus. Williams and Moore (1991) observed similar enhanced resistance in the progeny of immune wild bucks, and suggested that it reduces the effectiveness of myxomatosis in some wild rabbit populations.

7.10.4 Ambient temperature

Early observations by Parker and Thompson (1942) on the sparing effect of high environmental temperatures on rabbits infected with virulent myxoma virus were considerably amplified by Marshall (1959). The results with somewhat attenuated strains of virus were dramatic (Fig. 7.9 and Table 7.8). The severity and case-fatality rates were exacerbated by low and ameliorated by high environmental temperatures.

In experiments carried out with attenuated strains in unheated animal houses both Mykytowycz (1956) and Sobey *et al.* (1967) found that survival times were substantially shorter and mortality rates higher in winter than during the summer. In an attempt to promote recovery of enough rabbits infected with virulent myxoma virus to use this strain of virus in breeding experiments, Sobey *et al.* (1967) examined various regimes of exposure of infected rabbits to an ambient temperature of 85°F. They found such exposures had trivial effects on the disease process in genetically unselected rabbits, but that in genetically

Table 7.7 Genetic resistance to myxomatosis in wild rabbits in Britain[a]

Location	Year	Number of rabbits tested	Virulence grade of challenge virus[b]	Case-fatality rate (%)	Mean survival time of rabbits that died (days)
Norfolk	1966	41	III	90	26.7
	1967	34	III	94	23.2
	1968	71	III	86	25.5
	1969	74	III	84	25.9
	1970	27	III	59	24.8
	1974	15	III	13	—
	1974	11	I	100	14.1
	1975	11	I	100	17.1
	1976	63	III	21	35.9
Wiltshire	1978	71	III	45	28.6
	1979	53	II	45	29.4
	1980	44	I	56	22.1

[a] From Ross (1982).
[b] Checked on occasion of each test by inoculation into laboratory rabbits.

Fig. 7.9 The effect of ambient temperature on the response of rabbits to myxomatosis. All rabbits had been infected 20 days earlier with a small dose of virus of grade III virulence. A. Rabbit kept in cold room for 16 hours each day. B. Rabbit kept in hot room for 16 hours each day. Both were kept at ambient temperature (15°C) for the remaining 8 hours of each day. (From Marshall 1959.)

selected rabbits, twenty-four hours' exposure between the third and fifth days after infection prolonged survival and resulted in more recoveries.

Field observations (Williams *et al.* 1972b) support the view that low ambient temperatures probably have important effects in increasing the mortality in winter epizootics. One effect of flea transmission in Australia has been to move the onset of annual epizootics to spring rather than mid-summer, with resulting increases in case-fatality rates (Shepherd and Edmonds 1978).

In Britain, there seems to have been no detectable effect of ambient temperature on the mortality due to myxomatosis. During investigation of epizootics of myxomatosis at three study sites between 1971 and 1978, Ross *et al.* (1989) pointed out that on average the monthly numbers of myxomatous carcases found followed the same annual pattern as the monthly numbers of (live) infected rabbits seen, implying that the case-fatality rate did not vary during the year. This is not unexpected since summer temperatures in Britain are rarely as high as in many parts of Australia.

Table 7.8 Myxomatosis in rabbits held at room temperature (15–22°C) for 8 hours each day, and for the other 16 hours at low (0–4°C), room (15–22°C) or high (37–39°C) temperatures[a]

Virus strain		Cold room (0–4°C)	Mild room (15–22°C)	Hot room (37–39°C)
Moderately virulent (KM13)	Mortality	36/39	23/35	2/23[b]
	Clinical signs	Severe, progressive generalized disease	Variable	Mild with early demarcation of lesions
	Viraemia	High and prolonged	Moderately high and prolonged	Low and transient
	Antigen in serum[c]	Present in 4/5 from day 10 to day 15 (absent from one survivor)	Present in 2/4 on day 12	Absent
Highly virulent (Lausanne)	Mortality	—	6/6	6/6
	Survival time (days)	—	10, 11, 11, 11, 11, 13	9, 9, 10, 12, 12, 12
Nil (control rabbits)	Mortality	0/36	—	0/31

[a] Data from Marshall (1959).
[b] Seven deaths not due to myxomatosis omitted.
[c] Determined by complement fixation test.

In some parts of continental Europe, summer temperatures are comparable to those in Australia but there are no data for case-fatality rates in summer and winter outbreak in those areas.

7.10.5 Predators

Although Australia is usually regarded as a land with few predatory animals, from the rabbit's point of view it is well stocked with efficient enemies: foxes, feral cats, diurnal and nocturnal birds of prey, the domestic dog, and man. Like the rabbit, all except the birds are exotics. As predators, they are non-specific, but over 60 per cent of the diet of foxes, feral cats, and the larger birds of prey consists of rabbit meat and they depend on rabbits for the maintenance of their population densities (Baker-Gabb 1984; Parer 1977). Rabbit kittens are especially vulnerable, but foxes and wedgetailed eagles, and to a lesser extent cats, also kill adult rabbits. When rabbits were very numerous, as they were before myxomatosis, these predators had a negligible effect on rabbit numbers (Calaby 1951). With the reduction in rabbit numbers due to myxomatosis their effects became apparent (Fenner and Ratcliffe 1965: 46), not only because the percentage kill was higher but also because myxomatosis itself rendered both adult and juvenile rabbits very vulnerable to predation (Williams *et al.* 1973). Field observations over the past decade have reinforced the view that predators are now an important influence in controlling rabbit densities in Australia, and that their impact is enhanced by myxomatosis (Chapter 5).

In Britain also, predators—foxes, feral cats, and mustelids particularly—probably have a significant effect on post-myxomatosis rabbit populations (Trout and Tittensor 1989) but it has not been measured. Lloyd (1970, 1981) pointed out that areas on which rabbit densities had remained low had relatively high densities of predators, and conversely areas with low predator densities had high rabbit densities.

7.11 Coevolution of viral virulence and rabbit resistance

The fully virulent strains of virus which initiated the outbreaks in Australia, France, and Britain were largely replaced by moderately virulent strains which have remained predominant since the early 1960s (see Section 7.9), with no further increases in the proportion of more attenuated strains. Fenner and Ratcliffe (1965) speculated that since selection depends on transmissibility, and too mild a disease as well as too severe a disease is associated with lowered transmissibility, one might expect increased genetic resistance to select for what in genetically unselected rabbits would be classed as more virulent strains of virus.

Data from Victoria (Table 7.9) support this view. In the Mallee region, where resistance to myxomatosis was higher than elsewhere in the state (Edmonds *et al.* 1975) there was an increase in the virulence of field strains of virus in the early 1970s, whereas there was no similar increase in virulence of field strains from other parts of Victoria. In Britain also (Table 7.4) there was a marked increase in the virulence of field strains between 1962 and 1975, roughly coinciding with the first evidence of developing resistance in wild rabbits, and a further increase between 1975 and 1981 as resistance developed further. The more marked increase in virulence of field strains in Britain may have occurred because resistance in wild rabbits in Britain has proceeded further than in Australia (see Section 7.10.2).

Anderson and May (1982) used data on the changes in virulence of myxoma virus in Australia to show that evolution of the relationship between a parasite and its host need not result in complete loss of virulence by the parasite, a concept widely believed by microbiologists. Using a simple mathematical model relating the epidemiological parameters of different virulence grades, they concluded that the basic reproductive rate of myxoma virus was at a maximum for strains of virulence grade IV. They pointed out that, because of the over-simplifications of the model, the prediction was somewhat different from the observed facts (virulence grade III predominates), but their analysis showed that an intermediate virulence grade would have a selective advantage. Seymour (1992) has reworked the Anderson–May model including interaction between virulence, resistance, and rabbit density (as determined by resource capacity of the local environment) to

Table 7.9 The virulence of strains of myxoma virus recovered from the field in the Mallee region and elsewhere in Victoria[a]

Virulence grade	I	II	III	IV	V	Number of samples
Case-fatality rate (%)	>99	95–99	70–95	50–70	<50	
Mean survival time (days)	⩽13	14–16	17–28	29–50	—	
Collection area: Mallee region						
1959–63	0.0	4.3	57.1	34.3	4.3	70
1964–66	2.0	0.0	64.7	31.3	2.0	51
1967–69	0.0	0.0	68.1	31.9	0.0	31
1970–74	1.0	6.9	77.5	14.7	0.0	102
1975–81	3.0	5.8	67.8	23.4	0.0	121
Collection area: Victoria less Mallee						
1959–63	2.1	12.4	61.2	19.5	4.7	379
1964–66	0.4	0.4	63.5	34.5	1.2	255
1967–69	0.0	0.0	61.6	36.4	2.0	198
1970–74	0.0	1.4	69.4	29.2	0.0	72
1975–81	0.0	0.0	65.8	34.2	0.0	91

[a] From Edmonds *et al.* (1975) and J. W. Edmonds (personal communication, 1982). Enzootics recurred almost every year in the Mallee region, but less frequently elsewhere in Victoria. The genetic resistance of rabbits is highest in the Mallee region (Table 7.6).

give a quasi-epidemiological model which yields a prediction closer to reality.

Using the limited data available on the development of genetic resistance in rabbits, modelling by Dwyer *et al.* (1990) predicted that increasing host resistance should lead to selection for increased virulence of virus strains, and that the observed levels of genetic resistance in Australia would lead to the selection of grade III strains. This was based on the assumption that the decline in case mortality was similar for all strains of virus. However, on examining data on resistance and virus grades for Britain they concluded that this approach may not be valid. For further predictive models to be developed, more data on the levels of transmission of field strains in genetically resistant rabbits are required.

7.12 Commercial rabbit breeding and vaccination

All over the world, *Oryctolagus cuniculus* is an important laboratory animal for biomedical research and medical laboratory practice. It is also bred for food and fur in some countries (notably France), both in large commercial rabbitries and in backyard hutches. Outbreaks of myxomatosis in rabbit colonies bred for the laboratory have been known in South America since 1896 and in Australia since 1952, and in 1954 it was estimated that 30–40 per cent of the 140 million domestic rabbits in France were killed. A vaccine was sought for both these groups of domesticated rabbits. Tens of millions of doses of fibroma vaccine were sold annually in France in the 1950s, but the vaccine is now used on a much smaller scale, primarily to protect breeding stock.

7.12.1 Vaccination with fibroma virus

Fibroma virus was discovered by Shope (1932) in tumours on *Sylvilagus floridanus*. It differs from the antigenically-related myxoma virus in that it causes only benign tumours, localized to the site of inoculation, in *Oryctolagus cuniculus*. Prior inoculation with fibroma virus protects rabbits against myxomatosis (Shope 1932; Fenner and Woodroofe 1954; Rowe *et al.* 1956), although protection is not absolute and declines fairly quickly, so that revaccination every six months was recommended if protection was to be maintained.

In Australia, vaccination with fibroma virus was at first used only to protect laboratory rabbits. With the decline in the numbers of wild rabbits due to myxomatosis and the consequent fall in the supplies of rabbit meat and fur, attempts were made to establish commercial breeding of rabbits. Commercial producers sought to use fibroma virus to protect their breeding stock against myxomatosis, and from 1957 onwards rabbit fibroma virus was supplied as a vaccine to registered rabbit breeders in New South Wales, but not in other states.

In 1962, at the instigation of Dame Jean Macnamara, Dr Richard E. Shope was invited to Australia by CSIRO and the governments of Victoria and New South Wales to report on the use of fibroma virus in commercial rabbitries in Australia. His report was critical of what he regarded as the failure of Australian scientists to exploit myxomatosis adequately, and he recommended strongly against the use of fibroma virus for vaccination of rabbits reared for commercial purposes (though not objecting to the use of the vaccine for the much smaller number of animals bred for medical and biological research). Following receipt of this report, the Standing Committee on Agriculture determined in October 1962 that '. . . any continued use of fibroma must be under closely prescribed conditions', and the Director-General of Health reapplied restrictions upon scientific laboratories studying fibroma and myxoma viruses, although not upon the distribution of myxoma virus for rabbit control.

In view of a revival of interest in a domestic rabbit industry in Australia, Sobey (1981) reopened the question of the use of fibroma virus, but in spite of the fact that the likelihood of the 'escape' of the virus to the wild rabbit population is now acknowledged to be remote, the ban still remains in force. Highly attenuated strains of myxoma virus are now available as vaccines (see below), but these have not been used in the domestic rabbit industry in Australia.

A fibroma virus vaccine was developed in Britain (Rowe *et al.* 1956), and although no further work has been published on its effectiveness, such a vaccine is still available to rabbit breeders.

In France, there has been much research on fibroma virus vaccines, which have been used to protect not only domestic and commercially-reared rabbits but also large numbers of wild rabbits. It has been shown that the protection given by fibroma virus vaccines is improved by the addition of BCG vaccine (Metianu 1977), kaolin (Durand *et al.* 1974) or cortisone (Jacotot *et al.* 1962), and the use of the dermojet injector has been recommended (Joubert and Brun 1976).

7.12.2 Vaccination with attenuated strains of myxoma virus

Because of the threat of myxomatosis to domestic rabbit production in California, Saito *et al.* (1964) developed a highly attenuated strain of Californian myxoma virus by serial passage in rabbit kidney cells for use as a vaccine. This was used to a limited extent in France, but it was found to recover some or all of its pathogenicity on passage in rabbits (Jacotot *et al.* 1967). It was subsequently replaced by the attenuated strain SG33, which was produced by serial passage of an attenuated field strain of virus in a rabbit kidney cell line, and in chick embryo fibroblasts at 33°C (Saurat *et al.* 1978). SG33 was reported to give effective protection without regaining virulence and with only very small lesions on the site of vaccination. However, the use of SG33 vaccine was found to have an initial immunosuppressive effect which, in commercial rabbit farms with poor standards of hygiene, led to secondary bacterial infections (Brun *et al.* 1981*a*). It is now recommended that fibroma vaccine should be given first, with a second vaccination using SG33 six to eight weeks later (Joubert *et al.* 1982). Successful results, with no problems, have been reported by Picavet *et al.* (1990).

Despite considerable efforts, all attempts to produce an effective inactivated myxoma virus vaccine have proved unsuccessful (Joubert *et al.* 1973). This is not surprising, in view of the failure to produce an effective inactivated smallpox vaccine (Fenner *et al.* 1988).

7.13 Summary and conclusions

Myxoma virus, a member of the genus *Leporipoxvirus*, family *Poxviridae*, subfamily *Chordopoxvirinae*, produces small localized fibromas in the skin of rabbits of the species *Sylvilagus brasiliensis* in South and Central America and *Sylvilagus bachmani* in California and Oregon. Its host range is narrow, but it produces a very severe generalized disease in European rabbits (*Oryctolagus cuniculus*). Initially this was a nuisance for users of laboratory rabbits and breeders of European rabbits in these parts of the Americas, but in 1950 its significance was transformed when it was successfully introduced into Australia as a method of controlling wild European rabbits—the major vertebrate pest in that country. Two years later another strain of the virus was introduced into wild rabbit populations of France, and spread from that country all over Europe. In *Oryctolagus*, as in *Sylvilagus*, transmission is mechanical, and mosquitoes and the rabbit flea (*Spilopsyllus cuniculi*) are important vectors; mosquitoes in Australia and to some extent in France, fleas in Britain and the colder parts of Europe.

In both Australia and Europe, myxomatosis had a dramatic effect on rabbit populations, leading within five years to reductions in population size approaching 90 per cent. In Australia the disease was hailed as a major addition to methods for rabbit control, and deliberate introduction of virulent strains by inoculation or the introduction of virus-contaminated fleas into rabbit burrows was practised on a large scale. In Britain, because of sentimental attachment to rabbits in the countryside, deliberate inoculation was declared unlawful, but farmers and foresters welcomed the disease and ensured its spread by carrying infected rabbits to new areas, since they regarded the rabbit as a pest. In France the public reaction was mixed. Foresters and some farmers welcomed the reduction in the numbers of wild rabbits; chasseurs and domestic rabbit breeders saw it as a disaster for rabbit hunting and the production of rabbit meat.

Soon after its introduction, attenuated variants of myxoma virus were recovered from naturally-infected rabbits. Studies in Australia and England over the last thirty years have shown that such attenuated variants (which caused case-fatality rates of 70–90 per cent in genetically unselected rabbits, compared with >99 per cent for the strains originally introduced) soon became dominant in the field. Laboratory and field experiments showed that this shift is due to natural selection; less virulent strains produce the kind of disease that is best transmitted by both mosquitoes and fleas.

Because the emergence of attenuated variants allowed some 10 per cent of infected rabbits to survive, the opportunity arose for natural selection for genetic resistance. In areas subject to annual epizootics, an initial increase in resistance developed after a few years, and then the level of resistance increased more slowly. Interactions between virulence of the virus and resistance of the rabbit illustrate that coevolution is continuing, and more virulent strains of the virus are being selected for survival as the genetic resistance of the rabbit increases.

Oryctolagus cuniculus is used for laboratory experiments in all parts of the world, and domestic rabbits are bred for food in many countries, especially France. Such animals initially suffered severely from myxomatosis. Later they were given protection by vaccination with either Shope's fibroma virus, or with attenuated vaccine strains of myxoma virus.

Myxomatosis still exerts a substantial effect on the densities of rabbit populations in both Australia and Europe. In Australia, the rabbit population, which had been greatly depleted in the 1950s, began to increase again in the 1960s, was reduced again when the rabbit flea was well established, but began to build up again in the 1980s. In Britain, rabbit numbers have increased steadily from the very low levels in the mid-1950s, but with large local variations. In France, the rabbit population has also increased, and again, much more so in some regions than in others.

In the long term, there are several possibilities for further evolutionary changes in the virus and the rabbit:

a. A localized fibroma may be produced in genetically highly resistant wild rabbits. If virus from such lesions were transferred to genetically unselected rabbits it would probably behave as a very virulent myxoma virus, thus mimicking the climax association which has developed in *Sylvi-*

lagus in South America and California. There is as yet no evidence of any such trend in myxomatosis of *Oryctolagus* in Australia or Europe, but the association of leporipoxviruses with *Sylvilagus* probably dates back to the early Pleistocene, whereas myxoma virus has been spreading among *Oryctolagus* for only 40 years.

b. Selection appears to be producing different degrees of resistance in Europe and Australia, probably because the founder effect ensures that the genetic base of rabbits in Australia is much narrower than that of European rabbits. Nevertheless, it is reasonable to postulate that there will not be continued progressive selection for genetic resistance in either continent, because after a certain degree of resistance has been acquired, no further rapid changes in virus or host may be possible which would result in either a transmissible fibroma (*a* above) or a generalized disease that was readily transmitted by biting arthopods. We could then envisage a climax association in which myxomatosis persisted and still caused a moderately severe disease with an appreciable mortality, much as smallpox used to persist, until very recently, in human communities in Asia. Data collected over the past decade suggest that this may be the short-term 'stable' situation; however it must be remembered that evolution involving mutations in the host animals (rather than selection acting on existing genetic variability) would become operative only over a time-scale much longer than a few decades. The reproductive capacity of the rabbit is such that this sort of disease would not seriously interfere with its survival in nature, although it should help to keep the population size far below the extremes reached in pre-myxomatosis days.

References

Allan, R. M. (1956). A study of the populations of the rabbit flea *Spilopsyllus cuniculi* (Dale) on the wild rabbit *Oryctolagus cuniculus* in north-east Scotland. *Proceedings of the Royal Entomological Society of London* (A), **31**, 145–52.

Anderson, R. M. and May, R. M. (1982). Co-evolution of hosts and parasites. *Parasitology*, **85**, 411–26.

Andrewes, C. H., Muirhead-Thomson, R. C., and Stevenson, J. P. (1956). Laboratory studies of *Anopheles atroparvus* in relation to myxomatosis. *Journal of Hygiene*, **54**, 478–86.

Andrewes, C. H., Thompson, H. V., and Mansi, W. (1959). Myxomatosis: present position and future prospects in Great Britain. *Nature*, **184**, 1179–80.

Aragão, H. de B. (1927). Myxoma of rabbits. *Memorias do Instituto Oswaldo Cruz (Rio de Janeiro)*, **20**, 225–47.

Aragão, H. de B. (1943). O virus do mixoma do coelho do mato (*Sylvilagus minensis*) sua transmissão pelos *Aedes scapularis* e *aegypti*. *Memorias do Instituto Oswaldo Cruz (Rio de Janeiro)*, **38**, 93–9.

Armour, C. J. and Thompson, H. V. (1955). Spread of myxomatosis in the first outbreak in Great Britain. *Annals of Applied Biology*, **43**, 511–17.

Arthur, C. P. and Louzis, C. (1988). Myxomatose du lapin en France: une revue. *Revue Scientifique et Technique. Office International des Épizooties*, **7**, 939–57.

Arthur, C. P., Chapuis, J. L., Pages, M. V., and Spitz, F. (1980). Enquêtes sur la situation et la repartition écologique du lapin de garenne en France. *Bulletin mensuel de l'Office National de la Chasse. Numero Special Scientifique et Technique*, 37–90.

Baker-Gabb, D. J. (1984). The breeding ecology of twelve species of diurnal raptor in north-western Victoria. *Australian Wildlife Research*, **11**, 145–70.

Basse, A. (1961). Rabbit myxomatosis in Denmark. *Medlemsblad for den Danske Dyrlaegeforening*, **44**, 61–6.

Blazek, K. and Kral, J. (1954). Incidence of myxomatosis in Czechoslovakia. *Veterinárstri*, **4**, 357–8.

Block, W., Upton, C., and McFadden, G. (1985). Tumorigenic poxviruses: genomic organisation of malignant rabbit virus, a recombinant between Shope fibroma virus and myxoma virus. *Virology*, **140**, 113–24.

Borg, K. (1962). Om myxomatos. [On myxomatosis.] *Medlemsblad för Sveriges Veterinäförbund*, **4**, 1–10.

Bouvier, G. (1954). Quelques remarques sur la myxomatose. *Bulletin de l'Office International des Épizooties*, **46**, 76–7.

Bouvier, G. (1967). Myxomatosis of rabbits in Switzerland during 1954–66. *Recueil de Médecine Vétérinaire*, **143**, 427–33.

Brown, P. W., Allan, R. M., and Shanks, P. L. (1956). Rabbits and myxomatosis in the North-East of Scotland. *Scottish Agriculture*, **35**, 204–7.

Brun, A., Godard, A., and Moreau, Y. (1981*a*). La vaccination contre la myxomatose: vaccins heterologues et homologues. *Bulletin de la Société des Sciences Vétérinaires et de Médecine Comparée de Lyon*, **83**, 251–4.

Brun, A., Saurat, P., Gilbert, Y., Godard, A., and Bouquet, J. F. (1981*b*). Données actuelles sur l'épidémiologie, la pathogénie et la symptomatologie de la myxomatose. *Revue de Médecine Vétérinaire*, **132**, 585–90.

Bull, L. B. and Dickinson, D. B. (1937). The specificity of the virus of rabbit myxomatosis. *Journal of the Council for Scientific and Industrial Research of Australia*, **10**, 291–4.

Bull, L. B. and Mules, M. W. (1944). An investigation of *Myxomatosis cuniculi* with special reference to the possible use of the disease to control rabbit populations in Australia. *Journal of the Council for Scientific and Industrial Research of Australia*, **17**, 79–93.

Calaby, J. (1951). Notes on the little eagle; with particular reference to rabbit predation. *Emu*, **51**, 33–456.

Chapple, P. J. and Lewis, N. D. (1964). An outbreak of myxomatosis caused by a moderately attenuated strain of myxomatosis. *Journal of Hygiene*, **62**, 433–41.

Chapple, P. J. and Lewis, N. D. (1965). Myxomatosis and the rabbit flea. *Nature*, **207**, 388–9.

Cooke, B. D. (1983). Changes in the age-structure and size of populations of wild rabbits in South Australia, following the introduction of European rabbit fleas, *Spilopsylus cuniculi* (Dale), as vectors of myxomatosis. *Australian Wildlife Research*, **10**, 105–20.

Cooke, B. D. (1984). Factors limiting the distribution of the European rabbit flea, *Spilopsyllus cuniculi* (Dale) (Siphonaptera) in inland South Australia. *Australian Journal of Zoology*, **32**, 493–506.

Corazzola, S. and Zanin, E. (1970). Observations on a centre of myxomatosis in Veneto. *Agricoltura della Venezie*, **24**, 310–14.

Dalmat, H. T. (1959). Arthropod transmission of rabbit fibromatosis (Shope). *Journal of Hygiene*, **57**, 1–30.

Dalmat, H. T. and Stanton, M. F. (1959). A comparative study of the Shope fibroma in rabbits in relation to transmissibility by mosquitoes. *Journal of the National Cancer Institute*, **22**, 593–615.

Day, M. F., Fenner, F., Woodroofe, G. M., and McIntyre, G. A. (1956). Further studies on the mechanism of mosquito transmission of myxomatosis in the European rabbit. *Journal of Hygiene*, **54**, 258–83.

Delille, P. F. A. (1953). Une methode nouvelle permettant à l'agriculture de lutter efficacement contre la pullulation du lapin. *Compte rendu hebdomadaires des Séances de l'Académie d'agriculture de France*, **39**, 638–9.

Devon, A., Viaene, N., and Staelens, M. (1963). myxomatosis in Belgium from 1953 to 1963. *Vlaamsch diergeneeskundig Tijdschrift*, **32**, 170–81.

Douglas, G. W. (1962). The Glenfield strain of myxoma virus. Its use in Victoria. *Journal of Agriculture*, **60**, 511–16.

Dunsmore, J. D., Williams, R. T., and Price, J. R. (1971). A winter epizootic of myxomatosis in subalpine south-eastern Australia. *Australian Journal of Zoology*, **19**, 275–86.

Durand, M., Ravon, D., Guerche, J., and Prunet, P. (1974). Étude d'un nouveau vaccin contre la myxomatose. *Recueil de Médecine Vétérinaire*, **150**, 527–34.

Duran-Reynals, F. (1940). Production of degenerative inflammatory or neoplastic effects in the newborn rabbit by the Shope fibroma virus. *Yale Journal of Biology and Medicine*, **13**, 99–110.

Dwyer, G., Levin, S. A., and Buttel, L. (1990). A simulation model of the population dynamics and evolution of myxomatosis. *Ecological Monographs*, **60**, 423–47.

Edmonds, J. W., Nolan, I. F., Shepherd, R. C. H., and Gocs, A. (1975). Myxomatosis: the virulence of field strains of myxoma virus in a population of wild rabbits (*Oryctolagus cuniculus* L.) with high resistance to myxomatosis. *Journal of Hygiene*, **74**, 417–18.

Fenner, F. (1983). The Florey Lecture, 1983. Biological control, as exemplified by smallpox eradication and myxomatosis. *Proceedings of the Royal Society of London*, **B218**, 259–85.

Fenner, F. and Chapple, P. J. (1965). Evolutionary

changes in myxoma viruses in Britain. An examination of 222 naturally occurring strains obtained from 80 counties during the period October–November 1962. *Journal of Hygiene*, **63**, 175–85.

Fenner, F. and Marshall, I. D. (1954). Passive immunity in myxomatosis of the European rabbit (*Oryctolagus cuniculus*): the protection conferred on kittens born by immune does. *Journal of Hygiene*, **52**, 321–36.

Fenner, F. and Marshall, I. D. (1957). A comparison of the virulence of European rabbits (*Oryctolagus cuniculus*) of strains of myxoma virus recovered in the field in Australia, Europe and America. *Journal of Hygiene*, **55**, 149–91.

Fenner, F. and Ratcliffe, F. N. (1965). *Myxomatosis*. Cambridge University Press.

Fenner, F. and Woodroofe, G. M. (1953). The pathogenesis of infectious myxomatosis: The mechanism of infection and the immunological response in the European rabbit (*Oryctolagus cuniculus*). *British Journal of Experimental Pathology*, **34**, 400–11.

Fenner, F. and Woodroofe, G. M. (1954). Protection of laboratory rabbits against myxomatosis by vaccination with fibroma virus. *Australian Journal of Experimental Biology and Medical Science*, **32**, 653–67.

Fenner, F. and Woodroofe, G. M. (1965). Changes in the virulence and antigenic structure of strains of myxoma virus recovered from Australian wild rabbits between 1950 and 1964. *Australian Journal of Experimental Biology and Medical Science*, **43**, 359–70.

Fenner, F., Day, M. F., and Woodroofe, G. M. (1952). The mechanism of the transmission by myxomatosis in the European rabbit (*Oryctolagus cuniculus*) by the mosquito *Aedes aegypti*. *Australian Journal of Experimental Biology and Medical Science*, **30**, 139–52.

Fenner, F., Day, M. F., and Woodroofe, G. M. (1956). Epidemiological consequences of the mechanical transmission of myxomatosis by mosquitoes. *Journal of Hygiene*, 284–303.

Fenner, F., Poole, W. E., Marshall, I. D., and Dyce, A. L. (1957). Studies in the epidemiology of infectious myxomatosis of rabbits. VI. The experimental introduction of the European strain of myxoma virus into Australian wild rabbit populations. *Journal of Hygiene*, **55**, 192–206.

Fenner, F., Henderson, D. A., Arita, I., Jezek, Z., and Ladnyi, I. D. (1988). *Smallpox and its eradication*. World Health Organization, Geneva.

Fioretti, A. (1985). The problem of myxomatosis in Italy. *Rivista di Coniglicoltura*, **22**, 47–56.

Galaup, O. (1988). Contribution à l'étude sur la lutte anti-myxomatose à l'aide du vaccin SG33. Suivi de l'épidémiologie de la myxomatose sur la reserve de Donzere-Mondragon (Vaucluse). *Office National de la Chasse, Brevet d'Enseignement Professionnel Agricole Cytogénetique Promotion*, 1986–1988, 93pp.

Giban, J. (1956). Répercussion de la myxomatose sur les populations de lapin de garenne en France. *La Terre et la Vie*, **103**, 179–87.

Giban, J. (1959). Évolution de la myxomatose en France. *European Plant Protection Organization Report. International Conference on Harmful Mammals and Control, London 1958*, 67–71.

Heard, H. K., O'Connor, K., and Strayer, D. S. (1990). Molecular analysis of immunosuppression induced by viral replication in lymphocytes. *Journal of Immunology*, **144**, 3992–9.

Hudson, J. R., Thompson, H. V., and Mansi, W. (1955). Myxoma virus in Britain. *Nature*, **176**, 783.

Hurst, E. W. (1937). Myxoma and the Shope fibroma. II. The effect of intracerebral passage on the myxoma virus. *British Journal of Experimental Pathology*, **18**, 15–22.

Jacotot, H. and Vallée, A. (1953). Un foyer de myxome infectieux chez des lapins de garenne dans la region de Rambouillet. *Annales de l'Institut Pasteur*, **84**, 448–50.

Jacotot, H., Vallée, A. and Virat, B. (1962). Influence de la cortisone sur l'immunization du lapin contre la myxomatose par inoculation de virus du fibrome. *Annales de l'Institut Pasteur*, **103**, 285–90.

Jacotot, H., Virat, B., Reculard, P., and Vallée, A. (1967). Étude d'une souche attenuée du virus du myxome infectieux obtenue par passages en cultures cellulaires (MacKercher et Saito, 1964). *Annales de l'Institut Pasteur*, **113**, 221–37.

Joubert, L. and Brun, A. (1976). Prophylaxie médicale à dessein cynégétique de la myxomatose des garennes. La fibromatisation auriculaire au Dermojet. *Bulletin de la Société des Sciences Vétérinaire et de Médecine Comparée de Lyon*, **78**, 101–7.

Joubert, L., Leftheriotis, E., and Mouchet, J. (1972, 1973). *La myxomatose*, 2 vols. L'Expansion, Paris.

Joubert, L., Duclos, Ph., and Tuaillon, P. (1982). La

myxomatose des garennes dans le Sud-Est. *Revue de Médecine Vétérinaire*, **133**, 739–53.

Kessel, J. F., Prouty, C. C., and Meyer, J. W. (1931). Occurrence of infectious myxomatosis in southern California. *Proceedings of the Society for Experimental Biology and Medicine*, **28**, 413–14. University of Toronto Press, Toronto.

Kessel, J. F., Fisk, R. T., and Prouty, C. C. (1934). Studies with the Californian strain of the virus of infectious myxomatosis. *Proceedings of the Fifth Pacific Science Congress*, Canada, 1933, **4**, 2927–39.

Kilham, L. and Dalmat, H. T. (1955). Host-virus-mosquito relations of Shope fibromas in cottontail rabbits. *American Journal of Hygiene*, **61**, 45–54.

Kilham, L. and Fisher, E. R. (1954). Pathogenesis of fibromas in cottontail rabbits. *American Journal of Hygiene*, **59**, 104–12.

King, D. R. and Wheeler, S. H. (1985). The European rabbit in Western Australia I. Study sites and population dynamics. *Australian Wildlife Research*, **12**, 183–96.

King, D. R., Oliver, A. J., and Wheeler, S. H. (1985). The European rabbit flea, *Spilopsyllus cuniculi*, in south-western Australia. *Australian Wildlife Research*, **12**, 227–36.

Kotsche, W. (1964). Epidemiology of myxomatosis in the German Democratic Republic in 1955–62. *Archiv für experimentelle Veterinärmedizin*, **18**, 427–47.

Lankamp, C. T. (1953). First outbreak of myxomatosis in domestic rabbits in the Netherlands. *Tijdschrift voor Diergeneeskunde*, **78**, 816.

Lloyd, H. G. (1970). Post-myxomatosis rabbit populations in England and Wales. *European Plant Protection Organization Publication Series*, No. 58, 197–215.

Lloyd, H. G. (1981). Biological observations on post-myxomatosis wild rabbit populations in Britain 1955–1979. *Proceedings of the World Lagomorph Conference* (1979) (ed. K. Myers and C. D. MacInnes), pp. 623–7. University of Guelph, Ontario.

Lockley, R. M. (1954). The European rabbit-flea, *Spilopsyllus cuniculi*, as a vector of myxomatosis in Britain. *Veterinary Record*, **66**, 434–5.

Lubke, H. (1968). Ten years of state control of myxomatosis in rabbits in Berlin. *Berliner und Münchener tierärztliche Wochenschrift*, **81**, 275–9.

Marshall, I. D. (1959). The influence of ambient temperature on the course of myxomatosis in the rabbit. *Journal of Hygiene*, **57**, 484–97.

Marshall, I. D. (1961). Myxomatosis investigations carried out in Central and South America. *Report to the Australian Wool Research Fund Committee*.

Marshall, I. D. and Douglas, G. W. (1961). Studies in the epidemiology of infectious myxomatosis of rabbits. VIII. Further observations on changes in the innate resistance of Australian wild rabbits exposed to myxomatosis. *Journal of Hygiene*, **59**, 117–22.

Marshall, I. D. and Fenner, F. (1958). Studies in the epidemiology of infectious myxomatosis of rabbits. V. Changes in the innate resistance of Australian wild rabbits exposed to myxomatosis. *Journal of Hygiene*, **56**, 288–302.

Marshall, I. D. and Fenner, F. (1960). Studies in the epidemiology of infectious myxomatosis of rabbits. VII. The virulence of strains of myxoma virus recovered from Australian wild rabbits between 1951 and 1959. *Journal of Hygiene*, **58**, 485–7.

Marshall, I. D. and Regnery, D. C. (1960). Myxomatosis in a Californian brush rabbit (*Sylvilagus bachmani*). *Nature*, **188**, 73–4.

Martin, C. J. (1936). Observations on *Myxomatosis cuniculi* (Sanarelli) made with a view to the use of the virus in the control of rabbit plagues. *Bulletin of the Council of Scientific and Industrial Research of Australia*, No. 96.

Mead-Briggs, A. R. (1964). Some experiments concerning the interchange of rabbit-fleas, *Spilopsyllus cuniculi* (Dale) between living rabbit hosts. *Journal of Animal Ecology*, **33**, 13–26.

Mead-Briggs, A. R. and Rudge, A. J. B. (1960). Breeding of the rabbit flea, *Spilopsyllus cuniculi* (Dale): requirement of a 'factor' from a pregnant rabbit for ovarian maturation. *Nature*, **187**, 1136–7.

Mead-Briggs, A. R. and Vaughan, J. A. (1969). Some requirements for mating in the rabbit flea, *Spilopsyllus cuniculi* (Dale). *Journal of Experimental Biology*, **51**, 495–511.

Mead-Briggs, A. R. and Vaughan, J. A. (1975). The differential transmissibility of myxoma virus strains of differing virulence by the rabbit flea *Spilopsyllus cuniculi* (Dale). *Journal of Hygiene*, **75**, 237–47.

Metianu, T. (1977). Réinforcement du pouvoir immunisant du vaccin contre la myxomatose du lapin, à base de virus de Shope, par l'adjonction du BCG. *Recueil de Médecine Vétérinaire*, **153**, 573–7.

Moses, A. (1911). O Virus do mixoma does coelhos. *Memorias do Instituto Oswaldo Crus (Rio de Janeiro)*, **3**, 46–53.

Muirhead-Thomson, R. C. (1956). The part played by woodland mosquitoes of the genus *Aedes* in the transmission of myxomatosis in England. *Journal of Hygiene*, **54**, 461–71.

Müller-Using, D. (1953). Thoughts on the appearance of myxomatosis in Germany. *Zentralblatt für Veterinärmedizin*, **21/22**, 398–400.

Myers, K. (1954). Studies in the epidemiology of infectious myxomatosis of rabbits. II. Field experiments, August–November 1950, and the first epizootic of myxomatosis in the riverine plain of south-eastern Australia. *Journal of Hygiene*, **52**, 47–59.

Myers, K. (1970). The rabbit in Australia. In *Dynamics of populations* (ed. P. J. den Boer and G. R. Gradwell), pp. 478–506. Centre for Agricultural Publishing and Documentation, Wageningen.

Myers, K., Marshall, I. D., and Fenner, F. (1954). Studies in the epidemiology of infectious myxomatosis of rabbits. III. Observations on two succeeding epizootics in Australian wild rabbits on the riverine plain of south-eastern Australia, 1951–1953. *Journal of Hygiene*, **52**, 337–60.

Mykytowycz, R. (1956). The effect of season and mode of transmission on the severity of myxomatosis due to an attenuated strain of the virus. *Australian Journal of Experimental Biology and Medical Science*, **34**, 121–32.

Parer, I. (1977). The population ecology of the wild rabbit, *Oryctolagus cuniculus* (L), in a Mediterranean-type climate in New South Wales. *Australian Wildlife Research*, **4**, 171–205.

Parer, I., Conolly, D., and Sobey, W. R. (1981). Myxomatosis: the introduction of a highly virulent strain of myxomatosis into a wild rabbit population at Urana in New South Wales. *Australian Wildlife Research*, **8**, 613–26.

Parer, I., Conolly, D., and Sobey, W. R. (1985). Myxomatosis: the effects of annual introductions of an immunizing strain and a highly virulent strain of myxoma virus into rabbit populations at Urana, N.S.W. *Australian Wildlife Research*, **12**, 407–23.

Parker, R. F. and Thompson, R. L. (1942). The effect of external temperature on the course of infectious myxomatosis of rabbits. *Journal of Experimental Medicine*, **75**, 567–73.

Picavet, D. P., Lebas, F., Gilbert, T. and Brignol, E. (1990). Immunisation of rabbits against myxomatosis with an homologous vaccine. *Revue de Médicine Vétérinaire*, **140**, 823–7.

Ratcliffe, F. N., Myers, K., Fennessy, B. V., and Calaby, J. H. (1952). Myxomatosis in Australia. A step towards the biological control of the rabbit. *Nature*, **170**, 7–11.

Regnery, D. C. and Miller, J. H. (1972). A myxoma virus epizootic in a brush rabbit population. *Journal of Wildlife Research*, **8**, 327–31.

Regnery, R. L. (1975). Preliminary studies on an unusual poxvirus of the western grey squirrel (*Sciurus griseus griseus*) of North America. *Intervirology*, **5**, 364–6.

Rendel, J. M. (1971). Myxomatosis in the Australian rabbit population. *Search*, **2**, 89–94.

Ross, J. (1982). Myxomatosis: the natural evolution of the disease. *Symposium of the Zoological Society, London*, **50**, 77–95.

Ross, J. and Sanders, M. F. (1977). Innate resistance to myxomatosis in wild rabbits in England. *Journal of Hygiene*, **29**, 411–15.

Ross, J. and Sanders, M. F. (1984). The development of genetic resistance to myxomatosis in wild rabbits in Britain. *Journal of Hygiene*, **92**, 255–61.

Ross, J. and Sanders, M. F. (1987). Changes in the virulence of myxoma virus strains in Britain. *Epidemiology and Infection*, **98**, 113–17.

Ross, J., Tittensor, A. M., Fox, A. P., and Sanders, M. F. (1989). Myxomatosis in farmland rabbit populations in England and Wales. *Epidemiology and Infection*, **103**, 333–57.

Rothschild, M. (1953). Notes on the European rabbit flea. Report to the Myxomatosis Advisory Committee, Ministry of Agriculture, Fisheries and Food, United Kingdom.

Rothschild, M. and Ford, B. (1964). Breeding of the rabbit flea, *Spilopsyllus cuniculi* (Dale) controlled by the reproductive hormones of the host. *Nature*, **201**, 103–4.

Rothschild, M. and Ford, B. (1966). Hormones of the vertebrate host controlling ovarian regression and copulation of the rabbit flea. *Nature*, **211**, 261–6.

Rowe, B., Mansi, W., and Hudson, J. R. (1956). The use of fibroma virus (Shope) for the protection of rabbits against myxomatosis. *Journal of Comparative Pathology*, **66**, 290–7.

Saito, J. K., McKercher, D. G., and Castrucci, G. (1964). Attenuation of the myxoma virus and use of the living attenuated virus as an immunizing agent for myxomatosis. *Journal of Infectious Diseases*, **114**, 417–27.

Sanarelli, G. (1898). Das myxomatogene Virus. Beitrag zum Stadium der Krankheitserreger ausserhalb des Sichtbaren. *Zentralblatt für Bakteriologie, Parasitenkunde, Infektionskrankheiten und Hygiene (Abt. 1)*, **23**, 865–73.

Sanchez, B. C., Arroyo, C., and Blanco, A. (1954). Myxomatosis in rabbits in Spain. *Revista del Patronato de Biologia Animal*, **1**, 75–7.

Saurat, P., Gilbert, Y., and Ganière, J.-P. (1978). Étude d'une souche de virus myxomateux modifié. *Revue de Médecine Vétérinaire*, **129**, 415–51.

Schortanner, M. (1976). La cicatrisation des forêts rhenanes. *Introduction à l'étude de la végétation pionniere des alluvions du Rhin*. DEA, Université de Louis Pasteur, Strasbourg.

Service, M. W. (1971). A reappraisal of the role of mosquitoes in the transmission of myxomatosis. *Journal of Hygiene*, **69**, 105–11.

Seymour, R. M. (1992). A study of the interaction of virulence, resistance and resource limitation in a model of myxomatosis mediated by the European rabbit flea *Spilopsyllus cuniculi* (Dale). *Ecological Modelling*, **60**, 281–308.

Shanks, P. L., Sharman, G. A. M., Allan, R., Donald, L. G., Young, S., and Marr, T. G. (1955). Experiments with myxomatosis in the Hebrides. *British Veterinary Journal*, **111**, 25–36.

Shepherd, R. C. H. and Edmonds, J. W. (1977). Myxomatosis: the transmission of a highly virulent strain of myxoma virus by the European rabbit flea *Spilopsyllus cuniculi* (Dale) in the Mallee region of Victoria. *Journal of Hygiene*, **79**, 405–9.

Shepherd, R. C. H. and Edmonds, J. W. (1978). Myxomatosis: changes in the epidemiology of myxomatosis coincident with the establishment of the European rabbit flea *Spilopsyllus cuniculi* (Dale) in the Mallee region of Victoria. *Journal of Hygiene*, **81**, 399–403.

Shepherd, R. C. H. and Edmonds, J. W. (1979). Myxomatosis: the release and spread of the European rabbit flea *Spilopsyllus cuniculi* (Dale) in the Central District of Victoria. *Journal of Hygiene*, **83**, 285–94.

Shope, R. E. (1932). A transmissible tumor-like condition in rabbits. A filtrable virus causing a tumor-like condition in rabbits and its relationship to virus myxomatosum. *Journal of Experimental Medicine*, **56**, 793–802.

Sobey, W. R. (1960). Myxomatosis: the virulence of the virus and its relation to genetic resistance in the rabbit. *Australian Journal of Science*, **23**, 53–5.

Sobey, W. R. (1969). Selection for resistance to myxomatosis in domestic rabbits (*Oryctolagus cuniculus*). *Journal of Hygiene*, **67**, 743–54.

Sobey, W. R. (1981). Protecting domestic rabbits against myxomatosis in Australia and the problems associated with it past and present. In *Recent advances in animal nutrition in Australia* (ed. D. J. Farrell), pp. 168–72. University of New England Publishing Unit, Armidale.

Sobey, W. R. and Conolly, D. (1971). Myxomatosis: the introduction of the European rabbit flea *Spilopsyllus cuniculi* (Dale) into wild rabbit populations in Australia. *Journal of Hygiene*, **69**, 331–46.

Sobey, W. R. and Conolly, D. (1975). Myxomatosis: passive immunity in the offspring of immune rabbits (*Oryctolagus cuniculus*) infested with fleas (*Spilopsyllus cuniculi* Dale) and exposed to myxoma virus. *Journal of Hygiene*, **74**, 43–55.

Sobey, W. R. and Conolly, D. (1986). Myxomatosis: non-genetic aspects of resistance to myxomatosis in rabbits, *Oryctolagus cuniculus*. *Australian Wildlife Research*, **13**, 177–87.

Sobey, W. R. and Menzies, W. (1969). Myxomatosis: the introduction of the European rabbit flea *Spilopsyllus cuniculi* (Dale) into Australia. *Australian Journal of Science*, **31**, 404–6.

Sobey, W. R. and Turnbull, K. (1956). Fertility in rabbits recovering from myxomatosis. *Australian Journal of Biological Science*, **9**, 455–61.

Sobey, W. R., Menzies, W., Conolly, D., and Adams, K. M. (1967). Myxomatosis: the effect of raised ambient temperature on survival time. *Australian Journal of Science*, **30**, 322–4.

Sobey, W. R., Conolly, D., and Menzies, W. (1977). Myxomatosis: breeding large numbers of rabbit fleas (*Spilopsyllus cuniculi* Dale). *Journal of Hygiene*, **78**, 349–53.

Steffen, J. (1971). Myxomatosis—a disease of rabbits. *Medycyna Weterynarytjna*, **27**, 65–7.

Strayer, D. S. and Sell, S. (1983). Immunohistology of malignant rabbit fibroma virus—a comparative study with rabbit myxoma virus. *Journal of the National Cancer Institute*, **71**, 105–16.

Strayer, D. S., Skaletsky, E., Cabirac, G. F., Sharp, P. A., Corbeil, L. B., Sell, S., and Liebowitz, J. L. (1983). Malignant rabbit fibroma virus causes secondary immunosuppression in rabbits. *Journal of Immunology*, **130**, 399–404.

Thompson, H. V. (1954). The rabbit disease: myxomatosis. *Annals of Applied Biology*, **41**, 358–66.

Thompson, H. V. and Worden, A. N. (1956). *The Rabbit*, New Naturalist Monograph, Collins, London.

Trout, R. C. and Tittensor, A. M. (1989). Can predators regulate wild rabbit (*Oryctolagus cuniculus*) population density in England and Wales? *Mammal Review*, **19**, 153–73.

Trout, R. C., Tapper, S. C., and Harradine, J. (1986). Recent trends in the rabbit population in Britain. *Mammal Review*, **16**, 117–23.

Trout, R. C., Ross, J., Tittensor, A. M., and Fox, A. P. (1993). The effect on a British wild rabbit population (*Oryctolagus cuniculus*) of manipulating myxomatosis. *Journal of Applied Ecology*, **29**, 679–86.

Vaughan, H. E. N. and Vaughan, J. A. (1968). Some aspects of the epizootiology of myxomatosis. *Symposium of the Zoological Society, London*, **24**, 289–309.

Waechter, A. (1983). Notes sur les mammifères d'Alsace. 3. Lagomorphs et Rongeurs. *Mammalia*, **47**, 573–82.

White, H. C. (1929). Observations on rabbit myxoma. *New South Wales Department of Agriculture Veterinary Research Report No. 5, 1927–1928*, pp. 45–7.

Williams, C. K. and Moore, R. J. (1991). Inheritance of acquired immunity to myxomatosis. *Australian Journal of Zoology*, **39**, 307–11.

Williams, C. K., Moore, R. J., and Robbins, S. J. (1990). Genetic resistance to myxomatosis in Australian wild rabbits, *Oryctolagus cuniculus* (L.). *Australian Journal of Zoology*, **38**, 697–703.

Williams, R. T. and Parer, I. (1971). Observations on the dispersal of the European rabbit flea, *Spilopsyllus cuniculi* (Dale) through a natural population of wild rabbits, *Oryctolagus cuniculus* (L.). *Australian Journal of Zoology*, **20**, 391–404.

Williams, R. T., Dunsmore, J. D., and Parer, I. (1972*a*). Evidence for the existence of latent myxoma virus in rabbits (*Oryctolagus cuniculus* L.). *Nature*, **238**, 99–101.

Williams, R. T., Fullagar, P. J., Davey, C. C., and Kogon, C. (1972*b*). Factors affecting the survival time of rabbits in a winter epizootic of myxomatosis at Canberra, Australia. *Journal of Applied Ecology*, **9**, 403–14.

Williams, R. T., Fullagar, P. J., Kogon, C., and Davey, C. C. (1973). Observations on a naturally occurring winter epizootic of myxomatosis at Canberra, Australia, in the presence of rabbit fleas, *Spilopsyllus cuniculi* (Dale) and virulent myxoma virus. *Journal of Applied Ecology*, **10**, 417–27.

Zwar, D. (1984). *The Dame. The life and times of Dame Jean Macnamara medical pioneer*. Macmillan, Melbourne.

Index

Subject references concern the European rabbit unless otherwise indicated.

Personal references list the people and/or the events they influenced that are not directly documented in the seven chapter bibliographies.